Biomedical Photoacoustics

Biomedical Photoacoustics

Sihua Yang
Da Xing

Published by

Jenny Stanford Publishing Pte. Ltd.
Level 34, Centennial Tower
3 Temasek Avenue
Singapore 039190

Email: editorial@jennystanford.com
Web: www.jennystanford.com

British Library Cataloguing-in-Publication Data
A catalogue record for this book is available from the British Library.

Biomedical Photoacoustics

ISBN 978-981-4774-58-1 (Hardcover)
ISBN 978-0-203-70365-6 (eBook)

Contents

Preface

This book can be read on two different levels. First, it may be read by general readers with a limited, if any, scientific background. Throughout, the book has been written with this audience in mind. The second group of readers will be represented by professionals from the biomedical industry, academia, and government agencies. We do not expect everybody in the scientific community to agree with the content and ideas put forth in this book. But we do hope that the information and knowledge presented will become a wake-up call for the general public, regulatory agencies, legislators, business leaders, and scientists coming to the realization that the current state of biomedical technology is not so satisfactory, to say the least, and it needs to be developed—urgently.

The book comprises seven chapters. Chapter 1 is critical for the overall understanding since it is a general introduction to photoacoustics. Chapter 2 presents the fast photoacoustic imaging technology, which refers to multi-element array photoacoustic detection technology, and its applications. Chapter 3 introduces photoacoustic microscopy, including optical-resolution photo-acoustic microscopy, acoustic-resolution photoacoustic microscopy, and single-wavelength excited photoacoustic-fluorescence microscopy. Chapter 4 introduces photoacoustic endoscopy, intravascular photoacoustic endoscopy, and their biomedical applications. Chapter 5 focuses on photoacoustic viscoelasticity imaging technique and its medical applications. Chapter 6 compiles the known progress in the all-optical photoacoustic technology. Chapter 7 provides examples of nanoprobes as contrast agents for biomedical photoacoustic imaging.

We hope this book is widely read. If we are to develop biomedical technology, then we need to open our minds and start benefiting from the knowledge base created by scientists. We did have this chance a decade ago. But now we have more.

Chapter 1

Fundamentals of Photoacoustics

1.1 Introduction

Photoacoustics has a relatively long history dating back to the 1880s, when Alexander Graham Bell first discovered the photoacoustic (PA) effect via observing the generation of sound by absorbing modulated sunlight [1]. Since then, relatively little technological development or scientific research took place in PA application before the invention of laser in the 1960s. Benefiting from the directionality, high peak power and spectral purity of laser, numerous PA applications in industrial and sensing fields began to emerge in the 1970s and 1980s [2]. In the early stage of photoacoustics, applications were mostly based on the indirect gas-phase detection of signal of solids based on a PA cell, where the acoustic waves generated by laser-induced surface heating propagate in the gas-filled PA cell and then are detected with a microphone. General classes of applications of gas-phase PA technique include PA spectroscopy, PA monitoring of deexcitation process, PA probing of thermoelastic and other physical properties of materials, and PA generation of mechanical motions [3].

Biomedical Photoacoustics
Sihua Yang and Da Xing

ISBN 978-981-4774-58-1 (Hardcover), 978-0-203-70365-6 (eBook)
www.jennystanford.com

In mid-1990s, the PA technique began to be a non-destructive detecting tool for biomedical applications and the first PA images appeared [4, 5]. The PA imaging provides high-spatial-resolution images at traditionally ultrasonic depths, which is a potential benefit for medical imaging. In 2000, the first truly compelling in vivo PA images were obtained [6]. After that, this field has witnessed major growth in the development of instrumentation, image reconstruction algorithms, functional imaging capabilities, and in vivo application of the PA technique in biological research [7–10].

In the PA imaging, usually lasers with wavelengths in the visible and near-infrared (NIR) spectrum between 550 and 1200 nm are used as the irradiation sources. The quick optical absorption by the specific tissue chromophores such as melanin, hemoglobin, lipid, and water induces thermal expansion and PA wave emission, which were then detected by single-element or multi-element ultrasound transducers. Thereafter, PA images reflecting tissue absorption then can be obtained through image reconstruction algorithms. The PA imaging shows great potential for detecting and monitoring diseases. The schematic for biomedical PA imaging is shown in Fig. 1.1.

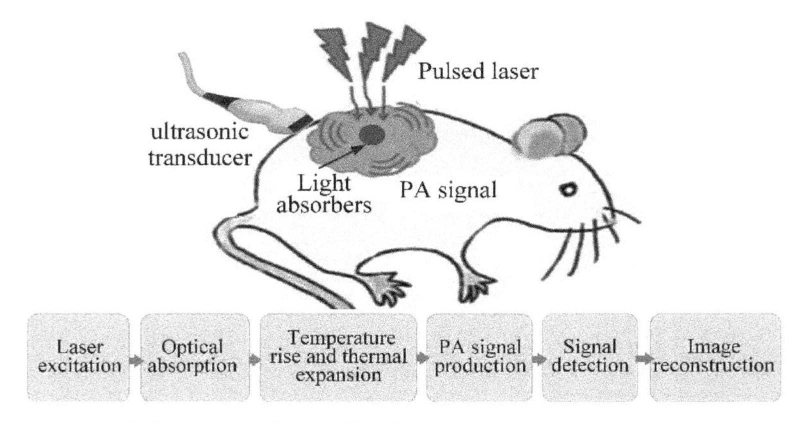

Figure 1.1 Schematic for biomedical PA imaging.

Absorption of photons by biomolecules thermoelastically induces PA waves through the PA effect. The conversion from absorbed optical energy to acoustic waves immediately brings unique advantages for biomedical applications. As a non-invasive imaging modality, PA imaging breaks through the optical diffusion limit by utilizing the relatively low acoustic scattering in biological tissues, which is several orders of magnitude lower than the optical scattering in tissue for certain transmission path [10]. Meanwhile, by switching excitation optical wavelengths, PA imaging is capable of revealing rich optical contrasts according to chemical composition of the tissues. Although both conventional ultrasound imaging and PA imaging are based on ultrasonic detection, ultrasound imaging only measures mechanical contrasts and PA imaging can provide optical and thermoelastic contrasts of tissues. Photoacoustic imaging has obvious advantages in the study of the structure, metabolism, and physiological characteristics of biological tissues.

Because the elegant marriage between light and sound enables the unique ability of scaling its spatial resolution and imaging depth across both optical and ultrasonic dimensions, PA imaging has the ability of providing multiscale multicontrast images of living biological structures ranging from organelles to organs [11]. As shown in Figure 2 of ref. [11], multiscale images of organelle, cell, tissues and organ in vivo can be achieved by PA technique with relatively high lateral and axial resolution, making PA technique a high-resolution modality across all four length scales.

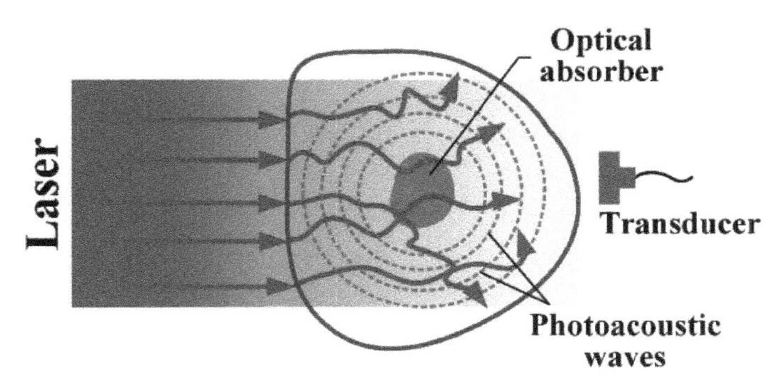

Figure 1.2 Schematic for thermoelastic regime. The red lines illustrate light propagation in tissue. Optical absorbers in tissue suffer thermal expansion after laser irradiation and emit PA waves.

At present, many groups are engaged in PA imaging and its biomedical applications: Da Xing, L. V. Wang, A. A. Oraevsky, S. Emelianov, R.A. Kruger, H. B. Jiang, V. P. Zharov, V. Ntziachristos,

P. C. Beard and A. A. Karabutov. PA technique has been developed as a useful tool for life science research and early disease detection.

1.2 Basic Theories for Photoacoustics

As tissue and materials [12] are irradiated by laser, PA signal can generate via a variety of mechanisms. Expect for the commonly used mechanisms—thermoelastic regime (thermal expansion) [1], other mechanism in PA application include:

- photochemical regime;
- cavitation, boiling, gas evolution;
- optical breakdown and plasma formation.

Different mechanisms possess their own advantages, and this is applicable for specific PA applications [5]. In this section, we present basic theories for these PA mechanisms to provide fundamental understanding of related PA application.

1.2.1 Thermoelastic Regime

In PA imaging, acoustic waves are excited by irradiating tissues with modulated laser, usually pulsed on a nanosecond or microsecond timescale. As the tissues absorb the irradiated laser energy, they suffer local temperature increase. In case the laser energy is moderate, temperature increase does not exceed the boiling temperature of the tissues, and quick thermal expansion of the heated area occurs followed by the emission of PA signal. In biological applications of PA imaging and sensing, thermal expansion is considered to be the only biologically safe mechanism. So thermal expansion is most commonly used in PA applications. In this section, we provide short-pulse excited PA theories for thermal expansion mechanism.

The basic theory for a point source was given by G. J. Diebold [13, 14], where PA pressure for other shapes of absorbers could be found through space integral. The production of ultrasound from absorbed electromagnetic radiation can be described as arising from a linear and nonlinear thermal expansion mechanism. The deposition of optical energy could cause heating and thermal expansion of the absorber, were the motion from expansion acts as a source for producing PA waves. Since PA signal is emitted wherever heat is deposited, the spatial and temporal character of the PA wave necessarily carries information about the optical properties and geometry of the absorber, which is the principle of PA imaging by image reconstruction program. There are a number of excellent reviews on the subject of basic photoacoustics. Here, the analytical theory for the thermal expansion-based PA effect generated by heat deposition at a point in space in an inviscid fluid is presented. Both linear and nonlinear thermal expansion effect are discussed below.

Assume that the heat capacity ratio of the irradiated point can be approximated as unity, the coupled differential equations for the PA pressure and temperature then uncouple, which gives a heat diffusion equation and a wave equation. The heat diffusion equation can be given for a spherical problem as [15]

$$\kappa\nabla^2 T(r,t) + H(r,t) = \rho c_{\mathrm{p}} \frac{\partial T(r,t)}{\partial t}, \tag{1.1}$$

where T is the temperature; H is the laser energy deposition per volume and time; κ, ρ, and c_p are the thermal conductivity, density and heat capacity, respectively; r is the radial coordinate; and t is the time. The wave equation is given in the form of displacement potential Φ as

$$\left(\nabla^2 - \frac{1}{c^2}\frac{\partial^2}{\partial t^2}\right)\Phi(r,t) = \beta T(r,t), \tag{1.2}$$

where c is the PA wave speed. Here, we discuss both linear and nonlinear thermal expansion mechanism by introducing $\beta = \beta_1 + \beta_2 T(r, t)$; thus β_1 and β_2 are the linear (first order) and nonlinear (second order) thermal expansion coefficients, respectively [16]. By utilizing Eq. (1.2), the PA pressure p can be found as

$$p = -\rho \frac{\partial^2 \Phi(r,t)}{\partial t^2}. \tag{1.3}$$

Meanwhile, the displacement due to thermal expansion can also be found as

$$u(r,t) = \nabla\Phi(r,t). \tag{1.4}$$

By transforming Φ into Ψ through the substitution $\Psi = r\Phi$, Eq. (1.1.2) can be transformed in to a one-dimensional equation of the form

$$\left(\frac{\partial^2}{\partial t^2} - c^2\frac{\partial^2}{\partial r^2}\right)\Psi = rQ(r,t). \tag{1.5}$$

Here, $Q(r, t)$ is the source function. Thus, Eq. (1.5) can be solved in the form of d'Alembert solution.

In order to obtain the analytical solution of Eq. (1.5), Eq. (1.2) should be solved. Considering a delta function optical heating pulse, according to the thermal conduction theory, the temperature distribution found for a source point in the case of linear thermal expansion ($\beta_2 = 0$) can be given as

$$T(r,t) = \frac{E_0\sigma}{8\rho c_{\mathrm{p}}}\frac{\exp(-r^2/4\chi t)}{(\pi\chi t)^{3/2}}, \tag{1.6}$$

where E_0 is the fluence of the incident laser, σ is the optical absorption cross-section, χ is the thermal diffusivity. Substitution of Eq. (1.6) into Eq. (1.1.2), we can obtain the solution of Eq. (1.5) for linear thermal expansion as

$$\Psi_1(r,t) = -\frac{E_0\sigma\beta_1 c}{16\rho c_{\mathrm{p}}(\pi\chi)^{3/2}}\int_0^t \frac{ds}{s^{3/2}}\int_{r-c(t-s)}^{r+c(t-s)} u\exp(-u^2/4\chi s)du. \tag{1.7}$$

By assuming that the term $(c^2t/\chi)^{1/2}$ is large enough, Eq. (1.7) can be analytically solved. Thus, we obtain the analytical solution for the displacement potential $\Phi(r, t)$ in the case of linear thermal expansion as

$$\Phi_1(r,t) = \frac{E_0\sigma\beta_1}{4\pi\rho c_{\mathrm{p}} r}\left(1-\exp\left(\hat{t}^-\right)\right)\left(1-\upsilon\left(\hat{t}^-\right)\right). \tag{1.8}$$

Here, υ is the Heaviside function, $\hat{t}^-$ is the dimensionless retarded time from the origin with $\hat{t}^- = \frac{c^2}{\chi}(t - \frac{r}{c})$ and $\hat{t}^+ = \frac{c^2}{\chi}(t + \frac{r}{c})$.

Thus, the PA pressure according to Eq. (1.1.3) in the case of linear thermal expansion can be obtained by combining Eq. (1.8) as

$$p_1 = \frac{E_0 \sigma \beta_1 c^4}{4\pi c_p \chi^2 r}\left(\exp(\hat{t}^-)\left(1 - \upsilon\left(\hat{t}^-\right)\right) - \delta\left(\hat{t}^-\right)\right). \tag{1.9}$$

The PA pressure is plotted in Fig. 1.3. As shown, it is a compressive, rising, exponential wave followed by a delta function rarefaction.

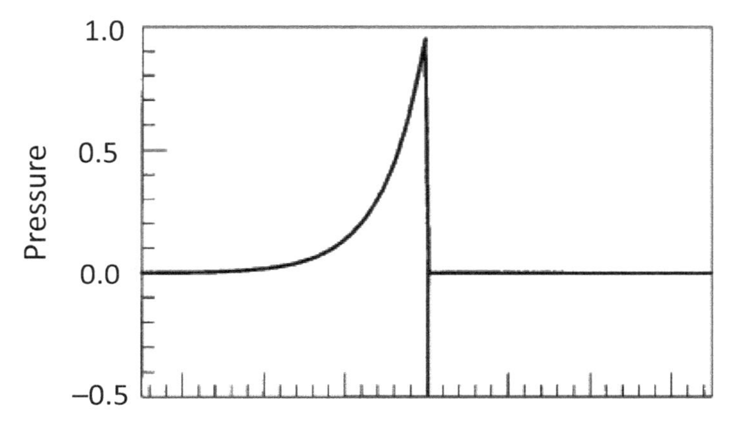

Figure 1.3 Photoacoustic pressure in arbitrary units versus retarded time from the origin from the linear temperature contribution. Reproduced with permission from Ref. [13].

Equation (1.9) gives the PA pressure response of a source point under delta laser irradiation for linear thermal expansion mechanism. The rise time of the waveform predicted by Eq. (1.9) for common fluids or tissues are so short, e.g., 60 fs for H_2O, that recording of the waveform would present serious experimental difficulties. In practical applications, laser pulses with much longer width are commonly used. Thus, it is more useful to determine the limiting form of the wave when the laser pulse is long compared with χ/c^2. The PA response of long laser pulse can be obtained by convolution of the pressure response from a delta function heating pulse from Eq. (1.9) with the intensity profile of the exciting optical beam. Assume that the laser pulse intensity is in the form of $I(t)=\frac{E_0}{\theta}f(t/\theta)$, the PA response can be obtained for a long laser pulse irradiation as

$$p=\frac{E_0\sigma\beta_1}{4\pi\theta^2 c_p r}\frac{d}{d\hat{\tau}}f(\hat{\tau}). \tag{1.10}$$

Here, $\hat{\tau}=(t-r/c)/\theta$. The corresponding PA response for a long Gaussian heating pulse irradiation is presented in Fig. 1.4.

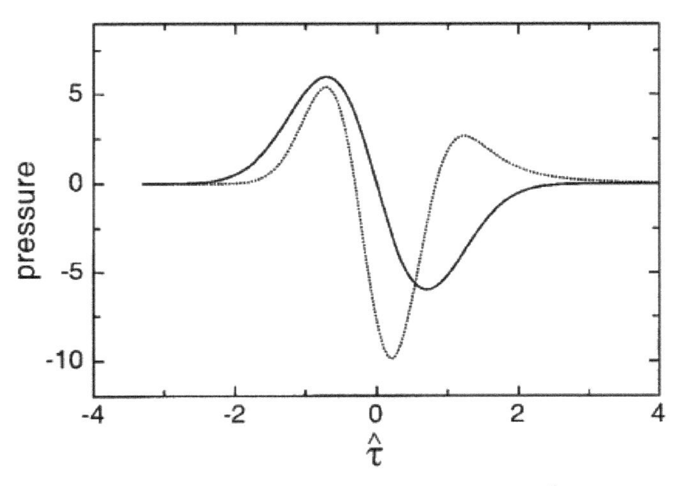

Figure 1.4 Photoacoustic pressure versus retarded time $\hat{t}$ from the origin for long, Gaussian heating pulses from (–) the linear source, and (···) the nonlinear source. The latter is found by numerical differentiation of the displacement potential. The pressure minimum is found numerically to be at $\hat{t}$ = 0.291. Reproduced with permission from Ref. [13].

The discussion above is based on the linear thermal expansion by assuming that the thermal expansion coefficient is constant during the PA signal production. However, in most cases, the thermal expansion coefficient is a function of temperature. Thus, the discussion of the nonlinear response of thermal expansion and PA signal production is necessary, especially for those applications with relatively high energy input per volume. Consider the solution to the wave equation with the nonlinear source term for long laser pulses as $\beta = \beta_1 + \beta_1 T(r, t)$. The corresponding PA pressure for nonlinear thermal expansion for short laser pulse irradiation can be solved as

$$P_1 = \frac{3E_0^2\sigma^2\beta_2 c^7}{64c_p^2\pi^3\rho\chi^5 r}\left((1/|\hat{t}^-|^4) - \frac{2}{3}(\delta(\hat{t}^-)/|\hat{t}^-|^3)\right). \quad (1.11)$$

The corresponding PA pressure is plotted in Fig. 1.3, which is a sharply rising wave with a discontinuity at zero point.

For a long Gaussian pulse where $f(t/\theta) = \pi^{-1/2}\exp(-(t/\theta)^2)$, the temperature function for nonlinear thermal expansion can be given as

$$T(r,t) = \frac{E_0\sigma}{8\pi^2\rho c_p\theta\chi^{3/2}}\int_0^\infty\left(\frac{\exp(-r^2/2\chi\zeta)}{\zeta^{3/2}}\exp(-(t-\zeta/\theta)^2)\right)d\zeta \quad (1.12)$$

Substituting Eq. (1.12) into Eq. (1.1.2) and then combining the equation of PA pressure with displacement potential, the PA pressure for the nonlinear thermal expansion for long pulse can be found as

$$p = \left(\frac{\beta_2 E_0^2\sigma^2}{128\pi^3 c_p^2\chi^{3/2}\theta^{7/2}\rho r}\right)\frac{\partial^2}{\partial\hat{\tau}^2}\exp\left(-2\hat{\tau}^2\right)$$
$$\times\int_0^\infty\left(\frac{erf(\zeta/\sqrt{2})}{\zeta^{3/2}}\right)\exp\left(-\zeta^2/2 - 2\zeta\hat{\tau}\right)d\zeta. \quad (1.13)$$

As shown in Fig. 1.5, the nonlinear long pulse PA response of a point source is an asymmetric tripolar wave. The waveforms here given by G. J. Diebold necessarily reflect the characteristics of the heat diffusion equation. Most notably, a rapid temperature rise that extends throughout space immediately after the laser pulse begins, which manifests itself in the acoustic wave as a failure to obey causality. As shown by Eq. (1.13), the PA pressure generated a long light pulse for the linear problem and is proportional to the first time derivative of the optical intensity.

Actually, in the thermoelastic regime, the PA pulse waveform is determined by both the optical absorption and pulse duration. We have already discussed a delta function optical heating pulse. Now we discuss the pulse duration on the PA signal production. In the case of linear regime, the equation of thermoelastic sound excitation can be given as [17]

$$\frac{\partial^2\phi}{\partial t^2} - c^2\Delta\phi = \left(\frac{\beta c^2}{\rho c_p}\right)\text{div}S,$$

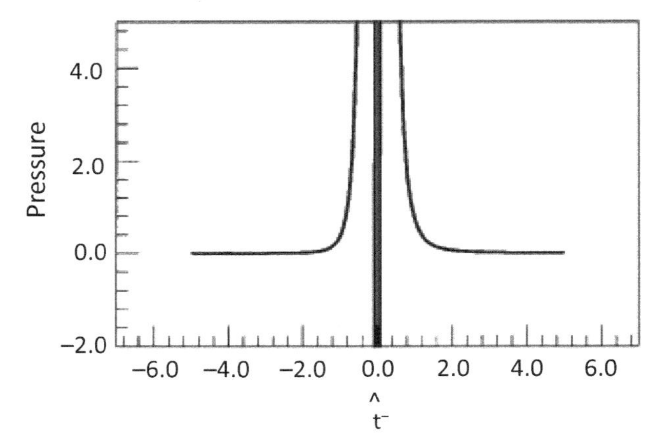

Figure 1.5 PA pressure in arbitrary units versus retarded time from the origin from the nonlinear temperature contribution, to the wave equation for a delta function heating pulse. Reproduced with permission from Ref. [13].

where $\phi(t, z)$ is the scalar potential of velocity field and S is the Poynting vector of the incident wave. For simplicity, one-dimensional problem is considered. Assuming that a homogeneous thin layer is illuminated by a relatively long optical pulse ($\mu_a c\tau_m >> 1$) with a pulse duration of τ_m, the Poynting vector of the medium can be determined as $S = I_0 \exp(-\mu_a z) f(t) n_z$, where I_0 is the optical intensity at the surface of the sample, μ_a is the absorption coefficient, z is the depth, $f(t)$ is the time function of the incident laser, n_z is the unit vector on the z axis. Then the PA signal can be described by

$$\frac{\partial^2\phi}{\partial t^2} - c^2\frac{\partial^2\phi}{\partial t^2} = -\left(\frac{\mu_a c^2\beta}{\rho c_p}\right)I_0 f(t)\exp(-\mu_a z).$$

By Fourier transform on each side of the equation, the equation can be solved using spectral method. Consequently, the Gaussian laser pulse excited PA signal profiles then can be described by

$$P(\tau,t) = -\frac{\mu_a I_0 \beta c^2}{2c_p} \int_{-\infty}^{\infty} f(t)\exp(-\mu_a c|\tau - t|)\text{sgn}(\tau - t)dt.$$

Figure 1.6 provides temporal profiles of PA/TA signals excited by Gaussian microwave pulse with various pulse durations and identical per pulse energy. It is shown that the amplitude of PA signals increases with a decreasing pulse duration. Therefore, by reducing the incident laser or microwave pulse width, improved PA/TA conversion efficiency can be obtained.

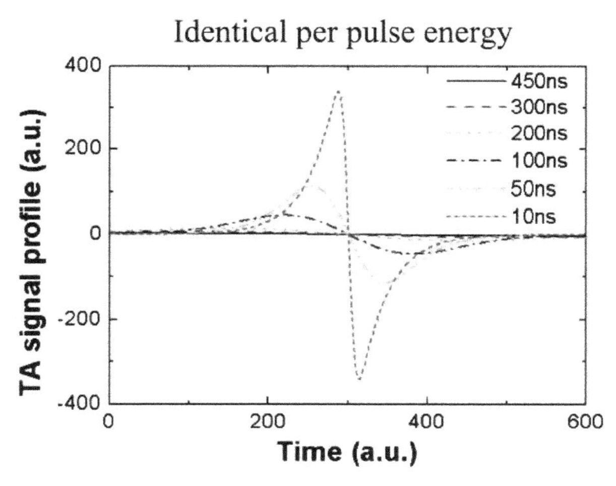

Figure 1.6 Simulated PA/TA waveform with various pulse duration and identical energy per pulse. Reproduced with permission from Ref. [17].

Benefit from its non-ionized and non-invasive characteristics, thermal expansion mechanism is widely applied in PA applications. However, thermal expansion is one of the least efficient mechanisms of light-sound energy conversion, which produces PA waves with relatively low amplitude.

1.2.2 Photochemical Process

PA waves can also be produced through the photochemical process. Huxiong Chen [18] reported that an anomalous PA effect is produced when a suspension of carbon particles in water is irradiated by a high-power, pulsed laser. The PA amplitude is on the order of 2000 times that produced by a dye solution with an equivalent absorption coefficient, where the effect is thought to originate in high-temperature chemical reactions between the surface carbon and the surrounding water, which is considered to be

$$\mathrm{C + H_2O} \text{ ® } \mathrm{CO + H_2}$$

Different from thermal expansion mechanism, a model to describe the PA effect that includes chemical reaction must include the consumption or liberation of thermal energy and volume changes produced on the conversion of reactants to products. As the carbon particles are irradiated by high-power pulsed laser, they absorb energy and cause the surrounding water to vaporization. In the case of the steam-carbon reaction, the volume change is substantial because of the great disparity between the molar volume of gases and liquid water. Consideration of the effects of energy consumption and volume change in the linearized hydrodynamic equations for a fluid gives the acoustic density δ and temperature T as solutions to the coupled equations [19]

$$\left(\nabla^2 - \frac{\gamma}{c^2}\frac{\partial^2}{\partial t^2}\right)\delta = -\rho\beta\nabla^2 T - \rho\beta_c\nabla^2 n, \tag{1.14}$$

and

$$\lambda\nabla^2 T - \rho C_V\frac{\partial T}{\partial t} = -H + \rho\mu^+\frac{\partial n}{\partial t} - \frac{C_V(\gamma-1)}{\beta}\frac{\partial\delta}{\partial t}, \tag{1.15}$$

where γ is the heat capacity ratio, c is the speed of sound, t is time, ρ is the ambient density, β is the thermal expansion coefficient, β_c is a "chemical" expansion coefficient, and n is the differential concentration of new chemical species formed. As shown in Eq. (1.14), two source terms on its right side indicate that changes in either the temperature (the first term) or the chemical composition (the second term) act identically to launch acoustic waves. λ is the thermal conductivity, C_V is the heat capacity at constant volume, H is the "heating function" (the energy per volume and time delivered to the fluid by the light beam), and μ^+

is a chemical potential. The chemical expansion coefficient is a measure of the change in fluid density at constant pressure and temperature on conversion of reactants to products, defined by

$$\beta_c = -\frac{1}{\rho}\left(\frac{\partial \rho}{\partial N}\right)_{P,T}, \tag{1.16}$$

where N is the concentration of chemical species reacted. The chemical potential is the amount of internal energy per mass stored per concentration of chemical species reacted at constant density and temperature, which is defined as

$$\mu^+ = \left(\frac{\partial \varepsilon}{\partial N}\right)_{P,T}. \tag{1.17}$$

Here, ε is the internal energy per mass of the fluid. The term in Eq. (1.16) contains the chemical potential acts as an energy source.

The model given here for chemical generation of the PA signal should be rigorously correct for differential changes in the state variables. However, the application of chemical mechanism is rather complicated, because high-temperature reactions and gas expansion are involved. Except for chemical mechanism, other mechanisms such as temperature-dependent thermal expansion, nonlinear heat conduction, vaporization of fluid around the perimeter of the particles, or even shock wave formation are possibly involved in the generation of the PA production process. Compared to thermal expansion mechanism, the photochemical process could produce much higher PA signal. However, the needed laser energy for chemical mechanism is much higher than that of thermal expansion, and may induce PA shock waves. Thus, it is not a biologically safe mechanism.

1.2.3 Cavitation, Gas Evolution, and Boiling

PA signal can also be produced by the high-power laser irradiation on a small region of tissues or liquids, where volume change undergone by transforming solution phase reactants to products gives a PA effect. In many PA applications, the interaction of laser radiation and tissue is initiated by the heating of small absorbing regions that are surrounded by less absorbing tissue [20]. Such absorbing sites are for example blood capillaries, tattoo pigments, or melanosomes in the retinal pigmented epithelium (RPE) of the eye. The strong laser energy absorption of these small absorbing sites may lead to the overheating of the sites to the boiling temperature. In this case, transit phase transition occurs which induce huge PA wave emission. Cavitation, gas evolution, or boiling-based PA effect have found important application as a means of high-sensitivity imaging, thermotherapy, and determination of thermodynamic reaction parameters. PA damage is caused by tensile stress waves with an amplitude that exceeds the tensile strength of the material in which they propagate.

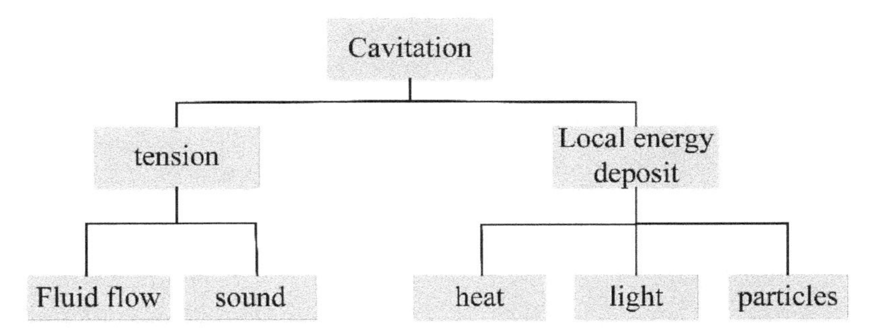

Figure 1.7 Systemic diagram of cavitation induced.

To describe the process of laser-induced cavitation and the PA signal production from cavitation, we give a basic theoretical model that describes the production of PA signal. As the absorber is irradiated by laser, it absorbs laser energy and converts light into heat. The temperature of the absorber can be described by Eq. (1.1). For high-dose laser irradiation, the temperature of the absorber rises quickly, which may exceed its supercritical temperature. Thus, cavitation occurs and bubble forms in the absorber, where the initial pressure in the newly formed vapor bubble can be predicted by Clausius–Clapeyron equation [21]:

$$p_{\text{sat}}(T) = p_0 \exp\left(\frac{H_{\text{vap}}}{R}\left(\frac{1}{T_0} - \frac{1}{T}\right)\right). \tag{1.18}$$

where p_0 = 101 KPa and T_0 = 273.15 K are the ambient pressure and temperature in our simulation, respectively. H_{vap} is the latent heat of vaporization, and R is the gas constant. The calculated initial pressure in water is about 100 atm. The formed bubble is in the non-equilibrium state with relatively high pressure and temperature. As the expansion goes on, the bubble dissipates energy by means of doing work to the external water accompanied by PA wave emission. The bubble dynamic can be described by the Rayleigh–Plesset equation [22]:

$$\rho\left(R\frac{d^2R}{dt^2}+\frac{3}{2}\left(\frac{dR}{dt}\right)^2\right)=P_{\text{bubble}}-P_0-\frac{2\sigma}{R}-\frac{4\mu}{R}\frac{dR}{dt}, \tag{1.19}$$

where ρ is liquid density, R is bubble radius, μ is viscosity, and P_{bubble} is the pressure in the bubble. By assuming the bubble to be ideal gas, the bubble pressure is estimated according to the equation-of-state equation [23]:

$$P_{\text{bubble}}=(\gamma-1)\rho_{\text{bubble}}C_{\text{bubble},V}T_{\text{bubble}}. \tag{1.20}$$

Here, $\gamma = C_{\text{bubble},p}/C_{\text{bubble},V}$, with $C_{\text{bubble},p}$ and $C_{\text{bubble},V}$ being specific heat capacity at constant pressure and constant volume, respectively. In the surrounding liquid, the motion of the fluid can be described by [24]

$$\rho\left(\frac{\partial v}{\partial t}+v\cdot\nabla v\right)=-\nabla P_{\text{liquid}}+\mu\nabla^2 v. \tag{1.21}$$

Through bubble expansion and collapse, the absorbed energy is converted into PA waves, which is described by the relationship between the PA pressure and the bubble-induced displacement [25]:

$$\nabla P_{\text{PA}}=-\rho\frac{\partial^2 u}{\partial t^2}, \tag{1.22}$$

where P_{PA} and u are the PA wave pressure and the bubble wall displacement. Equation (1.22) can be transformed into $P_{\text{PA}} = \rho c_{\text{L}}\partial u/\partial t$, with c_{L} to be the liquid sound velocity.

Compared with thermoelastic regime, the cavitation-induced PA effect possesses much higher signal amplitudes and conversion efficiency from absorbed photons to acoustic waves, thus overcoming the fundamental limits of thermoelastic regime. Cavitation/phase transition-induced PA effect has been widely applied in exogenous contrast agents medicated PA imaging monitored PA therapy.

1.2.4 Optical Breakdown or Plasma Formation

When extremely high laser radiation in focused in the interior of tissues, optical breakdown accompanied by PA shock waves may occur [26, 27]. Microexplosions take place in the laser focal region and cavities filled with a luminous plasma are formed. Laser radiation is absorbed in a dense plasma and this deposits additional energy in the cavities. The high pressure expands the cavities, which generates a shock wave. At the end of a laser pulse and energy evolution, the gas in the plasma cavities cools, the emission of radiation from the plasma stops, and a bubble which undergoes several pulsations is formed. An important feature of the optical breakdown is its threshold nature. The threshold intensity of laser depends on the properties of the tissue. It has been established that the threshold intensity is determined by the presence of solid microparticles, particularly for particles of soot always present in the atmosphere and therefore present in liquids (water) in contact with the atmosphere. Such solid particles absorb light and are heated to temperatures of the order of 10^4 K, which corresponds to the first ionization of the atoms and formation of a dense plasma. Light is absorbed strongly in the plasma, which causes further heating and creates a plasma cavity.

Even though the PA shock wave with high signal amplitudes can be obtained by optical breakdown or plasma formation, we should notice that this regime is not suitable for noninvasive biomedical PA applications, because the incident laser energy may exceed the safety threshold of tissues. The optical breakdown regime may be useful in PA therapy applications.

1.3 Basic Theory for Nanoprobe-Based Photoacoustics

As a fast-growing imaging technology, PA imaging offers extraordinary opportunities for detecting and monitoring disease pathophysiology, which is owing to its capacity for high-resolution sensing of rich optical contrast at depths beyond the optical transport mean-free-paths. With the aid of exogenous contrast agents, PA imaging has been extended to functional and molecular imaging for early detecting serious diseases [28–30]. Considering the basic thermal expansion mechanism of the PA effect, the signal amplitude of a nanoprobe is proportional to both its optical absorption and its PA conversion efficiency, which represents the efficiency of the conversion of absorbed optical energy to acoustic waves. Nanoprobes have a unique micro-mechanism of PA energy conversion due to the size effect. Here, we give basic theories for nanoprobe-based photoacoustics [31].

When a nanoprobe is irradiated by pulsed laser, its temperature increases. The temperature field can be described by the heat diffusion equation. The thermal confinement time of the nanoprobe can be expressed as [32, 33]

$$\tau_{\mathrm{th}} = \frac{D^2}{4\alpha_{\mathrm{th}}}, \tag{1.23}$$

where D is the diameter of the nanoprobe and α_{th} is the thermal diffusivity of the tissue surrounding the nanoprobe. Owing to the small size of the nanoprobes, the thermal confinement condition is no longer applicable, when the laser pulse width is about 10 ns. Therefore, heat cannot be completely confined inside the nanoprobe, and quick heat diffusion occurs from the sphere to the surrounding medium as soon as the laser pulse is applied.

To demonstrate this idea, quantitative simulation on the temperature field of a water-immersed gold nanosphere based on the finite element analysis (FEA) method is presented in Fig. 1.8. The heating pulse fluence is 6 mW/μm^2. The heating function is

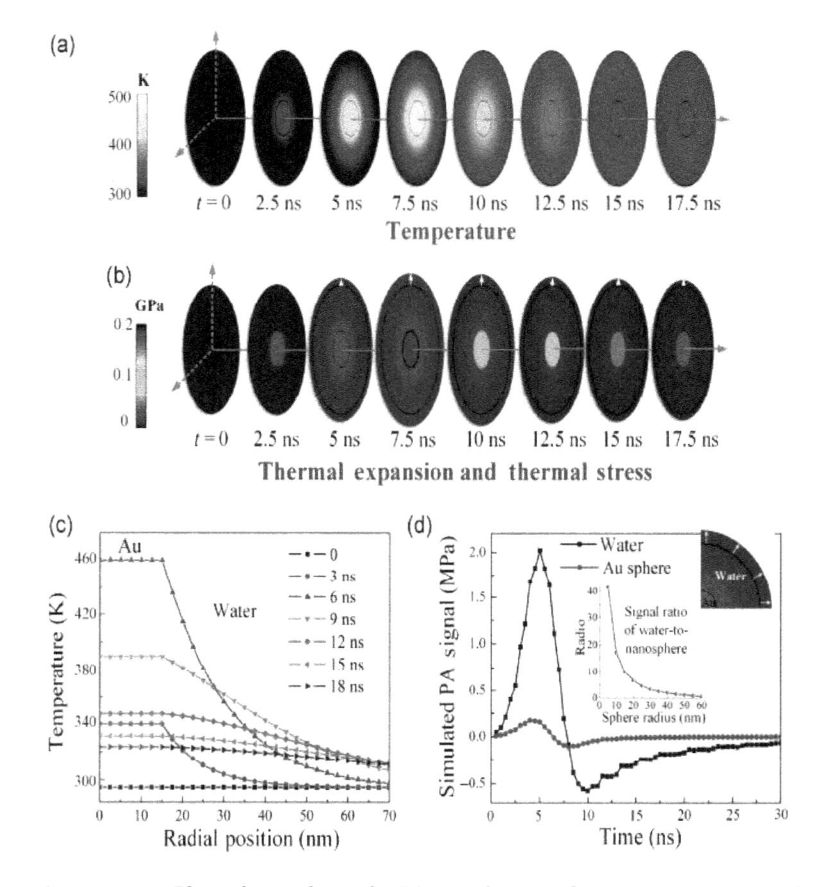

Figure 1.8 Photothermal and PA analysis of a water-immersed nanosphere (15 nm). (a) The temperature field in the sphere and in its surrounding water as a function of time. The laser is a temporal Gauss function with a peak value at 7.5 ns. (b) Thermal expansion–induced stress in the sphere and its surrounding water. (c) Quantitative temperature profiles at different times along the radial position. (d) The resulting PA signal from the sphere and its surrounding heated water. Reproduced with permission from Ref. [31].

$$f(t)=\frac{1}{\sqrt{2\pi}\xi}\exp\left(-\frac{(t-\tau/2)^2}{2\xi^2}\right), \tag{1.24}$$

where, ξ = 10 ns is the laser pulse width. The length of the heated liquid layer (thermal diffusion length) is a function of the thermal diffusivity of the liquid and the time, defined as $\chi = \sqrt{\alpha t}$ $(t \leq \tau)$ [32]. As shown in Fig. 1.8b, although both the sphere and the surrounding water are heated, the thermal expansion in the water is much more obvious than that of the sphere. This is mainly because of the high contrast between the thermoelastic properties of the water and the sphere. Simulation results for the PA signal from the sphere and water are presented in Fig. 1.8d, where the signal amplitude from water is several times that from the sphere. The produced PA signal that originates from the expansion-induced total displacement can be calculated by Eq. (1.22). By requiring temperature and displacement continuity at the sphere–water interface, the PA pressure obtained from the sphere can be obtained as [31]

$$P = -Q_{\mathrm{abs}} \frac{\rho_{\mathrm{H_2O}}}{4\pi r} \frac{\beta_{\mathrm{Au}} R^3/3 + \beta_{\mathrm{H_2O}} \chi_{\mathrm{H_2O}} R(R + \chi_{\mathrm{H_2O}})}{\rho_{\mathrm{Au}} C_{\mathrm{pAu}} R^3/3 + \rho_{\mathrm{H_2O}} C_{\mathrm{pH_2O}} R \chi_{\mathrm{H_2O}} (R + \chi_{\mathrm{H_2O}})}, \tag{1.25}$$

where, Q_{abs} is the absorbed power, ρ is density, r is radial coordinate, β is thermal expansion, and R is radius of the sphere. Equation (1.25) also shows that the produced PA signal is a sum contribution of the thermal expansion from the surrounding sphere and the surrounding heated medium.

In order to find the dominating parameters for nanoprobe-based PA conversion efficiency, two limiting cases of Eq. (1.25) are examined. For the case that the thermal diffusion length in the surrounding liquid is much smaller than the sphere radius, which means that the absorbed energy is mostly deposited in the sphere, Eq. (1.25) reduced to $P = -Q_{\mathrm{abs}} \frac{\rho_{\mathrm{H_2O}}}{4\pi r} (\beta/\rho C_{\mathrm{p}})_{\mathrm{Au}}$. Otherwise, $P = -Q_{\mathrm{abs}} \frac{\rho_{\mathrm{H_2O}}}{4\pi r} (\beta/\rho C_{\mathrm{p}})_{\mathrm{H_2O}}$ is obtained. Thus, the parameter $\Gamma = \beta/\rho C_{\mathrm{p}}$ known as the Grüneisen parameter then could be used to characterize the thermal-acoustic conversion capability of the materials [34].

However, the PA conversion problem is quite complex for liquid-immersed nanoprobes, because the total PA conversion efficiency is determined by both the thermal-acoustic conversion capability and the thermal energy distribution in the heated areas. Here, we define the produced PA signal pressure as a sum of conversion efficiency of the heated area:

$$P_{\text{total}} = \sum_i \Gamma_i W_i. \tag{1.26}$$

Here, W_i is the thermal energy deposited in domain i with Grüneisen parameter Γ_i. Much sharper PA signal can be obtained when more thermal energy is deposited in the thermal-acoustic efficient areas.

The thermal diffusivity of the liquid is one of the crucial factors that influence the thermal energy distribution between the absorbing nanoparticle and surrounding heated liquid. For liquids with large thermal diffusivity, energy can diffuse out of the nanoparticle quickly. Therefore, less energy is maintained in the nanoprobes. As shown in Fig. 1.9a, we simulated a 15 nm water-immersed sphere with different thermal diffusivities. With the thermal diffusivity increases, the thermal diffusion length becomes larger, leading to a sharp decrease of thermal energy in the sphere and an increase of thermal energy in the surrounding liquid. At the same time, with the expansion of the heated areas for larger thermal diffusivity, their average temperature decreases, as shown in the right temperature map for τ = 5 ns. Moreover, another factor affecting the thermal distribution is the size of the nanoparticles. Because larger nanoparticles can maintain more energy in their interiors, less energy diffuses into the surrounding liquid. In Fig. 1.9b, we simulated the relative thermal energy in water as a function of laser pulse width for spheres with different radii. When the laser pulse width increases, the thermal energy diffused out of the sphere increases for all spheres, while the sphere with smallest radius shows a much sharper increase. As shown in the temperature maps for τ = 5 ns, spheres with different radii have the same thermal diffusion length in the liquid (water).

Efficient probes/contrast agents are highly desirable for good-performance photoacoustic (PA) imaging, where the PA signal amplitude of a probe is dominated by both its optical absorption and the conversion efficiency from absorbed laser energy to acoustic waves. Nanoprobes have a unique micro-mechanism of PA energy conversion due to the size effect. Our work paves the way for the rational design and optimization of nanoprobes with improved conversion efficiency.

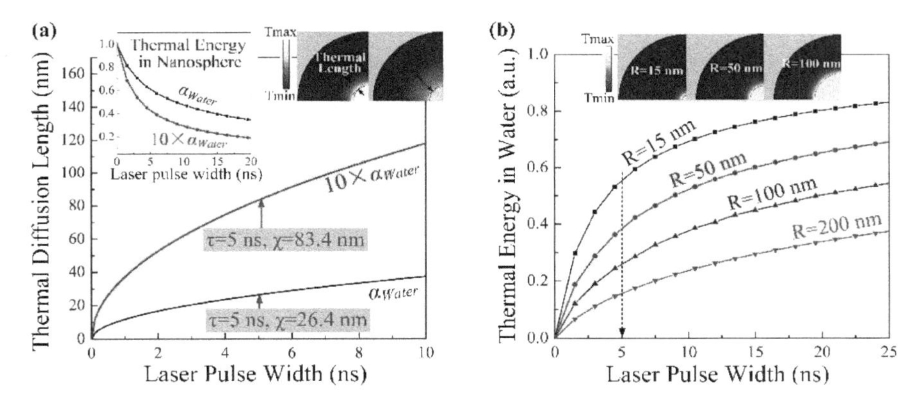

Figure 1.9 (a) Thermal diffusion length in water as a function of laser pulse width for different thermal diffusivities. (b) Thermal energy deposited in water as a function of laser pulse width for nanospheres with different radii. Reproduced with permission from Ref. [31].

1.4 Brief Introduction to Photoacoustic Techniques

PA imaging is a new biomedical imaging modality based on the use of laser-generated ultrasound that has emerged over the past decade. It is a hybrid modality, combining the high-contrast and spectroscopic-based specificity of optical imaging with the high spatial resolution of ultrasound imaging. PA imaging has been widely used in industrial and biomedical field, especially for biological applications of disease detection [11, 35]. The biomedical applications of PA technique in biology and medicine include imaging of angiogenesis [36], tumor microenvironments [37], microcirculation, gene activities [38], brain functions [39], and biomarkers and drug response [40]. Initial clinical applications include melanoma cancer imaging [41], gastrointestinal tract endoscopy, intravascular catheter imaging [42], neonatal brain imaging, breast cancer detection [43], prostate cancer detection [44], guided sentinel lymph node needle biopsy for cancer staging, early chemotherapeutic response imaging, dosimetry in thermal therapy [45], in vivo label-free histology, blood perfusion imaging, blood oxygenation imaging [46], and tissue metabolism imaging.

In this section, we provide a brief introduction to related PA techniques, including PA spectroscopy, PA microscopy, PA tomography, PA endoscopy, PA Doppler flowmetry, and PA thermometry.

1.4.1 PA Spectroscopy

A major advantage of PA technique is that it is intrinsically suited for spectroscopy detection [35, 46], where the imaging contrast can be selectively enhanced for specific components by tuning the excitation laser wavelength to the absorption features of their constituent chromophores. By obtaining images at multiple wavelengths with the usage of the spectroscopic analysis, the concentration of a specific chromophore is then quantified, which is similar to conventional optical spectroscopy. This technique is actually an "excitation spectrum" based on acoustic detection. A PA spectroscopy system is composed of three main parts: a laser irradiation source, an experimental chamber, and the data acquisition system. PA spectroscopy has found to be a very useful technique for research and analysis, not only in physics and chemistry, but also in biology and medicine.

PA spectroscopy is capable of providing functional information of tissues [46]. Here, we take the absorption of blood as example. The absorption spectrum of blood at visible and NIR wavelengths is strongly dependent on its oxygen saturation (sO_2), which is a consequence of the significant spectral differences between oxyhemoglobin (HbO_2) and de-oxyhemoglobin (Hb). Benefiting from the spectral differences, we are able to quantify the concentration of HbO_2 and Hb, as well as to estimate sO_2. These parameters are critical physiological parameters that intimately related to pathophysiological processes such as angiogenesis and tissue inflammatory processes. Assume that the blood can be considered as a mixture of Hb and HbO_2; thus the optical absorption coefficient of blood $\mu(\lambda)$ can be written as

$$\mu(\lambda) = A(\varepsilon_{\mathrm{Hb}}(\lambda)[\mathrm{Hb}] + \varepsilon_{\mathrm{HbO_2}}(\lambda)[\mathrm{HbO_2}]), \tag{1.27}$$

where λ is the wavelength, $\varepsilon_{\mathrm{Hb}}$ and $\varepsilon_{\mathrm{HbO2}}$ are the Hb and HbO_2 molar extinction coefficients ($cm^{-1}M^{-1}$), respectively, where [Hb] and $[HbO_2]$ are molar concentrations of two kinds of hemoglobin ($mol\ L^{-1}$), respectively. A is constant, which is usually set as 2.303. In physiology, the sO_2 is defined as

$$sO_2 = \frac{[HbO_2]}{[Hb] + [HbO_2]}. \quad (1.28)$$

sO_2 describes the relative concentration of the two kinds of hemoglobin. Thus, the sO_2 can be obtained.

1.4.2 PA Microscopy, Tomography, and Endoscopy

PA microscopy (PAM) refers to techniques in which a PA image is obtained by mechanically scanning either a focused laser beam or a focused ultrasound detector [11, 47]. The image is then reconstructed directly from the set of acquired A-line signals, without the aid of a reconstruction algorithm. If a focused laser beam is used in PAM, it is considered to be OR-PAM, where the spatial resolution in at least one plane (usually, the lateral) is defined by the spatial characteristics of a focused laser beam propagating in tissue. Otherwise, if a focused ultrasound detector is used, it is considered to be acoustic resolution PA microscopy (AR-PAM), where axial and lateral spatial resolution is defined by the physics of ultrasound propagation and detection. Different from optical microscopy, PAM has the ability of deep tissue imaging. For example, AR-PAM can be used to image to depths of several centimeters.

PA tomography (PAT) is usually considered as the traditional mode of PA imaging [11]. It is also the most general PA imaging technique with the fewest limitations on imaging performance imposed by its practical implementations. In PAT applications, full-field illumination is used, where a large diameter pulsed laser beam is irradiated on the surface of tissues. As laser penetrates deeply into tissues, the strong optical scattering of tissues results in a relatively large tissue volume becoming "bathed" in diffuse light. Then, tissue chromophores absorb incident laser radiation, which leads to impulsive heating of the irradiated tissue followed by the rapid generation of PA waves. The PA waves propagate to the tissue surface where they are detected by array of receivers or a mechanically scanned ultrasound receiver. The time-varying detected PA signals can then, with knowledge of the sound speed, be spatially resolved and back-projected to reconstruct a three-dimensional PA image.

PA endoscopy (PAE) has been intensively investigated in recent years as a means of imaging internal organs such as cardiovascular, esophagus and colon [48, 49]. A representative PAE design is shown in Fig. 1 of ref. [11]. A laser irradiates light which then is delivered by a multimode optical fiber placed in the central hole of a ring transducer. An optically and acoustically reflective mirror, driven by a micromotor through coupled magnets, rotates both the optical illumination and the acoustic detection for circumferential cross-sectional scanning. A linear motor pulls back the entire probe for volumetric imaging. In contrast to conventional optical microscopy/endoscopy, which has an imaging depth within the optical diffusion limit, PAE has shown a several-millimeter imaging depth in the dorsal region of a rat model. There are several potential clinical applications in which the target tissue can only be accessed by introducing a miniature endoscopic probe percutaneously or through a natural orifice. Among these are the clinical assessment of coronary artery disease, prostate cancer and gastrointestinal pathologies. A number of prototype PA endoscopic or intravascular devices, conceptually similar to conventional US probes, have been developed for these applications.

1.4.3 PA Doppler Flowmetry

The PA technique is also capable of providing blood flow velocity using Doppler flowmetry method, which is proposed by Hui Fang et al. [50] PA Doppler flowmetry (PADF) can be used to study the blood flow, where the flow velocity is potential signs for pathological changes. Compared to conventional Doppler flowmetry, PADF has much lower background noise, where it can potentially quantify tissue blood flow noninvasively because red blood cells are dominant endogenous light-absorbing tracer particles and can absorb light 100 times more than the background at certain wavelengths.

Assume that the particle is suspended in a liquid with velocity V. As irradiated by an amplitude-modulated continuous-wave laser with frequency f_0 and 100% modulation, continuous PA wave is produced. The PA signal intensity as a function of time t can be written as

$$I = \frac{I_0}{2}(1 + \cos(2\pi f_0 t)), \tag{1.29}$$

where, I_0 is the peak intensity. If the particle is not in motion, the frequency of the PA wave is f_0. Otherwise, it will subject to a Doppler shift. As shown in Fig. 1.10, the detected PA signal frequency is related to the velocity of the particle and the direction angles. Assume that the particle velocity is much smaller than the sound speed, the Doppler shift can be written as $-f_0 \frac{V}{c_0}\cos\alpha + f_0 \frac{V}{c_A}\cos\theta$, where c_0 and c_A are the speeds of light and sound, respectively. As c_0 is much larger than c_A, only the second term is detectable. Therefore, the Doppler shift can be written as

$$f_{\text{PADF}} = f_0 \frac{V}{c_A}\cos\theta \tag{1.30}$$

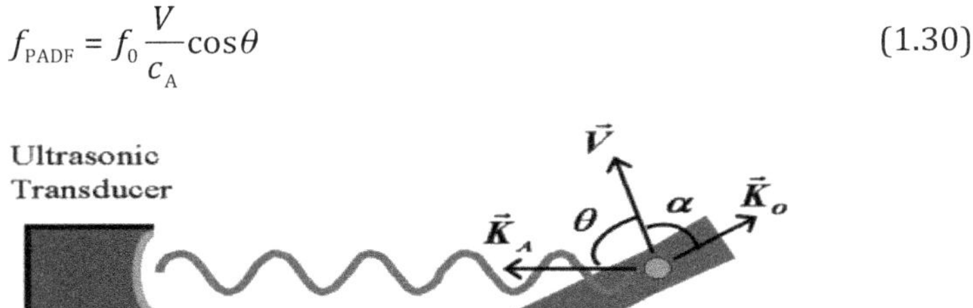
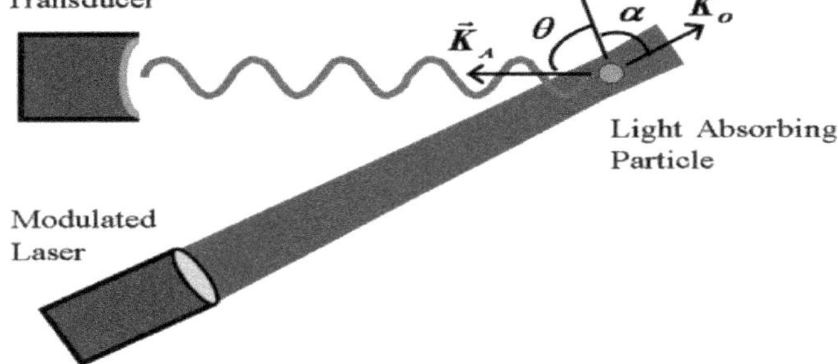

Figure 1.10 Schematic for photoacoustic Doppler shift. The small light-absorbing particle moving along velocity vector is illuminated by modulated continuous-wave light.

By detecting the Doppler shift of the PA signal, the velocity of the moving particles then can be obtained. This shift equals half of the shift in pulse-echo Doppler ultrasound and does not depend on the direction of the laser illumination. The PADF method is proved to capable of measuring low-speed blood flow and useful for measuring blood flow in microcirculation, which has average velocities from a fraction of mm/s in capillaries to tens of mm/s in small veins and arterials.

1.4.4 PA Thermometry

The temperature dependence of PA signal amplitude provides a means of non-invasively obtaining maps of temperature distributions in tissues. The PA thermometry (PATM) [51, 52] is applicable in temperature monitoring during photothermal cancer therapy.

When tissue is irradiated with a short laser pulse, the PA pressure can be estimated to be

$$P = \Gamma(T) \cdot \mu_{\alpha} \cdot F, \qquad \Gamma(T) = \frac{\beta \cdot c^2}{C_{\mathrm{P}}}. \tag{1.31}$$

Here, Γ is the Grüneisen parameter that is dependent on the temperature, μ_{α} is the absorption coefficient, F is the laser fluence, β is the thermal expansion coefficient, c is the speed of sound, and C_{p} is the heat capacity at constant pressure. For biological tissues, it can be assumed that the Grüneisen parameter is a linear function of temperature, which can be described as

$$\Gamma(T) = \Gamma_0 + \zeta \Delta T, \tag{1.32}$$

where ζ is constant for specific tissues, and ΔT is the temperature increment. Thus, the PA signal amplitude is a linear function of the temperature. Note that here we refer to the base temperature of the object, rather than the change in temperature due to the laser heating. The instantaneous temperature increase in the object due to the laser pulse heating is on the order of milliKelvin and its effect on the Grüneisen parameter is negligible. The base temperature of the object is a slowly varying parameter compared to the transient temperature increase induced by a laser pulse.

References

1. Bell, A. G. (1880). On the production and reproduction of sound by light. *Am. J. Sci.*, **118**, pp. 305–324.
2. Beard, P. (2011). Biomedical photoacoustic imaging. *Interface Focus*, **4**, p. 602.
3. Tam, A. C. (1986). Applications of photoacoustic sensing techniques. *Rev. Mod. Phys.*, **2**, p. 381.
4. Kruger, R. A., Liu, P., Fang, Y., Appledorn, C. R. (1995). Photoacoustic ultrasound (PAUS)—reconstruction tomography. *Med. Phys.*, **10**, pp. 1605–1609.
5. Hoelen, C. G. A., De Mul, F. F. M., Pongers, R., Dekker, A. (1998). Three-dimensional photoacoustic imaging of blood vessels in tissue. *Opt. Lett.*, **8**, pp. 648–650.
6. Ku, G., and Wang, L. V. (2000). Scanning thermoacoustic tomography in biological tissue. *Med. Phys.*, **5**, pp. 1195–1202.
7. Kolkman, R. G., Hondebrink, E., Steenbergen, W., de Mul, F. F. (2003). In vivo photoacoustic imaging of blood vessels using an extreme-narrow aperture sensor. *IEEE J. Sel. Top. Quant.*, **2**, pp. 343–346.
8. Zharov, V. P., Galanzha, E. I., Shashkov, E. V., Khlebtsov, N. G., Tuchin, V. V. (2006). In vivo photoacoustic flow cytometry for monitoring of circulating single cancer cells and contrast agents. *Opt. Lett.*, **24**, pp. 3623–3625.
9. Kim, C., Erpelding, T. N., Jankovic, L., Pashley, M. D., Wang, L. V. (2010). Deeply penetrating in vivo photoacoustic imaging using a clinical ultrasound array system. *Biomed. Opt. Express*, **1**, pp. 278–284.
10. Zhang, H. F., Maslov, K., Stoica, G., Wang, L. V. (2006). Functional photoacoustic microscopy for high-resolution and noninvasive in vivo imaging. *Nat. Biotechnol.*, **7**, pp. 848–851.
11. Wang, L. V., Hu, S. (2012). Photoacoustic tomography: In vivo imaging from organelles to organs. *Science*, **6075**, pp. 1458–1462.
12. Hutchins, D., Tam, A. C. (1986). Pulsed photoacoustic materials characterization. *IEEE Trans. Ultrason. Ferroelectr. Freq. Control*, **5**, pp. 429–449.
13. Calasso, I. G., Craig, W., Diebold, G. J. (2001). Photoacoustic point source. *Phys. Rev. Lett.*, **16**, p. 3550.
14. Diebold, G. J., Sun, T., and Khan, M. I. (1991). Photoacoustic monopole radiation in one, two, and three dimensions. *Phys. Rev. Lett.*, **24**, p. 3384.

15. Morse, P. M., Ingard, K. U., Stumpf, F. B. (1968). *Theoretical Acoustics.* (Princeton University Press, Princeton, NJ,).

16. Gusev, V. E., and Karabutov, A. A. (1991). Laser optoacoustics. NASA STI/Recon Technical Report A, 93.

17. Lou, C., Yang, S., Ji, Z., Chen, Q., Xing, D. (2012). Ultrashort microwave-induced thermoacoustic imaging: A breakthrough in excitation efficiency and spatial resolution. *Phys. Rev. Lett.*, **21**, p. 218101.

18. Chen, H., and Diebold, G. (1995). Chemical generation of acoustic waves: a giant photoacoustic effect. *Science*, **5238**, p. 963.

19. Skelton, E. A., and James, J. H. (1997). *Theoretical Acoustics of Underwater Structures.* World Scientific.

20. Paltauf, G., and Schmidt-Kloiber, H. (1999). Photoacoustic cavitation in spherical and cylindrical absorbers. *Appl. Phys. A Mater. Sci. Process.*, **5**, pp. 525–531.

21. Pan, H., Ritter, J. A., Balbuena, P. B. (1998). Examination of the approximations used in determining the isosteric heat of adsorption from the Clausius–Clapeyron equation. *Langmuir*, **21**, pp. 6323–6327.

22. Prosperetti, A. (1982). A generalization of the Rayleigh–Plesset equation of bubble dynamics. *Phys. Fluids*, **3**, pp. 409–410.

23. Furlani, E. P., Karampelas, I. H., and Xie, Q. (2012). Analysis of pulsed laser plasmon-assisted photothermal heating and bubble generation at the nanoscale. *Lab Chip*, **19**, pp. 3707–3719.

24. Chorin, A. J. (1968). Numerical solution of the Navier-Stokes equations. *Math. Comput.*, **104**, pp. 745–762.

25. Pelivanov, I. M., Kopylova, D. S., Podymova, N. B., and Karabutov, A. A. (2009). Optoacoustic method for determination of submicron metal coating properties: Theoretical consideration. *J. Appl. Phys.*, **1**, p. 013507.

26. Vogel, A., Busch, S., and Parlitz, U. (1996). Shock wave emission and cavitation bubble generation by picosecond and nanosecond optical breakdown in water. *J. Acoust. Soc. Am.*, **1**, pp. 148–165.

27. Lyamshev, L. M. (1981). Optoacoustic sources of sound. *Phys. Usp.*, **12**, pp. 977–995.

28. Roy, I., Shetty, D., Hota, R., Baek, K., Kim, J., Kim, C., Kim, K. (2015). A multifunctional subphthalocyanine nanosphere for targeting, labeling, and killing of antibiotic-resistant bacteria. *Angew. Chem.*, **50**, pp. 15367–15370.

29. Nie, L., Wang, S., Wang, X., Rong, P., Ma, Y., Liu, G., Chen, X. (2014). In vivo volumetric photoacoustic molecular angiography and therapeutic monitoring with targeted plasmonic nanostars. *Small*, **8**, pp. 1585–1593.

30. Li, P. C., Wang, C. R. C., Shieh, D. B., Wei, C. W., Liao, C. K., Poe, C., and Wu, Y. N. (2008). In vivo photoacoustic molecular imaging with simultaneous multiple selective targeting using antibody-conjugated gold nanorods. *Opt. Express*, **23**, pp. 18605–18615.

31. Shi, Y., Qin, H., Yang, S., and Xing, D. (2016). Thermally confined shell coating amplifies the photoacoustic conversion efficiency of nanoprobes. *Nano Res.*, **12**, pp. 3644–3655.

32. Rosencwaig, A., and Gersho, A. (1976). Theory of the photoacoustic effect with solids. *J. Appl. Phys.*, **1**, pp. 64–69.

33. Hinton, F. L. (1991). Thermal confinement bifurcation and the L- to H-mode transition in tokamaks. *Phys. Fluids B Plasma Phys.*, **3**, pp. 696–704.

34. Brugger, K. (1965). Generalized Grüneisen parameters in the anisotropic Debye model. *Phys. Rev.*, **6A**, p. A1826.

35. Rosencwaig, A. (1973). Photoacoustic spectroscopy of solids. *Opt. Commun.*, **4**, pp. 305–308.

36. Siphanto, R. I., Thumma, K. K., Kolkman, R. G. M., Van Leeuwen, T. G., De Mul, F. F. M., Van Neck, J. W., Steenbergen, W. (2005). Serial noninvasive photoacoustic imaging of neovascularization in tumor angiogenesis. *Opt. Express*, **1**, pp. 89–95.

37. Laufer, J., Johnson, P., Zhang, E., Treeby, B., Cox, B., Pedley, B., and Beard, P. (2012). In vivo preclinical photoacoustic imaging of tumor vasculature development and therapy. *J. Biomed. Opt.*, **5**, pp. 0560161–0560168.

38. Li, L., Zemp, R. J., Lungu, G., Stoica, G., and Wang, L. V. (2007). Photoacoustic imaging of lacZ gene expression in vivo. *J. Biomed. Opt.*, **2**, pp. 020504–020504.

39. Wang, X., Pang, Y., Ku, G., Xie, X., Stoica, G., and Wang, L. V. (2003). Noninvasive laser-induced photoacoustic tomography for structural and functional in vivo imaging of the brain. *Nat. Biotechnol.*, **7**, pp. 803–806.

40. Razansky, D., Buehler, A., and Ntziachristos, V. (2011). Volumetric real-time multispectral optoacoustic tomography of biomarkers. *Nat. Protoc.*, **8**, pp. 1121–1129.

41. Oh, J. T., Li, M. L., Zhang, H. F., Maslov, K., Stoica, G., and Wang, L. V. (2006). Three-dimensional imaging of skin melanoma in vivo by dual-wavelength photoacoustic microscopy. *J. Biomed. Opt.*, **3**, pp. 034032–034032.

42. Sethuraman, S., Amirian, J. H., Litovsky, S. H., Smalling, R. W., and Emelianov, S. Y. (2008). Spectroscopic intravascular photoacoustic imaging to differentiate atherosclerotic plaques. *Opt. Express*, **5**, pp. 3362–3367.

43. Pramanik, M., Ku, G., Li, C., and Wang, L. V. (2008). Design and evaluation of a novel breast cancer detection system combining both thermoacoustic (TA) and photoacoustic (PA) tomography. *Med. Phys.*, **6**, pp. 2218–2223.

44. Agarwal, A., Huang, S. W., O'donnell, M., Day, K. C., Day, M., Kotov, N., and Ashkenazi, S. (2007). Targeted gold nanorod contrast agent for prostate cancer detection by photoacoustic imaging. *J. Appl. Phys.*, **6**, p. 064701.

45. Bailey, R. T., Bernegger, S., Bicanic, D., Bijnen, F., Blom, C. W., Cruickshank, F. R., and Jalink, H. (2012). *Photoacoustic, Photothermal and Photochemical Processes in Gases,* vol. 46. Springer Science & Business Media.

46. Laufer, J., Delpy, D., Elwell, C., and Beard, P. (2006). Quantitative spatially resolved measurement of tissue chromophore concentrations using photoacoustic spectroscopy: Application to the measurement of blood oxygenation and haemoglobin concentration. *Phys. Med. Biol.*, **1**, p. 141.

47. Wang, L. V. (2009). Multiscale photoacoustic microscopy and computed tomography. *Nat. Photonics*, **3**(9), pp. 503–509.

48. Yang, J. M., Maslov, K., Yang, H. C., Zhou, Q., Shung, K. K., and Wang, L. V. (2009). Photoacoustic endoscopy. *Opt. Lett.*, **34**(10), pp. 1591–1593.

49. Yang, J. M., Favazza, C., Chen, R., Yao, J., Cai, X., Maslov, K., and Wang, L. V. (2012). Simultaneous functional photoacoustic and ultrasonic endoscopy of internal organs in vivo. *Nat. Med.*, **18**(8), pp. 1297–1302.

50. Fang, H., Maslov, K., and Wang, L. V. (2007). Photoacoustic Doppler effect from flowing small light-absorbing particles. *Phys. Rev. Lett.*, 99(18), p. 184501.

51. Pramanik, M., and Wang, L. V. (2009). Thermoacoustic and photoacoustic sensing of temperature. *J. Biomed. Opt.*, **14**(5), pp. 054024–054024.

52. Shah, J., Park, S., Aglyamov, S., Larson, T., Ma, L., Sokolov, K., and Emelianov, S. Y. (2008). Photoacoustic imaging and temperature measurement for photothermal cancer therapy. *J. Biomed. Opt.*, **13**(3), pp. 034024–034024.

Chapter 2

Fast Photoacoustic Imaging Technology

2.1 Multi-Element Array Photoacoustic Detection Technology

2.1.1 The Concept of Single-Element Photoacoustic Detection

Thermoacoustic (TA) signals can be detected using related technologies and equipment. Ultrasonic detectors are divided into single and multiple detectors by the number of elements. A single-element detector is designed to perform signal acquisition by scanning multiple locations (typically dozens to hundreds of locations). Although the images reconstructed can well reflect the electromagnetic wave absorption distribution of biological tissues, the experimental apparatus is complicated and the data acquisition requires a long period of scanning time (about 30 min). The detected TA signals are one-dimensional signal, to get a two-dimensional image by scanning linearly. The specific embodiments are:

As shown in Fig. 2.1, a two-dimensional array $z(i, j)$ is created. The discrete signal $u_0(y)$ detected by the detector is given

Biomedical Photoacoustics
Sihua Yang and Da Xing

ISBN 978-981-4774-58-1 (Hardcover), 978-0-203-70365-6 (eBook)
www.jennystanford.com

to $z(0, j)$ in the initial position of the probe, then using a mechanical or electronic means to move the position of detector, and the detected signal $u_1(y)$ is given to $z(1, j)$. After moving $i - 1$ positions, a complete two-dimensional array $z(i, j)$ is acquired to build a two-dimensional image. The structure of this scanning method, which can basically reflect the shape of the object, is simple and has a fast speed of acquisition and image processing.

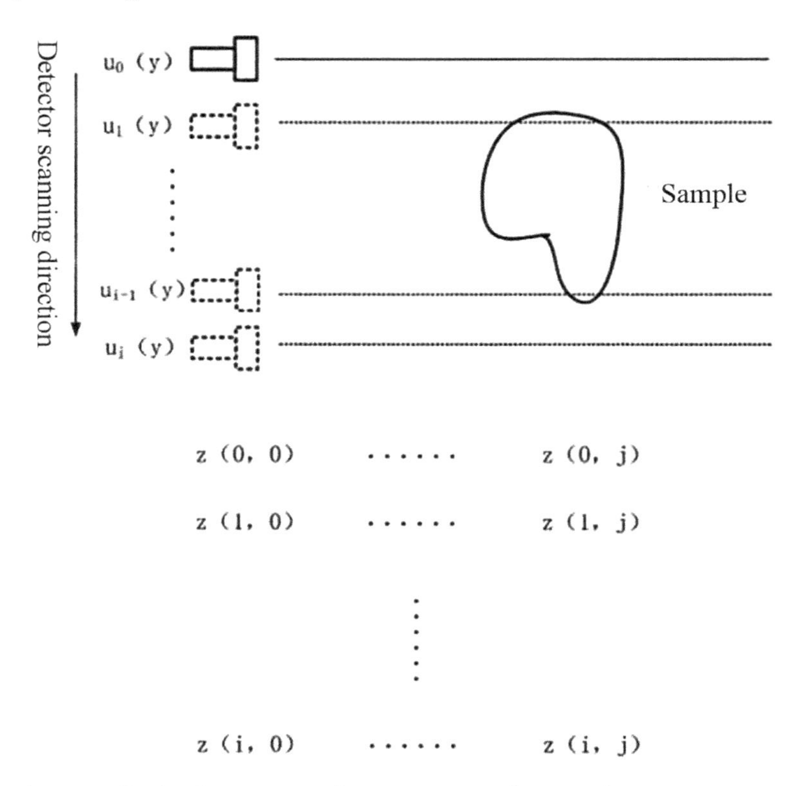

Figure 2.1 Single-element transducer scanning schematic diagram.

The linear-scanning modes are usually used in traditional TA imaging systems, which requires a single-element transducer rotating around the sample in circular fashion to acquire the complete information about the target. Therefore, the data acquisition requires a long scanning time, and acquiring the information for imaging needs dozens or hundreds of spatial positions. In order to reduce the imaging time, multi-element transducers have been introduced for acquiring more information at the same time [1].

2.1.2 The Concept of Multi-Element Photoacoustic Detection

A multi-element linear transducer array system (MLTAS) is developed for 2D tissue photoacoustic imaging (PAI). Due to the angular response property of the transducer array, a phase-control technique to detect time-domain signals is applied. We converted the signals into a one-dimensional (1-D) image along the acoustic axis. Multiple 1-D images acquired are combined with subgroups of transducers from the linear array, and we achieved a 2-D cross-sectional tomographic image of the sample. Compared to a single transducer (Hoelen et al., 1998, Hoelen and Mul, 2000, Wang et al., 2003, Kruger et al., 1995, Ku and Wang, 2001), MLTAS eliminated the mechanical shift or rotation of the detector, thus, reduced the time required for data acquisition. Taking the above advantages of MLTAS into consideration, it may provide a more convenient technique for future in vivo noninvasive imaging of tissues and clinic diagnosis.

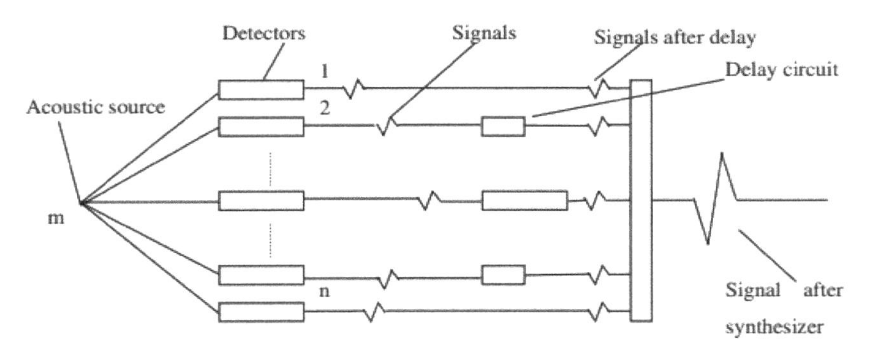

Figure 2.2 Basic principle of the phase-controlled focus algorithm: *n* detectors accept the acoustic pressure waves at the same time and the PA signals are synthesized, after a predetermined delay, for a coherent output [2].

As shown in Fig. 2.2, with a PA signal S_m originated from position m, the corresponding signal $S_n(t)$, for each detector, must be weighted for both the detector direction and the signal distance. The sum of $S_n(t)$ for all the detectors, S_m, is considered to be a PA signal from the corresponding reconstructed region m:

$$S_m = \sum_{n=1}^{N} \lambda_n S_n(t_{m1} + \tau_n), \tag{2.1}$$

where m is the position of a PA source, n is the position of a detector, N is the number of detectors and λ_n is an amplitude-weighting factor for each detector. $t_{m1} = r_{m1}/v$, t_{m1} is the time required for the PA signal to travel from the position m to the first detector; $\tau_n = (r_{mn} - r_{m1})/v$, is the time difference between the signals arriving at the detector n and the first detector, r_{mn} is the distance from the position m to the detector n, r_{m1} is the distance from the position m to the first detector, v is the average velocity of the acoustic wave in the coupling medium. From Eq. (2.1), the signal intensity is directly proportional to the number of active transducers. With τ_n calculated and the corresponding delay set in circuit for every signal, we can maximize the amplitude of the signal after the synthesizer, thus, a phase-controlled focusing is achieved. While an acoustic source of the target is outside the focus area, the signals gathered by the detectors are incoherent. Due to phase cancellation, the signal amplitude after the synthesizer is reduced, which is compared to that from the focus area. Hence, the signals collected by the transducer array are mainly from the acoustic source within the focus area [2].

Multi-element PAI can eliminate the mechanical shift or rotation of the transducer. Given the advantages of these, the time and laser energy required for data acquisition can be effectively reduced, and the reliability and convenience of PAI can be improved. However, the lateral resolution obtained by the system is not very well. There is a receiving solid angle along the acoustic axis of the array due to the directivity of the phase-controlled transducer array, within which PA signals can be gathered [3].

2.2 Multi-Element Array Photoacoustic Detection Algorithm

2.2.1 Filter Back-Projection Algorithm (Circular Scanning)

A phase-controlled technique is used to detect time-trace signals, and the phase-controlled algorithm can be used to reconstruct the image, but the system lateral resolution is not very good. The limited-field-filtered back-projection algorithm can be used to reconstruct the optical absorption distribution. The experimental results show that limited-field filtered back-projection algorithm within a certain range can greatly improve the lateral resolution of MLTAS.

The relationship between photoacoustic (PA) pressure and the distribution of optical absorption can be expressed as

$$P(r,t) \approx -\frac{\beta I_0 \tau c_0}{4\pi c_\mathrm{p}} \cdot \frac{d}{dt} \oiint_{|r'-r_0|=c_0 t} A(r') \frac{dr'}{c_0 t}, \tag{2.2}$$

where $A(r')$ is the fractional energy-absorption per-unit volume of soft tissue; β is the isobaric volume expansion coefficient; C_0 is the sound velocity in the target; C_p is the specific heat; $p(r, t)$ is the PA pressure; τ is the pulse width of pulse laser. Thus, the PA pressure recorded at position r and time t is the sum of radius $|r' - r_0|$ of the induction of all the pressure wave groups on the surface of the sphere.

Because subgroup of linear sensor array has a special impulse response, $p(r, t)$ in the detector position cannot be recorded directly. In practice, the measured signals $p(r, t)$ are convolution of the PA pressure and the impulse response function $h(t)$, so we can compute $p(r, t)$ as

$$P'(r,t) = P(r,t) \times h(t). \tag{2.3}$$

Then $p(r, t)$ can be obtained by measuring PA signal of impulse response.

$$P(r,t) \approx IFFT\left[\frac{j\omega p'(\omega)}{I(\omega)}\right]\left[1+\cos\left(\frac{\pi\omega}{\omega_c}\right)\right], \tag{2.4}$$

where $p'(\omega)$ and $I(\omega)$ are the Fourier transform of the PA signal $p'(r, t)$ and $h(t)$, respectively, the impulse response of the detector, $\omega_c\sqrt{2}$ is the cut-off frequency, and the apodizing function $1 + \cos(\pi\omega/\omega_c)$ is the window function used to band-limit the signals to ω_c.

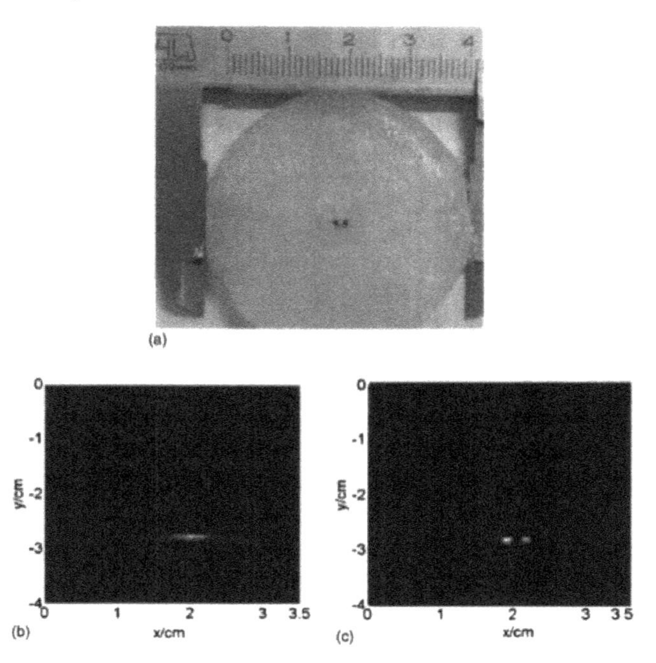

Figure 2.3 Reconstructed image of the two-graphite-rod phantom. (a) The picture of phantom. (b) The reconstructed image with the phase-controlled algorithm. (c) The reconstructed image with the limited-field filtered back-projection algorithm. Reproduced with permission from Ref. [6].

Figure 2.3a [6] shows the physical dimensions of the test sample. Figure 2.3c shows the reconstructed two-dimensional image with the limited-field-filtered back-projection algorithm, and Fig. 2.3b shows the reconstructed image with phase-controlled algorithm, respectively. Reconstruction results clearly show two pictures of good contrast. Reason is the use of phase-controlled technique to PA signal collection, signal strength of each location is all from different position sensor signal, signal combined with collecting from N different position from N sample average sampling signal has the same effect; thus, the signal-to-noise ratio is high. Figure 2.3c shows that the relative locations and sizes of the two-PA sources are resolved and matched to the original ones. But in Fig. 2.3b, two sources cannot clearly distinguish, with the imaging objects as a single source. It shows that the lateral resolution of the reconstruction image in Fig. 2.3c is better than Fig. 2.3b.

Based on filtered back-projection process of PAI, optical absorption images of two-dimensional distribution in the tissue is reconstructed. Here, we briefly introduce the PA theory. In a relatively short laser pulse radiation stress constraint, due to the energy deposition, the absorption of tissue temperature rises fast, leading to the expansion of the region. The laser pulse is short enough to ignore thermal diffusion, so it can produce acoustic waves according to the thermal elastic expansion effect. The acoustic waves $p_r(t)$, which reach a detector at position r and time t, can be express as

$$P_r(t) \approx \frac{\beta I_0 v_s}{4\pi C} \tau \frac{d}{dt} \oiint_{|r-r'|=v_s t} A(r') \frac{dr'}{v_s t}, \tag{2.5}$$

where β is the volume expansion coefficient; C is the specific heat; v_S is the speed of sound in the medium, τ is the pulse width, $A(r')$ is the fractional spatial absorption at position r', and I_0 is the energy of the light source.

The aim of photoacoustic tomography (PAT) is to reconstruct the absorption distribution from a set of recorded PA signals. To this end, Eq. (2.5) can be rewritten into

$$F_r(t) \approx \frac{4\pi C}{\beta I_0 \tau} t \int_0^t p_r(t')dt' = \oiint_{|r-r'|/v_s=t} A(r')dr'. \tag{2.6}$$

Taking the second derivative of $F_r(t)$ with respect to distance $x(x \equiv |r - r'| = v_S t)$

$$\frac{\partial^2 F_r(t)}{\partial t^2} = \frac{4\pi C}{\beta I_0 \tau}\left[t \frac{\partial p_r(t)}{\partial t} + 2p_r(t)\right] \approx K\left[t \frac{\partial p_r(t)}{\partial t}\right], \tag{2.7}$$

where K is a constant, $k \equiv 4\pi C/\beta I_0\tau$. So, we can compute $t\frac{\partial p_r(t)}{\partial t}$ from the recorded signals and then through the back-project to rebuild the optical absorption distribution.

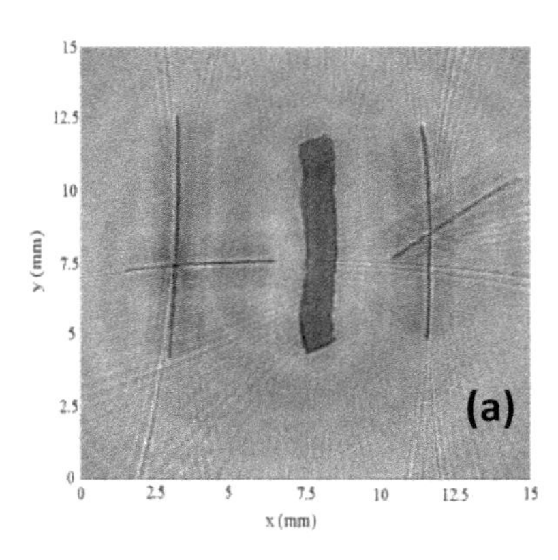

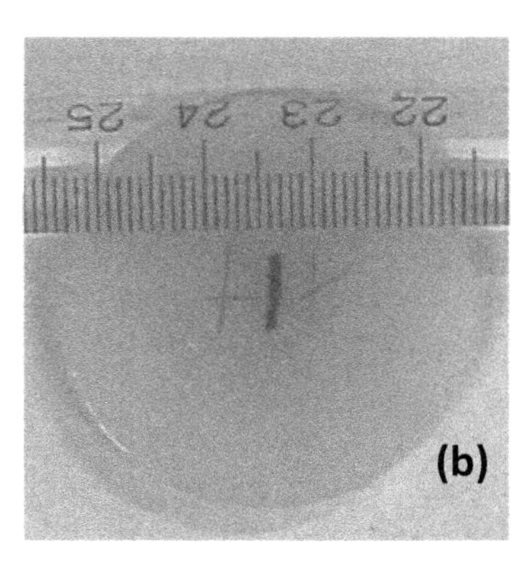

Figure 2.4 (a) PAT image of the phantoms; (b) photograph of the phantoms by removing the above 6 mm turbid media. Reproduced with permission from Ref. [7].

Experiments show that the reconstructed image and the original phantom in the PAT system is the same. The reconstructed image of the phantom is shown in Fig. 2.4a [7]. For comparison, Fig. 2.4b shows the phantom photo section is placed where the absorber. It can be found that the reconstructed image has a good correspondence with the original phantom. From the image result (Fig. 2.4a), there is a strong tail trace, especially the trace of the hair samples. This artifact is associated with the back-projection reconstruction method. In the weak absorption background can be seen that a strong linear absorption, tail tracking will strongly affect the image quality.

2.2.2 The Adaptive Projection Algorithm

As shown in Fig. 2.5 [4], object after microwave radiation, TA waves can be in the form of approximate spherical wave propagation. The TA waves arrive at element k at the time $t_k = R_k/c$, where R_k is the distance between element and the point object, and c is the acoustic velocity, which is assumed as constant. The back-projection formula can be written as

$$P(i,j) = \sum_N g_k(t)|R_k = \sum_N g_k\left(\sqrt{(i-x_k)^2+(j-x_k)^2}/c\right), \qquad (2.8)$$

where (i, j) is the 2D spatial coordinate of the point object, $g_k(t)$ is the time-domain signal received at element k, (x_k, y_k) is the spatial coordinate of the element k, and N is the total number of elements. The TA images can be found in suspicious areas of all discrete points for reconstruction. All in all, this algorithm is a transformation of the unfiltered (or simple) back-projection algorithm, and the difference is the feedback of coordinates of a group of elements when the transducer adapts the geometrical shape of the sample. This algorithm, although not theoretically true, is numerically equivalent to the delay-and-sum beam forming method used in ultrasound imaging systems. Therefore, for ultrasound-TA imaging system, it is very easy to implement.

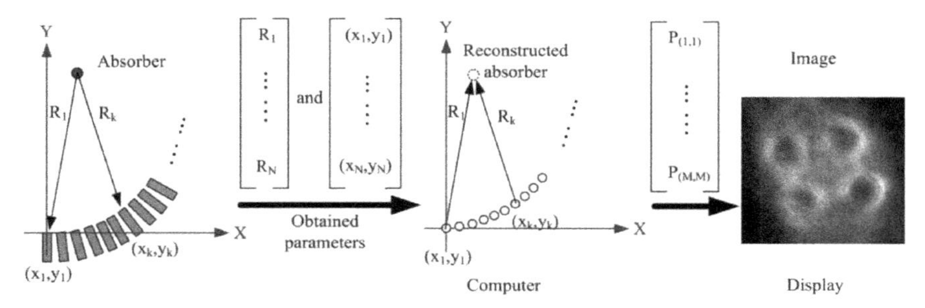

Figure 2.5 The process of adaptive back-projection algorithm. Reproduced with permission from Ref. [8].

2.3 Multi-Element Array Photoacoustic Imaging System

2.3.1 Multi-Element Date Collection System

In order to overcome the shortcomings of the unit detection system, the multi-element data acquisition system has become an inevitable choice. In fact, this is a strategy with space (device) for the time, a good solution to the problem. Our research group took the lead in 64-channel system in the international community. Furthermore, by establishing 384-multi-element data acquisition system and designing several kinds of multi-detector, the real-time PAI is realized.

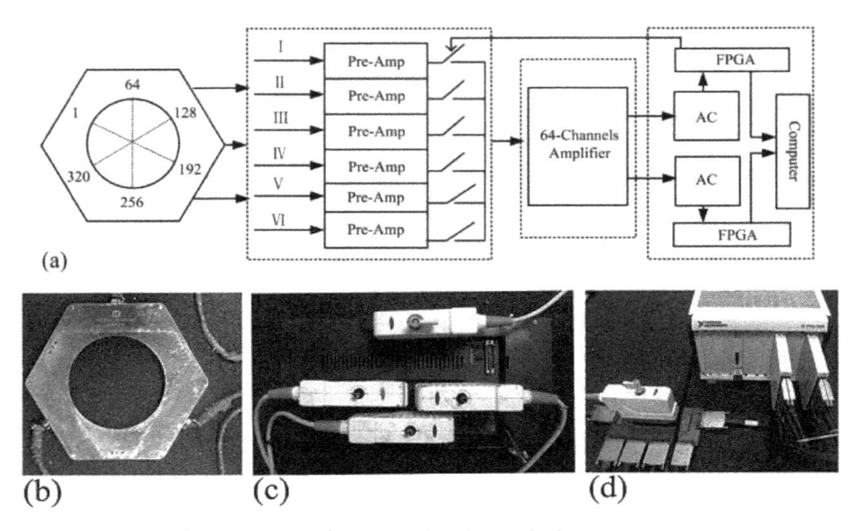

Figure 2.6 (a) Diagram of 384-multi-channel data acquisition system. Photographs of (b) the full ring transducer, (c) the 384–64-channel switch system, and (d) the 64-channel data acquisition system. Reproduced with permission from Ref. [9].

As shown in Fig. 2.6a, the diagram of 384-multi-channel data acquisition system, the front end is a 384-element detector (less than 384 elements can also be used), the full ring transducer (10C384-1.62 8-R100 AHA001, Doppler Ltd., China) consists of three parts, each containing 128 detectors. The central frequency of the transducer is 10 MHz and the bandwidth ranges from 64.5% to 92.4%. During the imaging process, the PA signal is detected simultaneously by 384 detectors as the laser pulse illuminates the object. The TA probe is converted to an electrical signal, the electrical signals enter into a high-speed digital switching system (actually a wavelength division multiplexer), which is grouped into six sets of 64-channel signals. Each set of 64-channel signals passes through the switching system in turn and is acquired by two 32-channel data acquisition boards (NI5752, NI Inc.USA) at a sampling rate of 50 MHz. After switching six times, all 384 signals are acquired. Finally, the 384-channel signals are used to reconstruct a 2-D image by a filtered back-projection algorithm. Figures 2.6b–d show the images of the full ring transducer, the 384–64-channel switch system, and the 64-channel data acquisition system, respectively [8, 9].

2.3.2 Multi-Element Linear Transducer

A linear array detector (Fig. 2.7) is commonly used for clinical ultrasound imaging; so it is referenced first by PAI technology. In conventional ultrasound imaging using linear array transducer, as illustrated in Fig. 2.8a, each transmitted ultrasound beam is focused at a specified depth and the received beam is dynamically focused at all ranges to form a high-quality ultrasound image. To interrogate the entire image, many ultrasound beams are required thus reducing the frame rate. However, if frame rate is critical, other ultrasound imaging techniques such as ultrafast ultrasound imaging [4, 5], parallel receive beamforming imaging [6], or explososcan [7] can be employed. Here the transmitted ultrasound beam is wider (slightly focused or even unfocused) and it surrounds the larger volume of tissue. The backscattered ultrasound signal, received on the desired elements of the array transducer at once, is then used to reconstruct the ultrasound image. For example, as shown in Fig. 2.8b, in ultrafast imaging plane wave is emitted by the whole aperture of the array transducer and every element of the array transducer aperture is used to receive the signals [8]. Most ultrasound imaging systems reconstruct the image via delay and sum beamforming [9]. The delay and summing beamforming takes into account the delay distance between the transducer elements and the reconstruction points in the image precisely. In order to improve the image quality, the directivity angle with dynamic receive focusing (fixed F-number) and apodization are considered [10, 11]. In general, delay-summation beamforming is the hardware implemented in an ultrasound system. It can be used to process signals and display images in real time.

In PAI, the tissue is irradiated with a short-pulse laser, and the light-absorbing body inside the tissue absorbs energy. The rapid thermoelastic expansion of the light absorber produces a PA pressure wave, which is detected by an ultrasonic transducer array. In contrast to conventional ultrasound imaging, however, in PAI, light is rapidly diffused throughout the tissue as the laser beam illuminates the tissue due to optical scattering. Therefore, laser beam instantly interrogates the entire volume of tissue and the PA response is simultaneously generated everywhere in tissue as depicted in Fig. 2.8c. Consequently, all transducer elements are used to receive these PA transients at once. Thus, in PAI there is no transmit focusing and only dynamic receive focusing can be used [3]. Apparently, PAI is similar to ultra-fast ultrasound imaging and therefore imaging techniques from ultra-fast ultrasound imaging may be employed. Generally, PA images can be reconstructed using simple back-projection [12], Fourier transform reconstruction [13], and delay-and-sum [14] image reconstruction techniques.

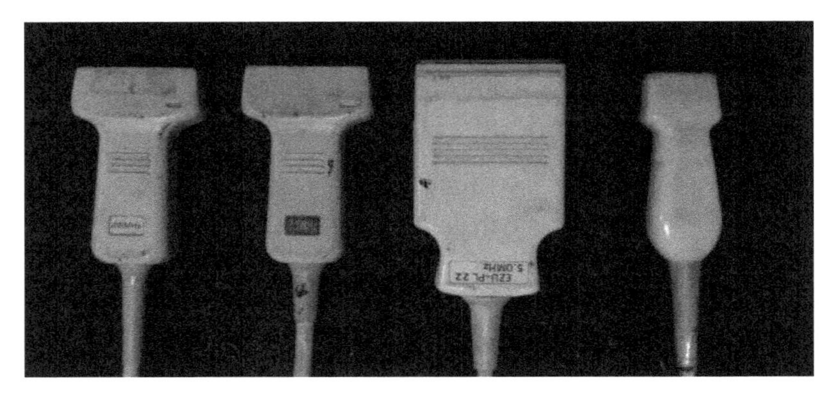

Figure 2.7 The photographs of multi-element linear array transducer.

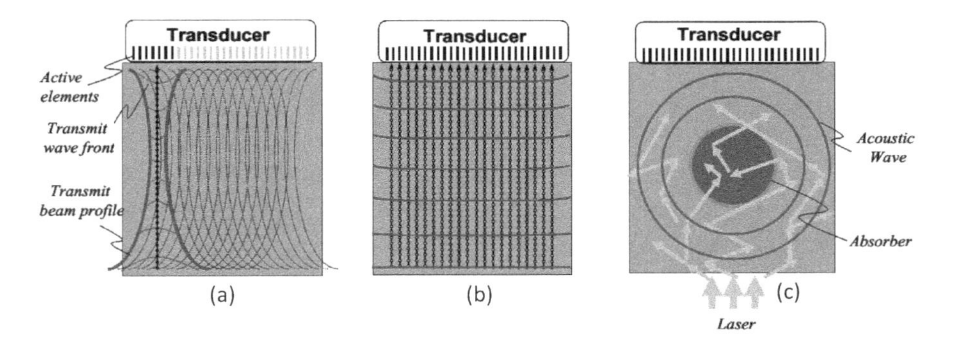

Figure 2.8 Schematic view of (a) conventional ultrasound, (b) ultrafast ultrasound, and (c) photoacoustic imaging modes. Reproduced with permission from Ref. [15].

In this chapter, we have developed a new PA tomographic system with a phase-controlled MLTAS. The schematic of the MLTAS is shown in Fig. 2.9. It consists of four parts: linear transducer array, pre-amplifier, phase adjustment module, and custom control circuit. The 30 Hz clock signal provided by the control circuit is used to excite the synchronous triggering of the laser, to select subgroups in the linear transducer array to capture the PA signal, and to trigger the data acquisition system (DAS) card. The linear transducer array (EZU-PL22, Hitachi, Japan) had 320 vertical transducers with a resonance frequency of 5 MHz and a scanning width of 56 mm. The transducer is divided into 80 blocks. Each subgroup can be selected by a control circuit and consists of four sensors. The signals from the transducers were imported into the DAS card (Compuscope12100, Gage Applied Co., Montreal, Quebec, Canada), after pre-amplification and phase adjustment. The card has a high-speed 12-bit analog-to-digital converter with a sampling rate of 100 MHz. A personal computer is used to control the operation system and data collection system.

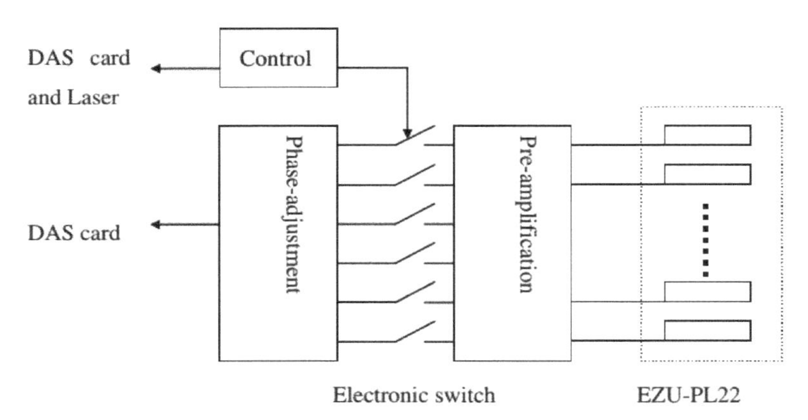

Figure 2.9 The schematic of the multi-element linear transducer array system (MLTAS).

The multi-element linear transducer array system provides an efficient and convenient approach for PAI. The technique can be used to distinguish optical absorption in a tissue phantom. It can provide a better way for noninvasive imaging and clinic diagnosis in vivo, compared to the existing PAI techniques [15].

2.3.3 Multi-Element Circular Transducer

Another type of detector that can be used for real-time monitoring is a circular array detector. Its basic idea is using space exchange for time, in order to increase the number of elements in exchange for the reduction of the scanning position. Figure 2.10 shows the photographs of the three ring-shaped array detector available in the laboratory. The main purpose of this section is to build data acquisition system based on multi-element circular transducer array.

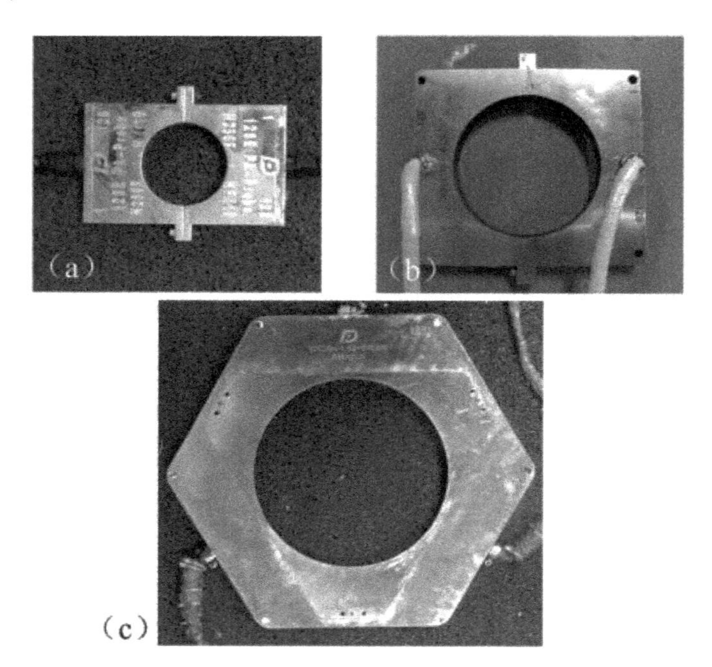

Figure 2.10 The photographs of multi-element circular array transducer.

The circular array detector shown in Fig. 2.10a has a dominant frequency of 7.5 MHz and a radius of 25 mm, the application is subject to great restrictions because of the small size. The circular array detector shown in Fig. 2.10b has a dominant frequency of 4 MHz and a radius of 50 mm. Due to process limitations, the array elements are not uniformly distributed at 360° and have no array elements at their junctions. The image reconstructed by this detector needs to be corrected.

The first two annular array detectors were defective in the test, and the circular array detector shown in Fig 2.10c was the most mature product today consisting of 384 elements, distributed along a 360°, with a radius of 100 mm. The center frequency of the detector is 10 MHz, and the relative bandwidth of each element is more than 60%. Each element in the array is 8 mm high and 1.62 mm wide, with an array spacing of 0.02 mm. In the imaging process, 384 elements are divided into 6 groups of 64 elements.

Figure 2.11 shows the schematic of the multi-element circular transducer array system. When the laser pulse arrives, only one set of signals enters the main amplifier, and the other group's signals remain off, controlled by a field-programmable gate array (FPGA, 7965R, NI). The main amplifier provides 30 dB of amplification, and the amplified signal is then sampled by two 32-channel data acquisition cards (NI5752, NI) at a sampling rate of 50 MHz.

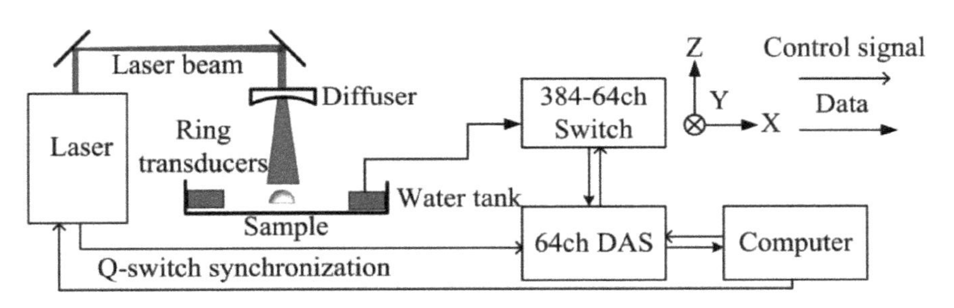

Figure 2.11 The schematic of the multi-element circular transducer array system.

The synchronization of the two acquisitions is controlled by two on-board FPGAs. After the arrival of the next laser pulse, the other groups of signals are sequentially amplified and collected, and then through the back-projection algorithm, the computer stored six sets of signals (384 channel signals) are used to reconstruct a two-dimensional image.

2.3.4 Multi-Element Flexible Transducer

In conventional PAI systems, the commonly used scanning methods include circular scanning (CS) or linear scanning (LS).

In the early decades of the PAI technique, the CS mode was used. To obtain complete information about the target, a single element transducer was required to rotate around the sample in a circular fashion. Because tens or even hundreds of spatial positions are required to obtain the information necessary for imaging. Therefore, the data acquisition requires a long scan time. In order to reduce the imaging time and to obtain more information simultaneously, linear, arc-shaped, and ring-shaped multi-element transducers have been introduced to PAI system. Similar to the B-scan in medical ultrasound imaging, the LS mode usually uses one-element transducers or linear transducers for one-dimensional scanning.

However, both CS and LS techniques have some disadvantages. As shown in Fig. 2.12a, the CS mode employs a fixed scanning radius to obtain information at different locations. However, for irregular samples, the scanning radius must be large enough to cover the longest axis of the imaging geometry. Therefore, a large amount of coupling liquid is required in order to fill in the gap between sample and transducer. Although this strategy is feasible in phantom study, it limits the practical application of PAI. On the other hand, medical ultrasound imaging for body scanning typically uses the LS mode, while it is not suitable for PAI of large objects. As shown in Fig. 2.12b, the LS technology enables seamless contact with the sample. However, due to the presence of idle elements, the utilization of the transducer elements is low. In addition, the relative position of the transducers and the PA waves in the transverse direction is difficult to detect, and therefore, some information required for imaging is lost. As a result, the reconstructed PA image is not completely correct, and is largely damaged due to distortion and other artifacts.

In this chapter, a sample-cling-scanning (SCS) model is proposed that combines the advantages of the CS and LS modes and provides a flexible multi-element transducer and adaptive back-projection algorithm to implement the model. As shown in Fig. 2.12c. Theoretically, it can accommodate any shape of the sample due to the controllable shape of the transducer; at the same time, most of the elements in the SCS model can be in intimate contact with the sample and therefore do not require coupling liquid. When the length of the flexible multi-element transducer is greater than the circumference of the sample, it can surround the sample so that relatively complete information can be obtained. Then the acquired information can be used to reconstruct the image by the adaptive back-projection algorithm.

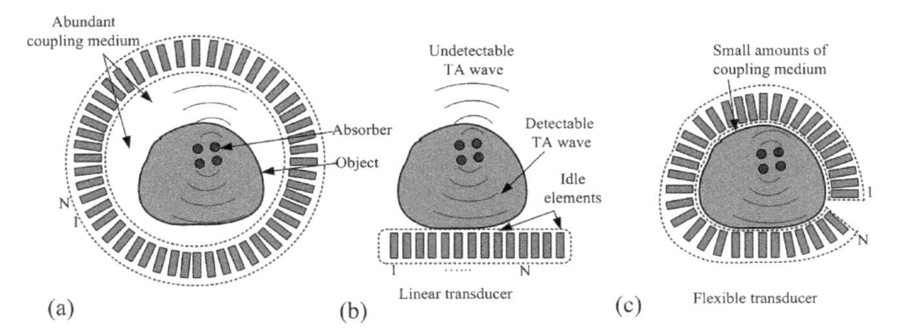

Figure 2.12 The demonstration of signal acquisition of CS (a), LS (b), and SCS (c) mode. Reproduced with permission from Ref. [8].

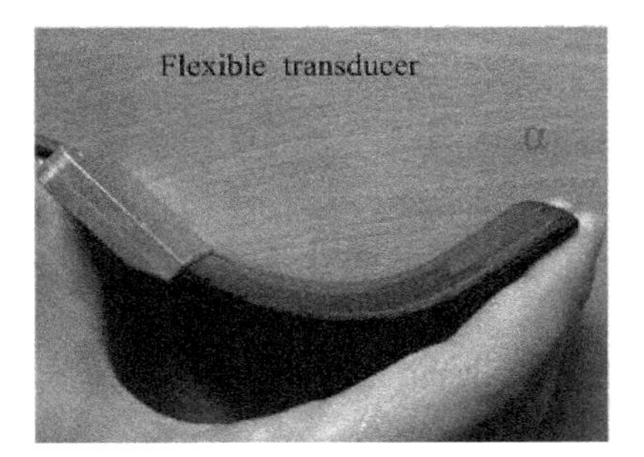

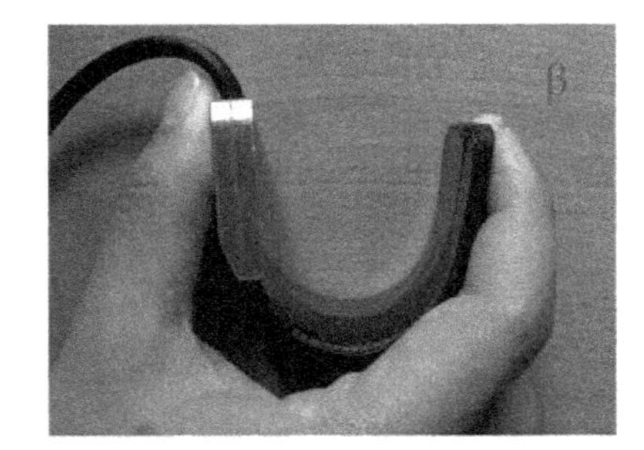

Figure 2.13 The photographs of flexible multi-element transducer with two different shapes.

A flexible 64-element transducer (7.5S64–0.5*10, Doppler Ltd., China) is used as energy converter from ultrasound to electrical signal with an operational frequency of 7.5 MHz and 70% bandwidth. The sensor is made of composite materials, including piezoelectric ceramics ($BaTiO_3$) and polyvinylidene fluoride. The underlayment is made of rubber; hence, this transducer has good flexibility, and its minimum curvature radius is about 5 mm. The transducer can endure bending at least 100,000 times. The effective length of the non-focusing transducer is 32 mm, and each element has a size of 0.5 × 10 mm^2. The relative echo sensitivity of each element is approximately 34.2 to 35.1 dB. Figure 2.13 shows the photographs of flexible multi-element transducer with two different shapes [16].

2.4 Multi-Element Array Photoacoustic Imaging Medical Applications

2.4.1 Photoacoustic Imaging of Breast Cancer

In women, the most frequently occurring malignancy and the leading cause of cancer death is breast cancer. In 2008, about 1.4 million women were diagnosed with breast cancer, which is 23% of all cancer cases. More important, in the same year, with 0.5 million deaths, the cause of 14% of all female cancer deaths was breast cancer.

PA (optoacoustic) imaging can visualize vasculature deep in tissue using the high contrast of hemoglobin to light, with the high-resolution possible with ultrasound detection [17]. Since angiogenesis, which is one of the hallmarks of cancer, leads to increased vascularity, as is shown, PA holds promise in imaging breast cancer in proof-of-principle studies [18]. For the first time, here we investigate if there are specific PA appearances of breast malignancies which can be related to the tumor vascularity, using an upgraded research imaging system. In addition to comparisons with x-ray and ultrasound images, in subsets of cases the PA images were compared with MR images as depicted in Fig. 2.14, and with vascular staining in histopathology. We could identify lesions in suspect breasts at the expected locations in 28 of 29 cases. We discovered that three types of PA appearances reminiscent of contrast enhancement types reported in MR imaging of breast malignancies, and first insights were gained into the relationship with tumor vascularity.

Between April 2012 and May 2013, the upgraded PAM was used to image suspect and highly suspect breasts of 29 patients. The results of those 29 patients' measurements gave rise to a research question which is further investigated in the 14 patients presented in this manuscript. Compared with previously reported

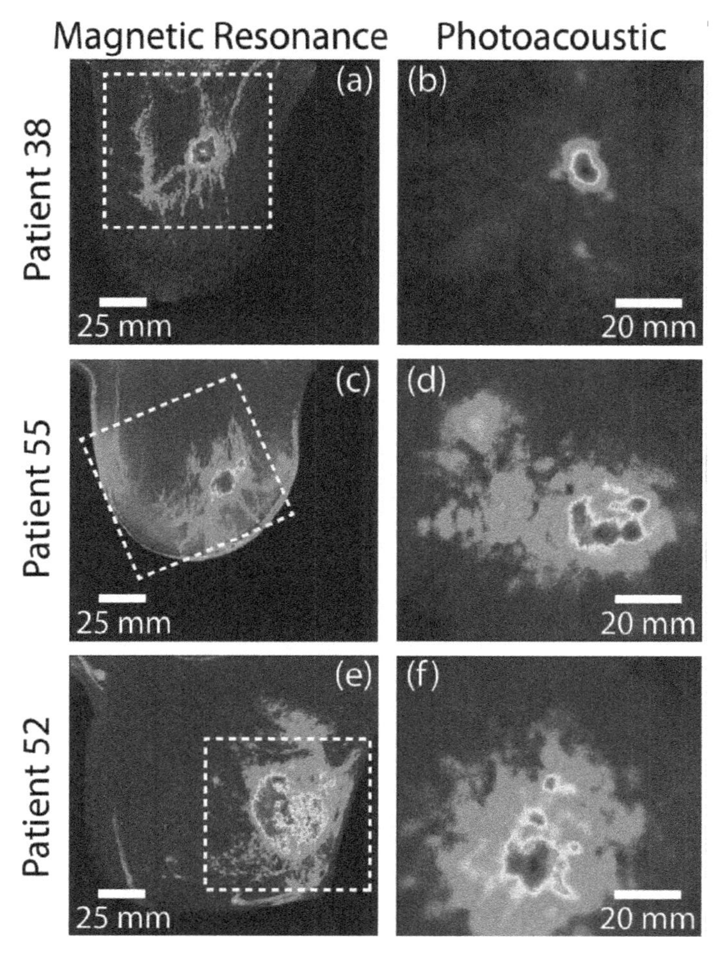

Figure 2.14 MR (left) and PA (right) cranio-caudal (CC) average intensity projections (AIP) of the breast for three cases. (a–b) Patient P38, having an IDC, grade 3 of 19 mm; (c–d) Patient P55, having an IDC, grade 2 of 34 mm; (e–f) Patient P52, having an IDC (unknown grade) of more than 50 mm. The dashed box in the MR image indicates the area from which the PA image is acquired. Reproduced with permission from Ref. [28].

clinical data, lesions not only presented as homogenous masses but also exhibited various types of appearances. We not only found that these patterns of PA intensity somewhat reminiscent of the contrast enhancement types generally reported in MRI of breast malignancies, but also we adopted the same descriptions as in the MR nomenclature. The following PA image patterns were identified generally:

(1) mass-like appearance consisting of a confined region of high intensity, which may be either homogenous or heterogeneous in spatial distribution, and regular or irregular in shape
(2) ring appearance consisting of a region of high intensity completely or partially surrounding an area with lower intensity
(3) non-mass or scattered appearance comprising scattered foci of high intensity within an extended region

This work indicates that PA breast imaging has potential in visualization of cancer [19]. For the first time, we observe that patterns in PA image intensities, which while not being completely clear-cut in categorization, can be described as mass-like, non-mass, and ring appearances. Comparing the image features in PA with those in contrast enhanced MR, and with vascular stained histopathology, led us to attribute PA intensities to the presence of vascularity. The contribution of water and fat to the PA intensities at 1064 nm is lower but will have to be confirmed by multiple wavelength studies. The elucidation of PA patterns of malignancy should be pursued in a large patient population. Such knowledge has the potential to contribute to the development of PA image descriptors, which may serve as diagnostic criteria for the technique in the long run.

2.4.2 Photoacoustic Imaging of Heart Lesion

PAI technology of in vitro rat cardiac lesion group applied in ex vivo organs of small animals [19]. We used PAT to visualize infarcted areas within mouse hearts and compared it to other imaging techniques (MRI and CT). In order to induce ischemia, an in vivo ligation of the Ramus interventricularis anterior (RIVA, left anterior descending, LAD) was performed on nine wild type C41 mice. After varying survival periods, the mice were sacrificed. The hearts were excised and immediately transferred into a formaldehyde solution for conservation. Various wavelengths in the visible and near infrared region (500–1000 nm) had been tested to find the best representation of the ischemic regions [19]. Samples were exposed by nanosecond laser pulses delivered by an Nd:YAG pumped optical parametric oscillator. Ultrasound detection was achieved by an optical Mach–Zehnder interferometer, working as an integrating line detector. The samples were located inside a water tank. The voxel data are computed from the measurement data by a Fourier domain–based reconstruction algorithm, followed by a sequence of inverse Radon transforms. Results clearly indicate the capability of PAI to detect pathological tissue and the possibility to produce three-dimensional images with resolutions below 100 μm. Different wavelengths allow the representation of structure inside an organ or on the surface even without contrast-enhancing tracers.

Scans of nine hearts from mice survived for three different periods of time in the visible and near infrared spectrum have been performed in order to find a wavelength which suits our demands [26]. For the detection of ischemic tissue in mouse hearts, a wavelength around 750 nm delivered best results. Slices which are shown in figure are at approximately the same imaging plane. The ischemic regions in the PA images which are marked by arrows in Figs. 2.15B, 2.16B, 2.17B are characterized by lower absorption compared to the surrounding healthy muscle tissue in the myocardial wall. In MR images, these parts are characterized by brighter shades of gray than unaffected myocardium, which appears almost black (marked by arrows in Figs. 2.15A, 2.16A, and 2.17A). Agar has a high water content, which depends on its concentration and is almost optically transparent. So, it is not visible in PA but in MR images (Figs. 2.15A, 2.16A, 2.17A). In consequence of soft tissue contrast, no details inside the organ are visible in μCT images (Figs. 2.15C, 2.16C, 2.17C). The white ring around the heart is the sample holder tube of the μCT system.

This study is confined to the detection of pathological tissue inside a mouse heart, but different wavelengths allow the representation of any desired structure inside an organ or on the surface. Changes in optical signals can be associated with the progress of disease or the functional status of tissues. The application of contrast-enhancing tracers is abandoned purposely by us, because it is impossible to ensure that the contrast agent is convenient for the target of interest and suitable for all imaging modalities applied in this study. It has been shown that the usage of contrast-enhancing agents can improve image significance and the field of application of PAI [21].

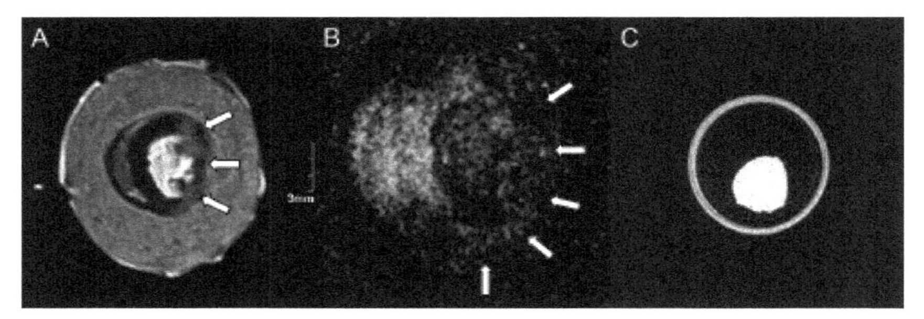

Figure 2.15 Slices of a mouse heart, mouse survived for 3 days. (A) MRI. (B) PAI. (C) μCT. Infarcted regions are marked by arrows. Reproduced with permission from Ref. [32].

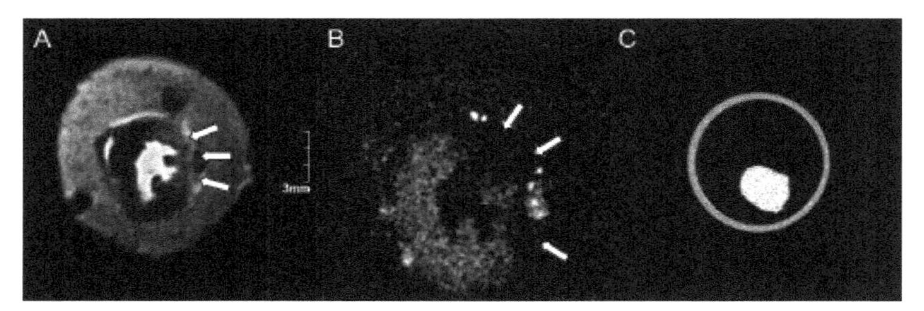

Figure 2.16 Slices of a mouse heart, mouse survived for 5 days. (A) MRI. (B) PAI. (C) μCT. Infarcted regions are marked by arrows. Reproduced with permission from Ref. [32].

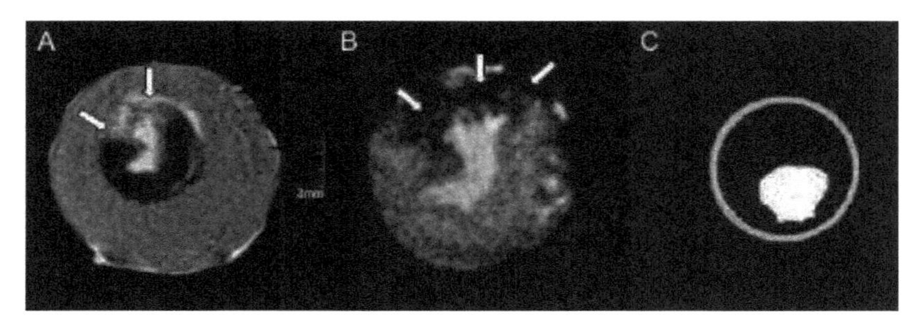

Figure 2.17 Slices of a mouse heart, mouse survived for 7 days. (A) MRI. (B) PAI. (C) μCT. Infarcted regions are marked by arrows. Reproduced with permission from Ref. [32].

2.4.3 Photoacoustic Imaging of Prostate Disease

Prostate cancer is the most frequently diagnosed type of cancer in men in the United States, with approximately 200,000 new cases of prostate cancer identified each year [22]. It is estimated that in 2009, 27,360 men in the United States will die from prostate cancer. In all over the world, prostate cancer is one of the most important causes of cancer fatalities among men, and it is the third leading cause worldwide. Several methods are proved to inspect for prostate cancer. Currently, blood screening for prostate-specific antigen (PSA), digital rectal examination (DRE),

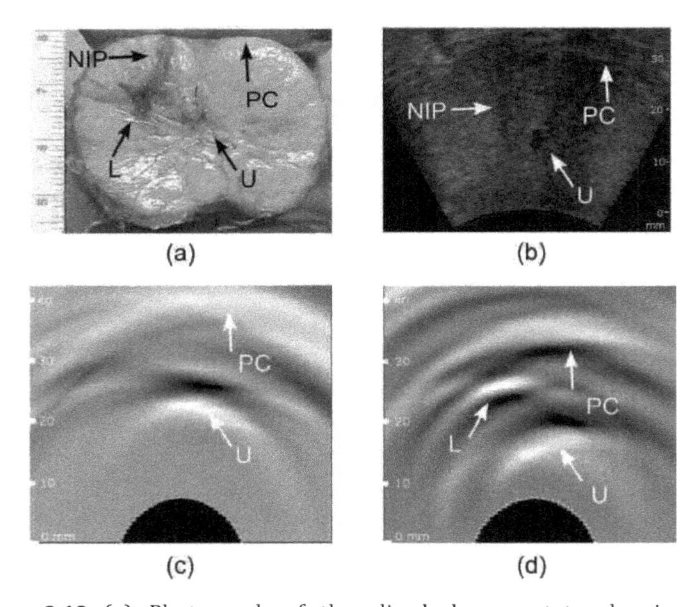

Figure 2.18 (a) Photograph of the sliced dog prostate showing the presence of the induced lesion with blood in the right lateral lobe, the urethra is visible but contracted after surgical excision; (b) ultrasonic image of the same dog prostate obtained in vivo after the surgery. OA images of the same dog prostate obtained in vivo (c) before and (d) after the lesion was induced. The induced bloody lesion can be seen in (a) and (d). The needle insertion path is visible in (a) and (b). Due to the acoustic ismmatch between tissue and air, the prostate capsule can be identified in OA images as a white band. Arrows indicate the prostate capsule (PC), urethra (U), needle insertion path (NIP), and lesion (L). The images are displayed using the standard 8-bit grayscale palette. Reproduced with permission from Ref. [40].

and tissue biopsy guided by transrectal ultrasound (TRUS) are the most widely accepted diagnostic methods for prostate cancer in clinics [23], as demonstrated in Fig. 2.18. But, these techniques all have limited ability to solve malignant tissue from benign enlarged prostate tissue [23]. Physicians typically overcome their limited sensitivity and specificity by using combinations of these methods to perform more accurate diagnoses. Although the entire prostate anatomy can be easily visualized by traditional TRUS imaging, malignant lesions are difficult to identify using TRUS. Evaluating other image factors such as prostate asymmetry and capsular distortion serve as the only indicator of cancer in TRUS images.

2.4.4 Photoacoustic Imaging of the Eye

Optical coherence tomography (OCT) and ultrasound (US) are methods widely used for diagnostic imaging of the eye. These techniques detect discontinuities in optical refractive index and acoustic impedance respectively. Because of these, both relate to variations in tissue density or composition, and OCT and US images share a qualitatively similar appearance. In PAI, short light pulses are directed at tissues, pressure is generated due to a rapid energy deposition in the tissue volume, and thermoelastic expansion results in generation of broadband US. PAI depicts optical absorption, which is independent of the tissue characteristics imaged by OCT or US. Our aim was to demonstrate the application of PAI in ocular tissues and to do so with lateral resolution comparable to OCT. We invented two PAI assemblies, both of which used single-element US transducers and lasers sharing a common focus. The first assembly had optical and 35 MHz US axes offset by a 30° angle. The second assembly consisted of a 20 MHz ring transducer with a coaxial optics. The laser emitted 5 ns pulses at either 532 or 1064 nm, with spot sizes at the focus of 35 μm for the angled probe and 20 μm for the coaxial probe. We compared lateral resolution by scanning 12.5 μm diameter wire targets with pulse/echo US and PAI at each wavelength. We then imaged the anterior segment in whole ex vivo pig eyes (shown in Fig. 2.19) and the choroid and ciliary body region in sectioned eyes (shown in Fig. 2.20).

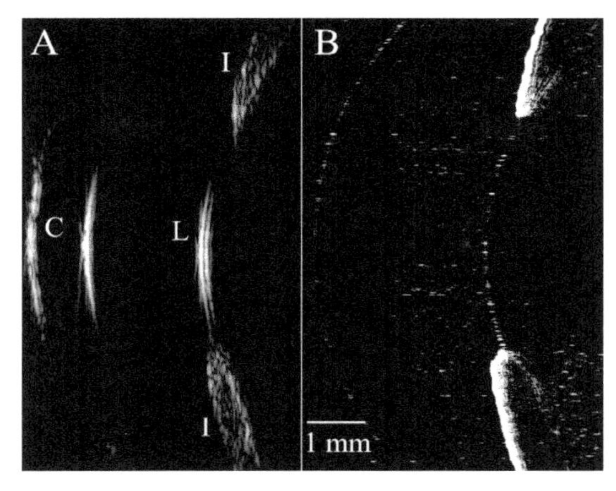

Figure 2.19 Pulse/echo (A) and 532 nm photoacoustic image (B) of anterior segment of ex vivo pig eye obtained using 20 MHz probe with coaxial optics. The pulse/echo image shows reflections from the specular corneal. Reproduced with permission from Ref. [44].

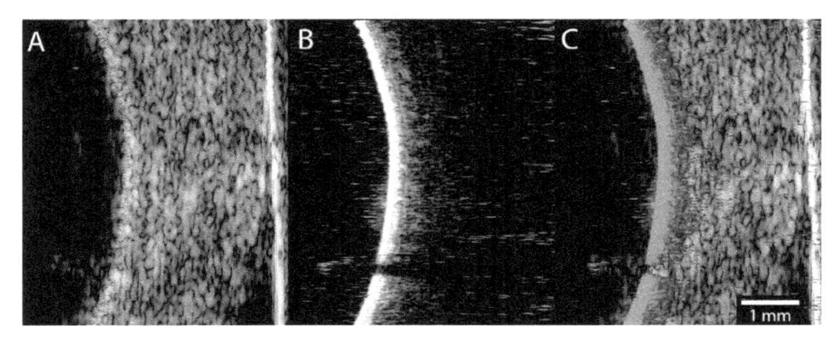

Figure 2.20 Pulse/echo (A) and photoacoustic (B) images of the choroid in an ex vivo pig eye obtained with the 20 MHz ring transducer using a 532 nm green laser. In pulse/echo mode, the choroidal surface and underlying sclera all produce echoes. In the photoacoustic image, a large signal is generated at the choroidal surface. Note the shadow from debris anterior to the choroidal surface (possibly detached retina). Acoustic signals appear in the photoacoustic image several microseconds after encountering strong absorbance by the choroid. In (C), the photoacoustic data (rendered in green) are superimposed on the grayscale pulse/echo data to form a fused image. Reproduced with permission from Ref. [44].

PAI data obtained at 1064 nm in the near infrared had higher penetration, but reduced signal amplitude compared to that obtained using the 532 nm green wavelength. Images were obtained of the iris, choroid and ciliary processes. The zonules and anterior cornea and lens surfaces were seen at 532 nm. Because the laser spot size was significantly smaller than the US beam width at the focus, PAI images had superior resolution than those obtained using conventional US.

We demonstrated high-resolution PAI of ex vivo eyes and found that images in the NIR provided better penetration than those obtained with the green (532 nm) laser, but with reduced sensitivity [24]. PAI images were distinctly different in appearance from those obtained with pulse ultrasound. In pulse ultrasound, significant reflections and backscatter occur in the sclera, iris, and choroid, and primarily surface reflections are from more homogeneous tissues such as the cornea and lens. In contrast, PA signals were generated primarily by pigmented tissues including the iris, ciliary body, and choroid.

PAI is an imaging modality that detects optical absorption and is thus independent of tissue properties detected by conventional pulse/echo ultrasound and OCT. Because PAI can be performed simultaneously with conventional pulse/echo ultrasound, these modalities can be readily combined in image fusion. For the thin tissue layers of the eye, the use of a focused laser may provide sufficient resolution to visualize structures such as the fovea that are beyond the resolution capabilities of pulse/echo ultrasound, and because of its lack of dependence on ballistic photons, it may offer improved penetration compared to OCT.

2.4.5 Photoacoustic Imaging of Tumor Angiogenesis

Vascularization plays an important part role in the development of tumors. Therefore, the most important technique in recognition of the prognostic value of tumor angiogenesis is monitoring angiogenesis, and to proposition of antiangiogenic therapies [25]. In this study, we have investigated whether PAI can be used to image vascularization in tumors. The PAI technique is based on the generation of sound waves by modulated or pulsed optical radiation. The efficiency of sound generation is higher for pulsed than for modulated PA generation. In pulsed PA, a short laser pulse heats absorbers inside the tissue, producing a temperature rise proportional to the deposited energy. The light pulse is so short that adiabatic heating of the absorber occurs, resulting in a pressure rise. The resulting pressure wave (acoustic wave) will propagate through the tissue and can be detected at the tissue surface. From the time, this pressure wave needs to reach the tissue surface (detector position), the position of the PA source can be determined. The temporal profile of the pressure transient will be dependent on the source geometry. For spherical and cylindrical sources, this temporal profile will have a typical bipolar shape, in which the distance between the two peaks is dependent on the size of the source.

PAI is a hybrid imaging modality that is based on the detection of acoustic waves generated by absorption of pulsed light by tissue chromophores such as hemoglobin in blood. Serial PAI has been performed over a 10-day period after subcutaneous inoculation of pancreatic tumor cells in a rat. The images were obtained from ultrasound generated by absorption in hemoglobin of short laser pulses at a wavelength of 1064 nm. The ultrasound signals were measured in reflection mode using a double-ring PA detector [26]. A correction algorithm has been developed to correct for scanning and movement artifacts during the measurements. Three-dimensional data visualize the development and quantify the extent of individual blood vessels around the growing tumor, blood concentration changes inside the tumor and growth in depth of the neovascularized region.

We have shown that we were able to noninvasively image the growth of a tumor in a rat model. The 3D reconstructions of the vasculature in the tumor show a growth of blood vessels. During the measurements movement, artifacts occurred due to the zigzag scanning pattern. Estimating the shift of the consecutive rows by from the cross-correlation of the adjacent image-planes successfully eliminated these artifacts. These results show that PAI is a promising tool to noninvasive visualization of tumor angiogenesis [25].

2.4.6 Photoacoustic Imaging of Brain Disease and Function

Team leader in the mice brain vascular blood oxygen hemoglobin concentration condition of high resolution imaging. They have presented a first-of-its kind real-time high-frequency PAI system

[26], have outlined our design including the hardware and software architecture, and have shown phantom and in vivo data. The B-scan acquisition rate is 50 frames/s, which is faster than the 30 Hz video rate [27]. It is noteworthy that this is the first article to our knowledge documenting real-time beamforming on a multi-core CPU desktop PC with high-level software, although others have certainly moved toward programmable hardware at a lower level. Because of the recent trend to multi-core processors, and leveraging Moore's Law, we may expect that our multi-threaded software approach may have considerable potential for PA systems, ultrasound systems, and other multi-channel systems requiring beamforming as the number of processor cores expands [30–32]. Having provided some discussion on the capabilities and limitations of our real-time PAI system, one question that remains to be seen is what biomedical applications can best make use of this emerging technology. With recent progress, real-time PAI is sure to find a niche in several biomedical applications. Now equipped with real-time imaging capabilities, PAI is ready to make seminal contributions in biological and clinical imaging.

Imaging techniques based on optical contrast analysis can be used to visualize dynamic and functional properties of the nervous system via optical signals resulting from changes in blood volume, oxygen consumption, and cellular swelling associated with brain physiology and pathology. Here we report in vivo noninvasive transdermal and transcranial imaging of the structure and function of rat brains by means of laser-induced PAT. The advantage of PAT over pure optical imaging is that it retains intrinsic optical contrast characteristics while taking advantage of the diffraction-limited high spatial resolution of ultrasound. We accurately mapped rat brain structures, with and without lesions, and functional cerebral hemodynamic changes in cortical blood vessels around the whisker-barrel cortex in response to whisker stimulation. We also imaged hyperoxia- and hypoxia-induced cerebral hemodynamic changes. This neuroimaging modality holds promise for applications in neurophysiology, neuropathology and neurotherapy.

The PAI system was used for noninvasive monitoring of traumatic mouse brain in vivo with high-quality reconstructed images [33]. Traumatic lesions accompanying the hemorrhage in the mouse cortical surface were accurately mapped, and foreign bodies of two small copper wires inserted in the mouse brain were also detected. Furthermore, the time course of morphological changes of cerebral blood during rehabilitation process of a mouse brain with traumatic brain injury was obtained using a series of PA images. Experimental results demonstrate that PA technique holds the potential for clinical applications in brain trauma and cerebrovascular disease detection [33, 36].

2.4.7 Diagnostic Photoacoustic Imaging

In our studies, anatomical, functional, and molecular information of in vivo mice bearing cancerous tumors was acquired using a combination of ultrasound and multi-wavelength PAI. All methods follow protocols approved by the Institutional Animal Care and Use Committee at the University of Texas at Austin. First, we developed a small-animal model of breast cancer, which consisted of two tumors established from human breast cancer cell lines with differential cell biomarker expression. For this small-animal model, we initiated tumors within the mammary fat pad using injections of BT-474 cancer cells, which over-express the cellular receptor HER2, and MDA-MB-231 cancer cells, which over-express $\alpha v\beta 3$ integrin, present on the cell surface of epithelial cells of neovasculature. Gold nanorods were chosen as the molecular contrast agent since, in addition to being highly optically absorbing due to surface plasmon resonance effects, their optical absorption spectra can be tuned by changing their aspect ratio [36]. By tuning nanorods to have different peak optical absorption wavelengths, it is possible to distinguish between multiple nanorod contrast agents through multi-wavelength PAI [37]. We coated the nanorods with amorphous silica, which improves the gold nanorod thermal stability [38] and the PA signal generation efficiency [39]. Silica-coated nanorods with two different aspect ratios were chemically modified to attach targeting antibodies, allowing the nanoparticle to be preferentially uptaken by tumor cells over-expressing the targeted cellular receptor [37]. The contrast agents were injected into the bloodstream of a mouse growing the tumors. The injected contrast agents distributed through the circulatory system, and infiltrated through the tissues of the mouse, including the cancerous tumors. The tumor region of the mouse was imaged using a Vevo 2100 high-frequency small-animal ultrasound scanner (VisualSonics Inc.), integrated with a SpectraPhysics QuantaRay Pro Nd:YAG nanosecond pulsed laser with a GWU PremiScan optical parametric oscillator (OPO) to tune the light wavelength. A fiber optic bundle was used to deliver the laser light on either side of a linear transducer array. The system was used to deliver 10–20 m J/cm^2 of light over a range of wavelengths from 680–930 nm. A 21 MHz ultrasound transducer (MS250, VisualSonics Inc.) collected the PA signals at each wavelength, followed by the transmission of ultrasound and the receiving of the reflected ultrasound by the same transducer, resulting in co-registered ultrasonic and PA images. A linear stepper motor was used to translate the transducer and fiber bundle construct, in steps of 150 μm, to acquire PA and ultrasound images in the third dimension, averaging four PA signals at each step. Three-dimensional (3D) ultrasound and PA images were captured in a volume surrounding the two tumors within the mammary fat pad of the mouse.

The amplitude of the PA signal received is dependent upon the wavelength of the laser light used to illuminate the sample. This dependence can be unmixed into the individual absorption spectra of the absorbers within the tissue—oxyhemoglobin, deoxyhemoglobin, nanoparticle 1 and nanoparticle 2. We used a linear least squares error spectral unmixing method where each voxel was assumed to contain a combination of the four optical absorbers [42]. In this method, the initial PA signal, located at a position within the image, depends on the concentration of the absorbers in the region, the laser fluence, and the Grüneisen parameter [40]. The optical absorption spectra of hemoglobin were obtained from the literature [41]. And the nanoparticle spectra were measured using UV-Vis spectroscopy (Synergy HT microplate reader, Biotek Instruments, Inc.). If the spectral unmixing problem is over-constrained, which is achieved by acquiring PA data at more wavelengths than the number of absorbers present in the tissue, then a minimum mean squared error estimate can be obtained for the absorber concentrations. The estimated concentrations of deoxyhemoglobin, oxyhemoglobin, nanoparticle 1, and nanoparticle 2 are displayed using blue, red, green, and violet colormaps, respectively. For spectral unmixing, the data was averaged into voxels of size 500 μm, 200 μm, 300 μm. The images of the optical absorbers were overlaid on co-registered ultrasound grayscale images to visualize the anatomy in the region.

High-resolution 3D anatomical images of the tumor region were generated by processing the ultrasound signals received by the integrated ultrasound and PAI system (Fig. 2.21). In Fig. 2.21a, a 3D image of a single wavelength PA image, overlaid on the ultrasound image, clearly shows the anatomy within the imaged region of the mouse, including the vasculature (in color, provided by the PA image), and the layer of skin and the femur within the leg (in grayscale, provided by the ultrasound image). In our mouse model, we can identify the tumor regions by locating two hypoechoic regions under the skin, as shown in the 2D ultrasound slice in Fig. 2.21b. From the PA signal (Fig. 2.21c,d), it is clear that there are light absorbers within the skin, light absorption from vasculature surrounding the tumors, and from the tumors themselves. While both ultrasound and PA images indicate the presence of tumors, the anatomical images alone do not provide sufficient information to diagnose malignancy, since the accuracy of ultrasound in the distinction of benign and malignant tumors is insufficient [42]. While the in vivo ultrasound and single wavelength PA images in Fig. 2.21 provide high-resolution anatomical information, the cellular composition of the tumor is unknown, and thus, at this stage of imaging, we are unable to identify functional characteristics or the cellular molecular expression, which might aid in the identification of a benign versus cancerous tumor. Functional PA images, for example of the blood oxygen saturation of the tumor, could provide additional criteria to assess tumor function. We acquired functional information by detecting PA signals at multiple laser wavelengths and spectrally unmixing the signals as described above. For functional imaging of the animal model described above, we acquired PA signals generated by laser light between 680–850 nm. We can observe the oxygen saturation of the blood within the "normal" tissue and within the tumor regions. The estimated concentrations of deoxyhemoglobin and oxyhemoglobin are shown using a color scale ranging from blue (0% oxygen saturation) to red (100% oxygen saturation).

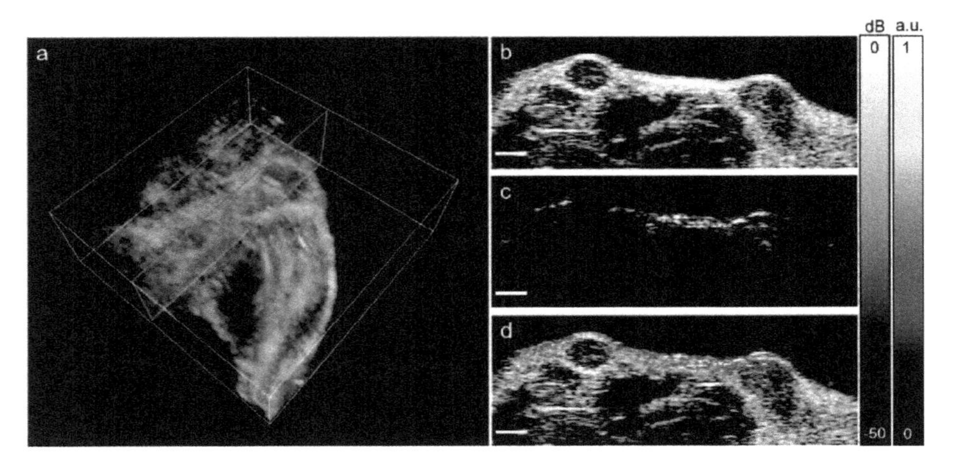

Figure 2.21 In vivo anatomical photoacoustic and ultrasound images. (a) 3D overlay of photoacoustic and ultrasound images of the upper leg/abdominal region. Blue box shows the 2D imaging plane of (b–d). (b) Ultrasound image showing two hypoechoic tumors. (c) Photoacoustic image, acquired using a laser wavelength of 850 nm. (d) Overlay of photoacoustic and ultrasound images. Scale bars = 2 mm. Reproduced with permission from Ref. [46].

The 3D image indicates distinction of venous flow from arterial flow within vasculature of the imaged tissue. While the study of vasculature has relevance to some niches of medical diagnostics, more significantly, we can use the functional information provided by the multiwavelength PAI to study the tumor blood oxygen saturation. Hypoxic malignant tumors have a worse prognosis, making the blood oxygen saturation useful for the assessment of therapy and therapy response [43]. As shown in Fig. 2.22b, there is a hypoxic region within the core of the imaged tumor. This lower oxygen saturation is likely due to the high metabolic activity of the cells located within that region, leading to insufficient oxygen delivery within the tumor. Functional PAI of the blood oxygen saturation could be used to study the response of the tumor to therapeutic treatment, noninvasively, in real time, and over long time periods of tumor growth. Finally, we can further process the PA images to view the signal that correlates to the optical absorption spectra of the injected contrast agents.

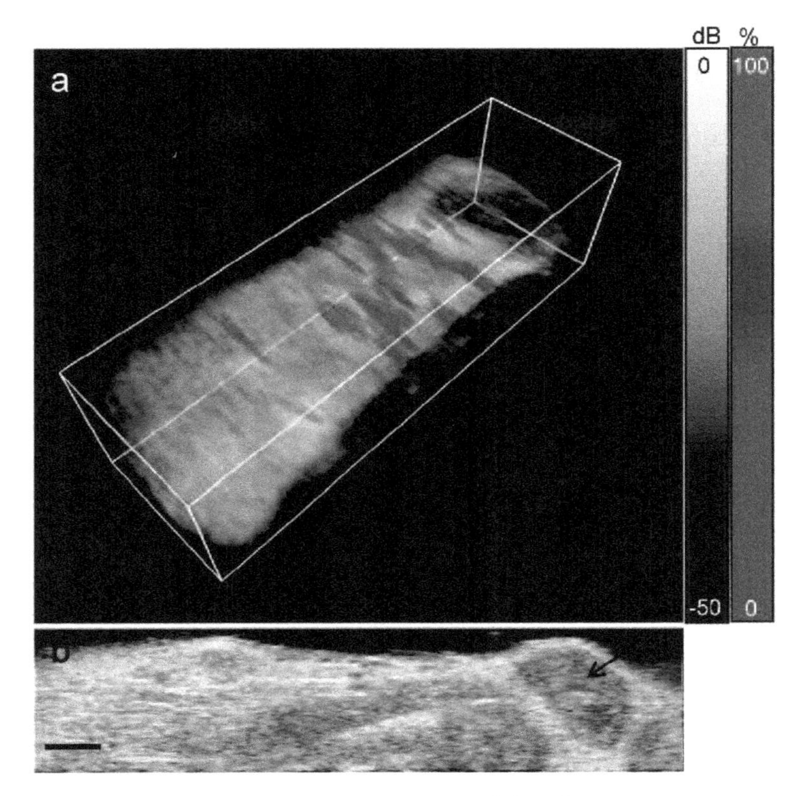

Figure 2.22 In vivo functional imaging of blood oxygen saturation. (a) 3D overlay of percent blood oxygen saturation, determined from spectral unmixing of the multiwavelength photoacoustic signal, and the ultrasound image. Volume is 23 mm (wide) × 6 mm (tall) × 7 mm (scanned direction). (b) 2D slice of the percent blood oxygen saturation and ultrasound signal. Black arrow indicates a hypoxic region within a tumor. Scale bars = 2 mm. Reproduced with permission from Ref. [47].

In the studies shown here, we have used silica-coated gold nanorods, but it important to note that a large variety of contrast agents can be successfully used for molecular PAI [44]; in vivo PAI with enhanced contrast has been demonstrated with the FDA approved dye methylene blue [45], with gold nanospheres or with silver nanoplates [46]. As shown in Fig. 2.23, within our mouse model, regions of nanoparticle accumulation correspond to the tumor regions identified with the ultrasound imaging and the functional PAI of the blood oxygen saturation. Over time, there is an increase in the signal attributed to the nanoparticle contrast agents within the tumor region (Fig. 2.23b,c). The accumulation of nanoparticles could be due to two effects. It is known that tumors have "leaky" vasculature, enabling the increased delivery of nano-sized particles through the enhanced permeability and retention effect. Additionally, the injected nanoparticles were bioconjugated to antibodies specific for cell receptors over-expressed on cell types used to initiate the tumors, thus it is likely that the active targeting improves the retention of the specific nanoparticles within the tumor region. This example demonstrates how the use of nanoparticles can provide molecular information about the cellular expression of the tumor. In this way, photoacoustics combines the benefits of high-contrast optical resolution of targetable light-interacting probes, with the imaging depth capabilities of ultrasound.

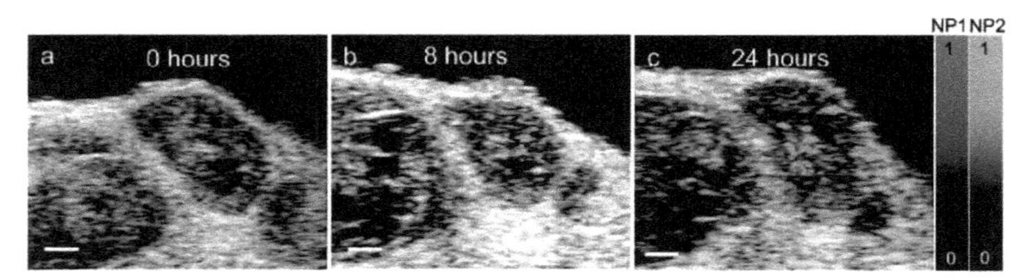

Figure 2.23 In vivo molecular imaging of contrast agent accumulation within a targeted tumor. The Multi-wavelength photoacoustic signal was unmixed into two components corresponding to two different contrast agents, nanoparticle 1 (NP1) and nanoparticle 2 (NP2). (a) 2D overlay of ultrasound image, distribution of NP1 and NP2 before the injection of the contrast agent, showing minimal background photoacoustic signal. Accumulation of NP1 and NP2 within the tumor is shown (b) 8 h and (c) 24 h after intravenous injection. Scale bars = 1 mm.

References

1. Ji, Z., Ding, W., Ye, F., Lou, C., and Xing, D. (2015). Shape-adapting thermoacoustic imaging system based on flexible multi-element transducer, *Appl. Phys. Lett.*, **107**, p. 094104.

2. Yin, B., Xing, D., Wang, Y., Zeng, Y., Tan, Y., and Chen, Q. (2004). Fast photoacoustic imaging system based on 320-element linear transducer array, *Phys. Med. Biol.*, **49**, pp. 1339–1346.

3. Yang, D., Xing, D., Gu, H., Tan, Y., and Zeng, L. (2005). Fast multielement phase-controlled photoacoustic imaging based on limited-field-filtered back-projection algorithm, *Appl. Phys. Lett.*, **87**, p. 194101.
4. Park, S., Mallidi, S., Karpiouk, A. B., Aglyamov, S., and Emelianov, S. Y. (2007). Photoacoustic imaging using array transducer, *Proc. SPIE Intl. Soc Opt Eng*, **6437**, pp. 643714–1.
5. Su, Y., Wang, R. K., Xu, K., Zhang, Fan, Xu, K., and Yao, J., (2005). Photoacoustic tomography imaging of biological tissues, *Opt. Health Care Biomed. Opt. Diagn. Treat. II*, **5630**, pp. 582–586.
6. Ye, F., Ji, Z., Ding, W., Lou, C., Yang, S., and Xing, D. (2016). Ultrashort microwave-pumped real-time thermoacoustic breast tumor imaging system, *IEEE Trans. Med. Imag.*, **35**, pp. 840–841.
7. Park, S., Shah, J., Aglyamov, S. R., Karpiouk, A., Mallidi, S., Gopal, A., Moon, H., Zhang, X., Scott, W. G., and Emelianov, S. Y. (2006). Integrated system for ultrasonic, photoacoustic, and elasticity imaging, *Proc. 2006 SPIE-Medical Imaging Symp.: Ultrasonic Imaging Signal Process.*, **6147**, pp. 61470H–1.
8. Delannoy, B., Torquet, R., Bruneel, C., Bridoux, E., Rouvaen, J. M., and LaSota, H. (1979). Acoustical image reconstruction in parallel-processing analog electronic systems, *J. Appl. Phys.*, **50**, pp. 3153–3159.
9. Fink, M., Sandrin, L., Tanter, M., Catheline, S., Chaffai, S., Bercoff, J., and Gennisson, J. L. (2002). Ultra high speed imaging of elasticity, *Proc. IEEE Ultrasonic Symposium*, **2**, pp. 1811–1820.
10. O'Donnell, M. (1990). Efficient parallel receive beam forming for phased array imaging using phase rotation, *Proc. IEEE Ultrasonic Symposium*, **3**, pp. 1495–1498.
11. Shattuck, D. P., Weinshenker, M. D., Smith, S. W., and Ramm von, O. T. (1984). Explososcan: A parallel processing technique for high speed ultrasound imaging with linear phased arrays, *J. Acoust. Soc. Amer.*, **75**, pp. 1273–1282.
12. Park, S., Aglyamov, S. R., Scott, W. G., and Emelianov, S. Y. (2007). Strain imaging using conventional and ultrafast ultrasound imaging: Numerical analysis, *IEEE Trans. Ultrason. Ferroelect. Freq. Cont.*, **54**, pp. 987–995.
13. Thomenius, K. E. (1996). Evolution of ultrasound beamformers, *Proc. IEEE Ultrasonic Symposium*, **2**, pp. 1615–1622.
14. Jensen, J. A. (1996). Field: A program for simulating ultrasound systems, *10th Nordicbaltic Conf. Biomed. Imaging*, **4**, pp. 351–353.

15. Christensen, D. A. (1988). *Ultrasonic Bioinstrumentation*, Wiley.

16. Norton, S. J., and Linzer, M. (1987). Backprojection reconstruction of random source distributions, *J. Acoust. Soc. Amer.*, **81**, pp. 977–985.

17. Kostli, K. P., and Beard, P. C. (2003). Two-dimensional photoacoustic imaging by use of Fourier-transform image reconstruction and a detector with an anisotropic response, *Appl. Opt.*, **42**, pp. 1899–1908.

18. Hoelen, C. G. A., and Mul, de F. F. M. (2000). Image reconstruction for photoacoustic scanning of tissue structures, *Appl. Opt.*, **39**, pp. 5872–5883.

19. Heijblom, M., Steenbergen, W., and Manohar, S. (2015). Clinical photoacoustic breast imaging: The Twente experience, *IEEE Pulse*, **6**, pp. 42–46.

20. Heijblom, M., Piras, D., Brinkhuis, M., Van Hespen, J. C. G., Van den Engh, F. M., Van der Schaaf, M., Klaase, J. M., Leeuwen van, T. G., Steenbergen, W., and Manohar, S. (2015). Photoacoustic image patterns of breast carcinoma and comparisons with Magnetic Resonance Imaging and vascular stained histopathology, *Sci. Rep.*, **5**, 11778.

21. Holotta, M., Grossauer, H., Kremser, C., Torbica, P., Völkl, J., Degenhart, G., Esterhammer, R., Nuster, R., Paltauf, G., and Jaschke, W. (2010). Photoacoustic tomography of pathological tissue in ex-vivo mouse hearts, *SPIE*, **7564**, pp. 75642X–5.

22. Liu, B., Kruger, R., Reinecke, D., and Stantz, K. M. (2010). Monitor hemoglobin concentration and oxygen saturation in living mouse tail using photoacoustic CT scanner, *SPIE*, **7564**, pp. 756439–1.

23. Stantz, K. M., Cao, M., Liu, B., Miller, K. D., and Guo, L. (2010). Molecular imaging of neutropilin-1 receptor using photoacoustic spectroscopy in reast tumors, *Proc. SPIE*, **7564**, pp. 756410–7.

24. Yaseen, M. A., Ermilov, S. A., Brecht, H. P., Su, R., Conjusteau, A., Fronheiser, M., Bell, B. A., Motamedi, M., and Oraevsky, A. A. (2010). Optoacoustic imaging of the prostate: Development toward imageguided biopsy, *J. Biomed. Opt.*, **15**, pp. 021310–8.

25. Wang, X., Roberts, W. W., Carson, P. L., Wood, D. P., and Fowlkes, J. B. (2010). Photoacoustic tomography: A potential new tool for prostate cancer, *Biomed. Opt. Express*, **1**, pp. 1117–1126.

26. Silverman, R. H., Kong, F., Chen, Y. C., Lloyd, H. O., Kim,H. H., Cannata, J. M., Shung, K. K., and Coleman, D. J. (2010). High-resolution photoacoustic imaging of ocular tissues, *Ultrasound Med. Biol.*, **36**, pp. 733–742.

27. Kolkman, R. G. M., Thumma, K. K., Ten Brinke, G. A., and Thumma, K. K. (2008). Photoacoustic imaging of tumor angiogenesis, *Proc. SPIE*, **6856**, pp. 685602-6.

28. Jankovic, L., Shahzad, K., Wang, Y., Burcher, M., Scholle, F. D., Hauff, P., Mofina, S., and Skobe, M. (2008). In vivo photoacoustic imaging of nude mice vasculature using a photoacoustic imaging system based on a commercial ultrasound scanner, *Proc. SPIE*, **6856**, pp. 68560N-12.

29. Zemp, R. J., Bitton, R., Li, M. L., Shung, K. K., Stoica, G., and Wang, L. V. (2007). Photoacoustic imaging of the high-frequency ultrasound array transducer, *J. Biomed. Opt.*, **12**, pp. 10501-3.

30. Zemp, R. J., Song, L., Bitton, R., et al. (2008). Real-time photoacoustic microscopy in vivo with a 30-MHz ultrasound array transducer, *Opt. Express*, **16**, pp. 7915-7928.

31. Lao, Y., Xing, D., Yang, S., and Xiang, L. (2008). Noninvasive photoacoustic imaging of the developing vasculature during early tumor growth, *Phys. Med. Biol.*, **15**, pp. 4203-4212.

32. Li, C., and Wang, L. V. (2009). Photoacoustic tomography of the mouse cerebral cortex with a high-numerical-aperture-based virtual point detector, *J. Biomed. Opt.*, **14**, pp. 024047-3.

33. Yang, S., Xing, D., Lao, Y., Zeng, L., Xiang, L., and Chen, W. R. (2007). Noninvasive monitoring of traumatic brain injury and post-traumatic rehabilitation with laser-induced photoacoustic imaging, *Appl. Phys. Lett.*, **90**, pp. 243902-243903.

34. Wang, X., Pang, Y., Ku, G., Xie, X., Stoica, G., and Wang, L. V. (2003). Noninvasive laser-induced photoacoustic tomography for structural and functional in vivo imaging of the brain, *Nat. Biotechnol.*, **21**, pp. 803-806.

35. Bayer, C. L., Luke, G. P., and Emelianov, S. Y. (2012). Photoacoustic imaging for medical diagnostics, *Acoust. Today*, **8**, pp. 15-23.

36. Jain, P. K., Lee, K. S., El Sayed, I. H., and El Sayed, M. A. (2006). Calculated absorption and scattering properties of gold nanoparticles of different size, shape, and composition: Applications in biological imaging and biomedicine, *J. Phys. Chem. B*, **110**, pp. 7238-7248.

37. Bayer, C. L., Chen, Y. S., Kim, S., Mallidi, S., Sokolov, K., and Emelianov, S. (2011). Multiplex photoacoustic molecular imaging using targeted silica-coated gold nanorods, *Biomed. Opt. Express*, **2**, pp. 1828-1835.

38. Chen, Y. S., Frey, W., Kim, S., Homan, K., Kruizinga, P., Sokolov, K., and Emelianov, S. (2010). Enhanced thermal stability of silica-coated gold nanorods for photoacoustic imaging and image-guided therapy, *Opt. Express*, **18**, pp. 8867–8878.

39. Chen, Y. S., Frey, W., Kim, S., Kruizinga, P., Homan, K., and Emelianov, S. (2011). Silica-coated gold nanorods as photoacoustic signal nanoamplifiers, *Nano. Lett.*, **11**, pp. 348–354.

40. Kim, S., Chen, Y. S., Luke, G. P., and Emelianov, S. Y. (2011). In vivo three-dimensional spectroscopic photoacoustic imaging for monitoring nanoparticle delivery, *Biomed. Opt. Express*, **2**, pp. 2540–2550.

41. Prahl, S. (1999). Optical absorption of hemoglobin, *Oregon Medical Laser Center*, http://omlc. ogi. edu/spectra/hemoglobin/index. html, **15**.

42. Sehgal, C. M., Weinstein, S. P., Arger, P. H., and Conant, E. F. (2006). A review of breast ultrasound, *J. Mammary Gland Biol. Neoplasia*, **11**, pp. 113–123.

43. Harris, A. L. (2002). Hypoxia: A key regulatory factor in tumour growth, *Nat. Rev. Cancer*, **2**, pp. 38–47.

44. Luke, G. P., Yeager, D., and Emelianov, S. Y. (2012). Biomedical applications of photoacoustic imaging with exogenous contrast agents, *Ann. Biomed. Eng.*, **40**, pp. 422–437.

45. Song, K. H., Stein, E. W., Margenthaler, J. A., and Wang, L. V. (2008). Noninvasive photoacoustic identification of sentinel lymph nodes containing methylene blue in vivo in a rat model, *J. Biomed. Opt.*, **13**, pp. 054033–6.

46. Homan, K. A., Souza, M., Truby, R., Luke, G. P., Green, C., Vreeland, E., and Emelianov, S. (2012). Silver nanoplate contrast agents for in vivo molecular photoacoustic imaging, *ACS Nano*, **6**, pp. 641–650.

Chapter 3

Photoacoustic Microscopy

The ability to in vivo image microstructures of biological tissue with high resolution at desired penetration depth is critically important for the early diagnosis of diseases. A promising technique to accomplish this objective is photoacoustic microscopy (PAM). Compared with current high-resolution optical imaging techniques such as confocal microscopy and optical coherence tomography (OCT), in which the imaging depth is limited to approximately one transport mean free path (1–2 mm) in biological tissues, PAM has the ability to image deeper because it utilizes not only ballistic but also diffusive photons. Thus, microstructures such as micrangium then can be imaged by PAM with high optical contrast and high-resolution.

3.1 Introduction to Photoacoustic Microscopy

In the applications of photoacoustics, biological tissues absorb pulsed laser energy and induce transient temperature rise, which further generate high-frequency ultrasonic waves. Photoacoustic (PA) imaging provides an exquisite way to resolve the optical absorption distribution in biological tissue ultrasonically, thus

Biomedical Photoacoustics
Sihua Yang and Da Xing

ISBN 978-981-4774-58-1 (Hardcover), 978-0-203-70365-6 (eBook)
www.jennystanford.com

leading to a new branch of photoacoustic microscopy—PAM, [1]. In PAM [2–4], through optically focusing and/or acoustic focusing, two kinds of microscopy have been proposed: optical-resolution PAM [5–9] (OR-PAM) and acoustic-resolution PAM [10, 11] (AR-PAM). The fundamental imaging systems are shown in Fig. 3.1 for OR-PAM (Fig. 3.1A) and AR-PAM (Fig. 3.1B). In OR-PAM, by focusing both optically and acoustically, high-resolution images can be achieved with lateral resolution up to several microns. As shown in Fig. 3.1C, single capillaries in an adult mouse brain through the intact skull can be imaged by OR-PAM clearly with lateral resolution of 5 μm.

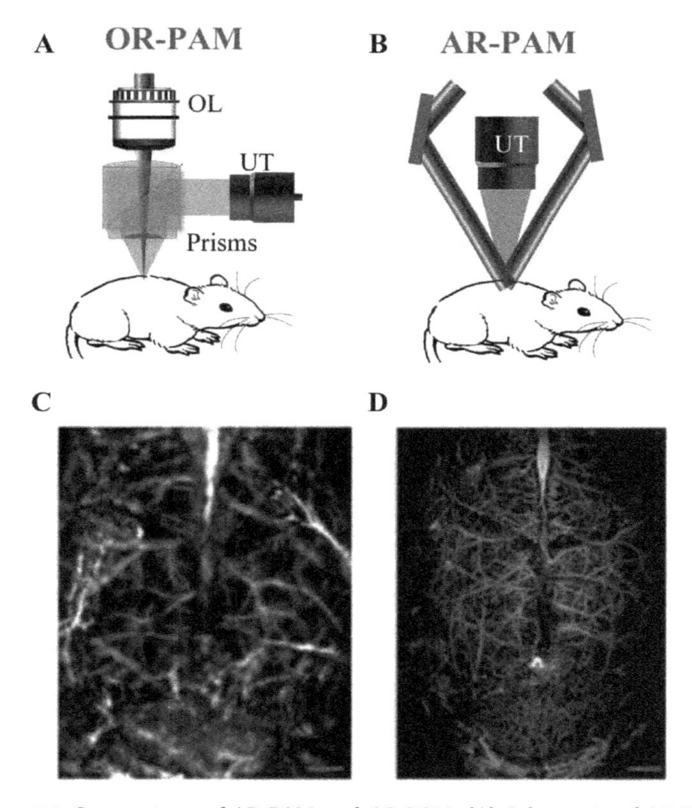

Figure 3.1 Comparison of AR-PAM and OR-PAM. (A) Schematic of AR-PAM. (B) Schematic of OR-PAM. (C) AR-PAM image of the cortical vasculature in a living adult mouse with both the scalp and the skull intact. (D) OR-PAM image of the cortical vasculature in a living adult mouse with the scalp removed and the skull intact. Scale bars: 1 mm.

However, the confocal of laser and sound limits the penetration depths compared with that of AR-PAM. Benefiting from the deep-focusing capability of ultrasound, AR-PAM with tight acoustic focusing and weak optical focusing are widely used in deep-seated tissues' imaging in the photon diffusive regime. Nevertheless, owing to the weak optical focusing, the lateral resolution of PA image obtained by AR-PAM is limited and inadequate for cellular and subcellular imaging. As shown in Fig. 3.1D, the lateral resolution for the mouse brain image obtained by OA-PAM is about 45 μm. We should notice that the weak ultrasonic scattering in tissues allows the AR-PAM to penetrate both the scalp and the skull of the mouse, while the OR-PAM requires the scalp to be removed during the imaging process.

3.2 Optical-Resolution Photoacoustic Microscopy

OR-PAM combines the advantages traditional optical imaging and ultrasound imaging, enabling in vivo detection of endogenous optical contrast with optical-resolution [12]. By fine optical focusing, the OR-PAM systems are capable of providing high lateral resolution, while the axial resolution is still derived from time-resolved ultrasonic detection. Although OR-PAM has comparable depth penetration limit with existing high-resolution optical imaging modalities such as OCT [13] and optical confocal microscopy [14], it is primarily sensitive to optical absorption rather than scattering, thus high contrast can be obtained by OR-PAM. In this section, we summarize fundamentals for OR-PAM, including the fundamental systems, multi-wavelength OR-PAM [15], and fast variable focus OR-PAM using an electrically tunable lens [16].

3.2.1 System for OR-PAM

An OR-PAM system employs optical focusing to achieve micrometer-level lateral resolution of imaging. Usually a pulsed laser (Nd:YLF laser) is used the irradiation source. As the laser pulse (pulse duration, 7 ns) is launched out, it is spatially filtered by a pinhole with a diameter of 25 μm, as shown in Fig. 3.2.

The pinhole is then imaged to a diffraction-limited focal spot with a diameter of 3.7 μm by an objective lens (NA, 0.1; depth of focus, 40 μm). The ultrasonic focusing is achieved through a plano-concave lens. The radius curvature of the plano-concave lens is about 5.2 mm with aperture of 6.4 mm and NA (in water) of 0.46. In this way, the optical objective lens and the ultrasound transducer with central frequency of 75 MHz can be axially and confocally configured.

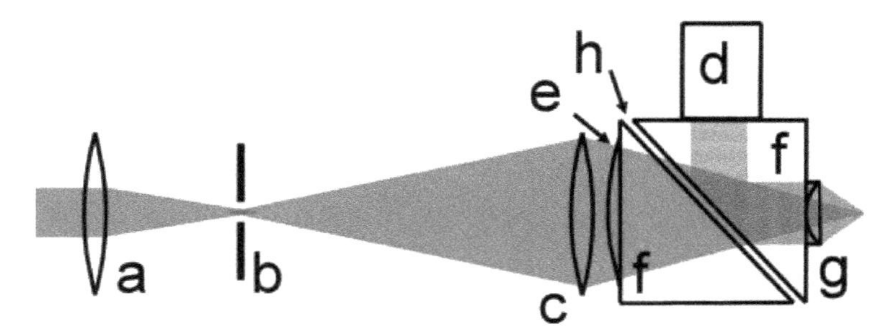

Figure 3.2 Schematic of the OR-PAM system. (a) Condenser lens; (b) pinhole; (c) microscope objective; (d) ultrasonic transducer; (e) correcting lens; (f) isosceles prism; (g) acoustic lens; (h) silicon oil. Reproduced with permission from Ref. [13].

Once the produced PA signal is detected by the ultrasound transducer, an image of a region of interest (ROI) can be obtained by two-dimensional scanning along the *x*–*y* plane, which can be viewed through direct volumetric rendering for a volumetric image, cross sectional (B-scan) images for a sectional image, or maximum amplitude projection images.

In order to obtain the imaging depth of the OR-PAM system, Lihong V. Wang et al. measured the largest imaging depth of two horse hairs with diameter of about 200 μm, which are separately placed above and below a piece of freshly harvested rat scalp. As shown in Fig. 3.3A, the two hairs can be clearly seen in the OR-PAM image, which was acquired at the optical wavelength of 630 nm with 32 times signal averaging. The B-scan image in Fig. 3.3B that the bottom hair was placed 700 μm deep in tissue. Result shows that the OR-PAM system has the imaging penetration ability of at least 700 μm.

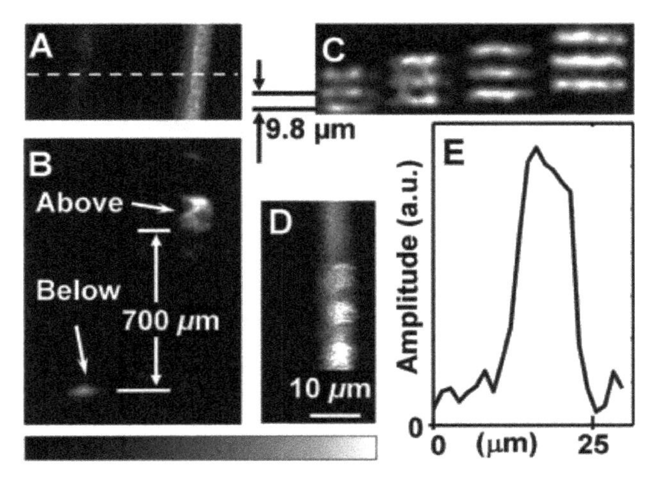

Figure 3.3 Imaging depth and lateral resolution of the ORPAM system. (A) MAP image of two horse hairs placed above and below a piece of rat skin; (B) B-scan image at the location marked by the dashed line in (A); (C) MAP image of a USAF resolution test target group 6; (D) MAP image of a 6 m diameter carbon fiber; (E) Cross-sectional profile of a 6 m diameter carbon fiber under a 0.2 mm-thick nude mouse ear. Reproduced with permission from Ref. [13].

The axial resolution of the OR-PAM system depends on the ultrasonic parameters, including the central frequency, bandwidth, and the NA. Here, by imaging a United States Air Force (USAF-1951, Edmund) resolution test target in an optically clear medium with laser wavelength of 590 nm, the axial resolution was estimated to be 15 μm based on the measured transducer bandwidth (100 MHz in receiving-only mode) and the speed of sound in tissue (1.5 mm/s). Figure 3.3C shows a PA image of the resolution test target. Nonlinearly fitting the modulation transfer function followed by extrapolation yields a lateral resolution of 5 μm, which is 30% greater than the diffraction limit of 3.7 μm. To further illustrate the lateral resolution, a 6 μm diameter carbon fiber immersed in water was imaged as shown in Fig. 3.3D. The mean full width at half-maximum (FWHM) value of the imaged fiber is estimated to be 9.8 μm, which is 3.8 μm wider than the fiber diameter and hence in agreement with the 5 μm resolution. As the lateral resolution is expected to deteriorate with an imaging depth owing to optical scattering, a 6 μm diameter carbon fiber covered by a 0.2 mm thick nude mouse ear was imaged, as shown in Fig. 3.3E, where the PA amplitude profile of a cross section of the carbon fiber is provided. The FWHM value of the imaged fiber is estimated to be 10 μm. Therefore, it can be concluded that the OR-PAM system maintains a 5 μm lateral resolution in tissue up to at least 0.2 mm in depth.

3.2.2 Multi-Wavelength OR-PAM

By replacing the single-wavelength pulsed laser with an optical parametric oscillator (OPO) laser (VIBRANT B 532I, OPOTEK), our group developed a multi-wavelength OR-PAM system for noninvasively detecting hemoglobin oxygen saturation (SO_2) and carboxyhemoglobin saturation (SCO) in subcutaneous microvasculature.

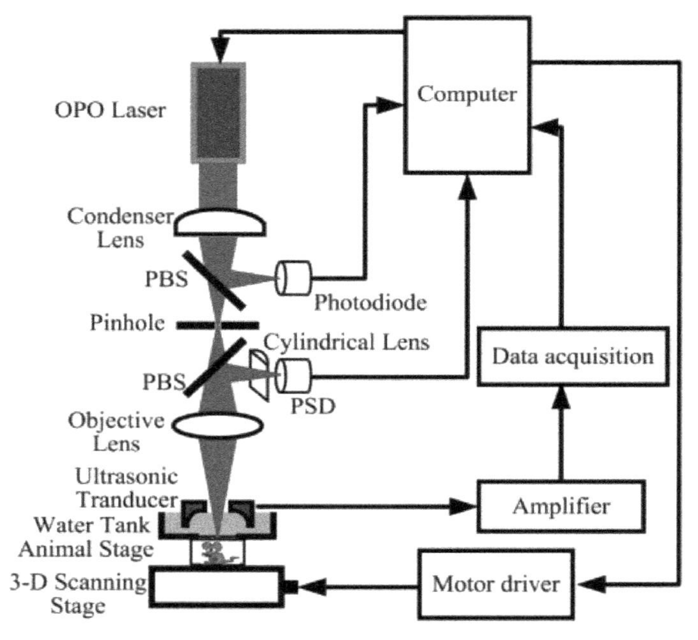

Figure 3.4 Scheme of the in vivo PAM system. The diameter of the pinhole is 100 μm. Reproduced with permission from Ref. [15].

The scheme of the multi-wavelength OR-PAM system [15] is shown in Fig. 3.4. The tuning range of the OPO laser ranges from 680 nm to 950 nm. A custom-made piezocomposite hollow focused ultrasound transducer (Doppler Electronic Technologies Co., Ltd., China) with focus length of 5 mm and center frequency of 15 MHz was used to detect the PA signals. The laser beam was focused by an objective lens (NA 0.1; WD 37.5 mm) through the small hole (diameter 2 mm) in the center of the hollow sensor. A data acquisition card (NI5124, National Instruments) received the amplified PA signals from the hollow sensor at a sampling rate of 200 M samples/s. Three-dimensional scanning stage was driven by computer-controlled stepper motors (MC600AS, Zolix). During the experiments, a silicon photodiode was used to monitor and calibrate the intensity and stability of the laser beam. The cylindrical lens and the position sensor device (PSD) comprised an auto focusing system.

In order to validate the imaging ability of the multi-wavelength OR-PAM system, the resulting PA signals are given for three wavelengths (760 nm, 850 nm, 900 nm) for mouse blood samples at 70.54% SO_2 and 28.1% SCO level, respectively, as shown in Fig. 3.5. According to the PA effect, the PA signals of different SO_2 and SCO levels are proportional to the absorptions of each blood sample, respectively. Results proved that the multi-wavelength OR-PAM system is capable of providing multi-wavelength PA imaging for biological applications, as shown in Fig. 3.5.

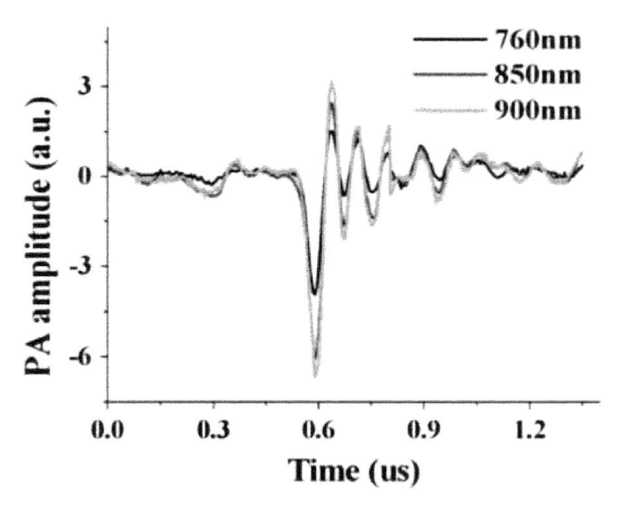

Figure 3.5 The PA signals of in vitro blood sample at 70.54% sO_2 and 28.1% SCO level. Reproduced with permission from Ref. [15].

3.2.3 Fast Variable Focus OR-PAM Using an Electrically Tunable Lens

Since OR-PAM can provide high-resolution and high-contrast images noninvasively, it has been widely applied in many biomedical research fields. However, due to a single depth of focus, only a narrow depth range in focus is acquired in OR-PAM. This limits the application of OR-PAM and cannot meet the requirement of depth focusing scan. Our group has developed a fast variable focus OR-PAM system (VF-PAM) using an electrically tunable lens (ETL), which provides an inertia-free focusing scan over a large depth range. The focus depths of the VF-PAM system can be varied as a function of the supplied electrical signal and shift continuously.

The schematic diagram of the fast VF-PAM system [16] is depicted in Fig. 3.6a. We used an Nd:YAG laser (Surelite II-20, Continuum, USA) operating 10 ns pulses and 532 nm wavelength with a repetition rate of 20 Hz to excite the PA signal. The laser pulse fluence is estimated about 13 mJ/cm^2, which is lower than the American National Standard Institute (ANSI) safety limit (20 mJ/cm^2). The laser beam was first passed through a spatial filter and scanned with a two-dimensional (2-D) scanning galvanometer (6231H, Cambridge Technology, Inc) for x–y plane scanning and then was reshaped by scanning lens, tube lens, and ETL (EL-10-30, Optotune AG, Switzerland) for z-axis scanning. The reshaped laser beam was focused by a plan microscope objective lens to irradiate the test sample. The 2-D scanning galvanometer controlled by a computer was triggered by the signals of the laser. During the experiments, a silicon photodiode was employed to monitor and calibrate the intensity and stability of the laser beam. A custom-made unfocused ultrasonic transducer with center frequency of 10 MHz and −6 dB bandwidth of 99.8% was used to detect the PA signals. The PA signals were first amplified with an amplifier (ZFL-500, Minicircuits), then digitized by a dual-channel data acquisition card (NI 5122, National Instrument, USA) at a sampling rate of 100 M samples/s, which then finally recorded in the computer for imaging reconstruction through a MATLAB program.

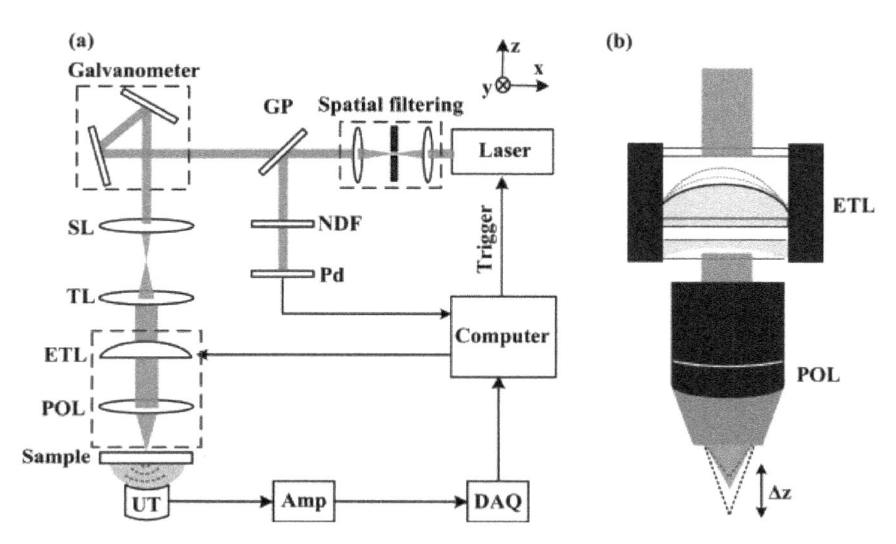

Figure 3.6 Schematic diagram of the VF-PAM system (a) and electrically tunable lens (ETL) placed on top of the plan microscopy objective lens (b). GP, glass plate; SL, scan lens; TL, tube lens; ETL, electrically tunable lens; POL, plan objective lens; UT, ultrasonic transducer; Amp, amplifier; DAQ, data acquisition system; NDF, neutral density filter; Pd, photodiode. Reproduced with permission from Ref. [16].

The detailed schematic of the fast variable focus system is presented in Fig. 3.6b (blue dashed box in Fig. 3.6a). The advantages of the ETL include high transmission, large aperture, high damage threshold, and fast response time. It consists of a container, which is filled with an optical fluid and sealed off with an elastic spherical polymer membrane. The radius of the membrane is proportional to the pressure exerted on its outer zone by an electromagnetic actuator. The ETL and plan microscope objective were connected together via an optical interface. The focal length of the ETL was controlled by a high-precision USB driver (EL-E-4, Optotune), which delivered a stable current output from 0 mA to 300 mA with a resolution of 0.1 mA. When using a 0.3 NA plan microscope objective lens, within the control current range, the maximum focus-shifting range was measured about 2.82 mm with a 1 μm continuously shifting accuracy. The shift speed of the depth refocusing of the excitation spot between two focal planes is faster than mechanical scan, which is well within 15 ms.

In order to obtain the focus-shifting range of the proposed VF-PAM system, several carbon fibers at different depths were imaged by focus shift. The lateral scan step was set to be 1.8 μm. The measurements showed that the maximum focus-shifting range was 2.82 mm. Figures 3.7a,b were PA images of the carbon fibers at upper and lower planes, respectively. The inset at the bottom right corner of Fig. 3.7b shows the schematic diagram of the two planes carbon fibers with a depth interval of 2.82 mm. The FWHM of the fibers along the dashed line in Figs. 3.7a,b were

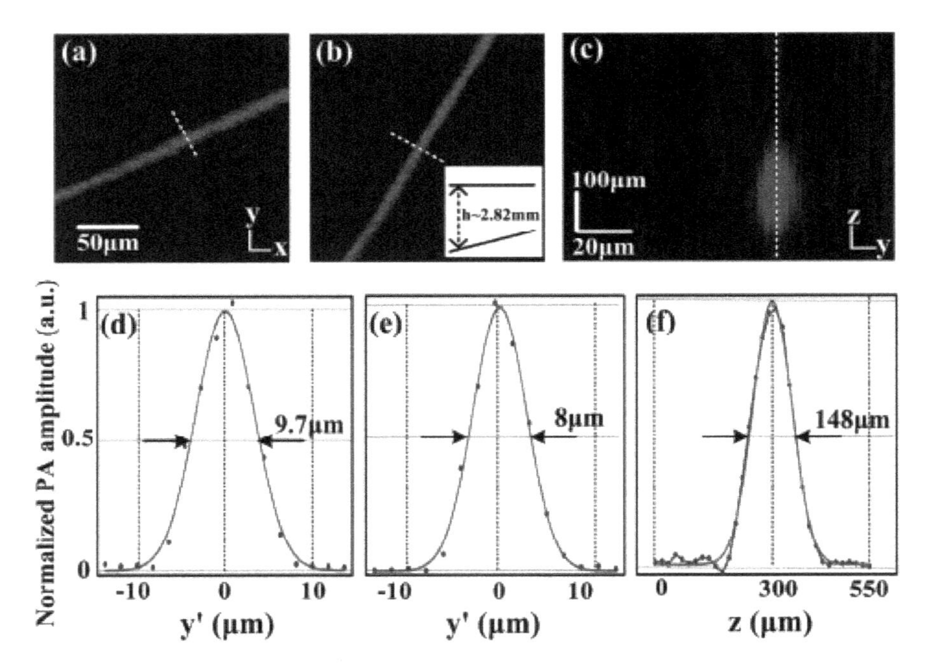

Figure 3.7 Focus-shifting range and axial resolution of the VF-PAM system. PA images of (a) the upper plane carbon fiber and (b) the lower plane carbon fiber. The inset at the bottom-right corner of (b) showed schematic diagram of the two planes carbon fibers, the depth interval between the two planes was 2.82 mm. (c) The axial section PA image of carbon fiber in (a). (d–e) Distribution of the PA amplitude (dots) along the dashed line in (a)-(b); the solid line is a Gaussian fitted curve. (f) Axial profile of the carbon fiber in (c); Blue line with dot: experimental signal; Red solid line: envelope of the experimental signal. Reproduced with permission from Ref. [16].

shown in Figs. 3.7d,e. The axial resolution of the VF-PAM system is determined by the bandwidth of the ultrasound transducer. Here, the theoretical axial resolution was calculated to be 132 μm. The experimental axial resolution was measured by using a 7 μm diameter carbon fiber, which is shown in Fig. 3.7c. Owing to the small diameter of the carbon fiber, the FWHM of the envelope of the axial spread profile could be regarded as the axial resolution of the imaging system. A PA axial spread profile of the carbon fiber along the dashed line is presented in Fig. 3.7f, which shows the axial resolution for the VF-PAM system was estimated to be 148 μm.

In order to further demonstrate the in vivo focusing scan ability of the VF-PAM system, microvasculature of the mouse ear was performed. As shown in Fig. 3.8a, PA images of a mouse ear microvasculature were obtained by focusing scan. The axial scanning range was taken from the tissues surface to approximately 225 μm below the surface. The depth interval was set to be 25 μm between two PA images, and the time for focus shift in depth direction was within 15 ms. Results showed that through focusing scan, more micro blood vessels in the focal plane can be seen in PA images. Three typical PA images at depths of 50 μm, 150 μm, and 225 μm in the mouse ear are displayed in Figs. 3.8b–d, respectively. As shown in Fig. 3.8c, many capillaries in the dashed circle and more micro blood vessels indicated by white arrows were imaged with good contrast, which were not viewed in Fig. 3.8b. However, since the capillaries were not in the depth of field, they vanished in Fig. 3.8d.

To further demonstrate that the VF-PAM system can achieve deeper focus scan, brain tissues at different depths of the mouse were performed. We obtained brain subcutaneous vessels (~200 μm) by PA imaging, as shown in Fig. 3.8f, brain cortex vessels (~400 μm) shown in Fig. 3.8g. The brain skull is from depth of 200 μm to 400 μm. Results show that the VF-PAM system can detect depth vessels in tissues through focusing scan. These PA images indicate the capability of providing many structural details of microvasculature and brain tissues by using the VF-PAM system.

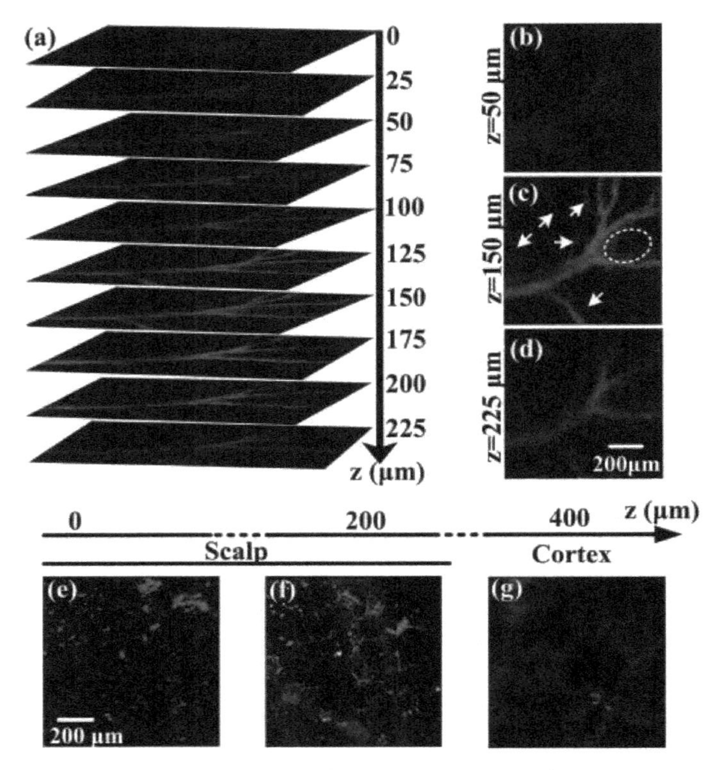

Figure 3.8 In vivo PA imaging of microvasculature of a mouse ear and brain tissues of mouse at different depths: (a) PA images of microvasculature of a mouse ear at different focus depths were acquired with the VF-PAM system, the depth interval was ~25 μm between PA images. (b–d) Three typical PA images of microvasculature, the focus depths were about 50 μm, 150 μm and 225 μm, respectively. (e) Epidermis. (f) Brain subcutaneous vessels (~200 μm). (g) Brain cortex vessels (~400 μm). Brain skull is located at the depth from 200 μm to 400 μm.

3.3 Acoustic-Resolution Photoacoustic Microscopy

In OR-PAM, the lateral spatial resolution is determined by tight optical focusing. However, the penetration depth is limited to one optical transport mean free path. Acoustic-resolution photoacoustic microscopy (AR-PAM) (Fig. 3.9) [10, 18] is a promising fusion imaging technique for deep imaging of biological tissues, where the excitation laser beam is loosely focused to the entire acoustic detection area. It combines the strong optical absorption contrast with the fine ultrasound spatial resolution, while overcoming the limited penetration depth (~1 mm) of optical imaging, and the weak imaging contrast and speckle artifacts of ultrasound. In this section, we introduce AR-PAM systems, including dark-field confocal AR-PAM system and multimode fiber bundle mediated AR-PAM system.

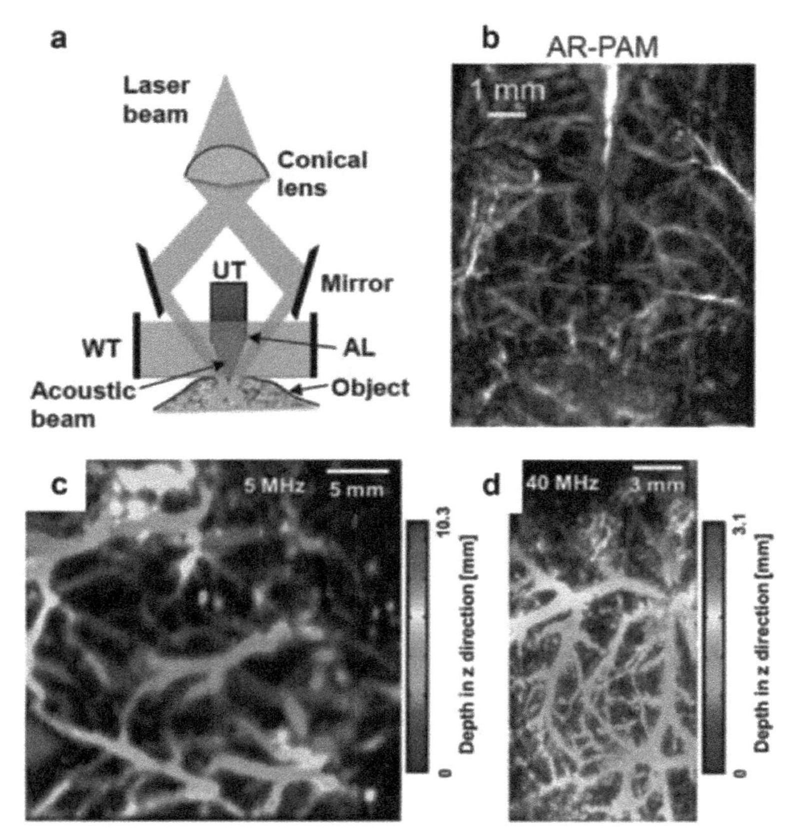

Figure 3.9 (a) Schematic of dark-field acoustic resolution photoacoustic microscopy (AR-PAM). (b) AR-PAM image of brain microvasculatures in a mouse with the intact skull but removed scalp. PA image of vasculature of rat's back region with ultrasound transducer of (c) 5 MHz and (d) 40 MHz. AL, acoustic lens; CorL, correction lens; RAP, right angled prism; RhP, rhomboid prism; SOL, silicon oil layer; UT, ultrasound transducer; WT, water tank. Reproduced with permission from Ref. [18].

3.3.1 Dark-Field Confocal AR-PAM

The system of dark-field confocal AR-PAM system [17] is shown in Fig. 3.10. A tunable OPO laser and Qswitched Nd:YAG pump laser with a pulse width of 4 ns and laser repetition rate of 10 Hz [19] are used in the system. The beam is split into two paths: One diverges directly at the output of the pump laser (Fig. 3.10a, green) and delivers λ = 532 or 1064 nm; the other is guided at the output of the OPO laser (Fig. 3.10a, red) and delivers the wavelength of light that the OPO laser provides (i.e., 680 ≤ λ ≤ 2,500 nm). Each beam path is guided by several prisms. A collimating lens is placed in the beam path from the OPO laser to reduce the divergence of the output beam. A spherical conical lens generates a ring-shaped light beam, so that the ultrasound transducer can be located at the center of the light without blocking beam propagation. The diverged ring-shaped beam is refocused by total internal reflection of the light in an optical condenser. The refocused light excites biological tissues, which generate PA waves. The generated PA waves are detected by an ultrasound transducer and the detected signals are amplified and saved in a data acquisition board. This dark-field co-axial geometry provides a deep imaging depth and high SNR when imaging biological tissues. The object biological tissues, especially small animals such as mice or rats, are positioned under a water tank that has a bottom opening that is sealed by a translucent membrane for light and ultrasound. The center frequency of the ultrasound transducer can be exchanged (e.g., 5 and 40 MHz), so the PA images can be acquired at different spatial resolutions and consequent imaging depths can be modulated. AR-PAM uses a point-by-point detection method. Thus, the PA waves are detected at each point of the area to be imaged. Therefore, to acquire a 3D PA image, it is necessary to apply raster scanning with a single element spherically focused ultrasound transducer. Because a single-element focused ultrasound transducer is used, reconstruction of the 3D image is not required a mathematical algorithm. Therefore, the resolution of the resulting image is much better than that acquired using an array transducer. Further, the use of raster scanning does not cause side lobes or image artifacts. Yet, the imaging acquisition rate is relatively slow since it depends on the mechanical scanning speed and laser repetition rate [20].

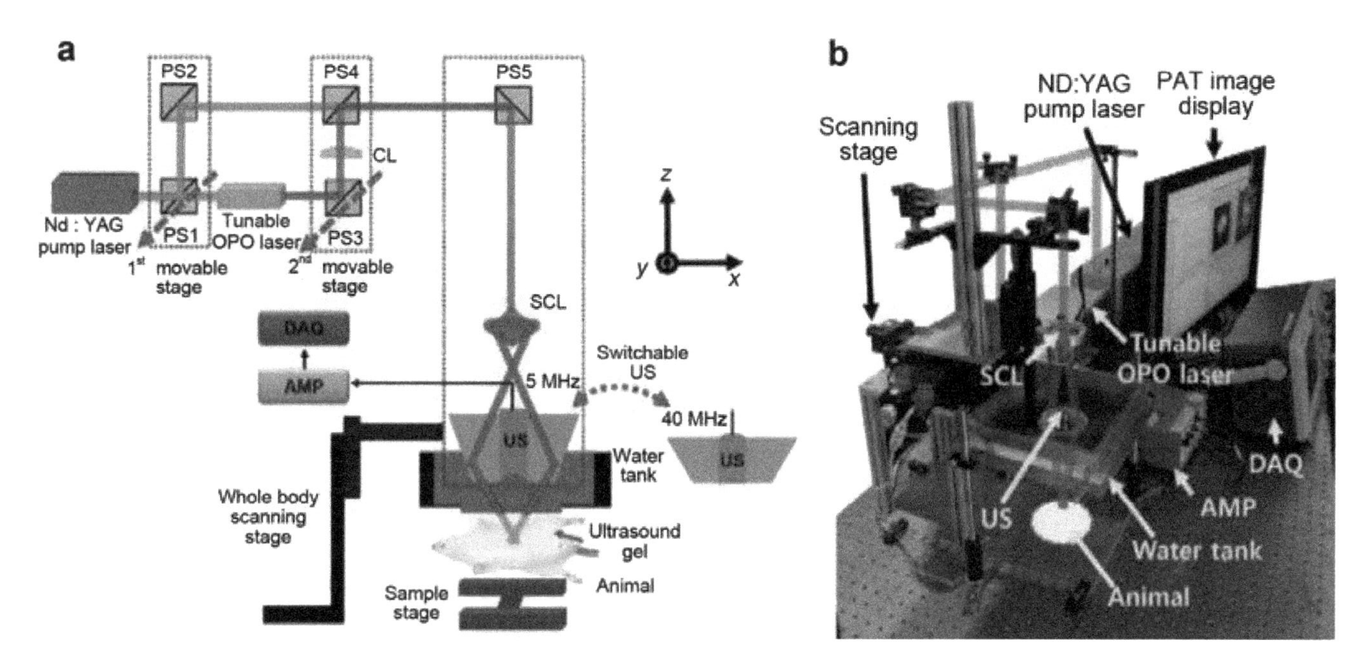

Figure 3.10 Dark field confocal AR-PAM system. (a) Schematic of the system. (b) Photograph of the system. OPO, optical parametric oscillator; PS, prism; CL, collimating lens; SCL, spherical conical lens; US, ultrasound transducer; AMP, amplifier; and DAQ, data acquisition. Reproduced with permission from Ref. [18].

3.3.2 AR-PAM Using Multimode Fiber Bundle

In OR-PAM, by tightly focusing the laser beam using high-NA optical objective lens, PA images with high lateral resolution can be obtained. However, the confocal of laser and sound limits the penetration depths. AR-PAM is a promising technique for deep imaging compared to OR-PAM, where the excitation laser beam is loosely focused to the entire acoustic detection area. Since AR-PAM and OR-PAM systems have complementary properties, both systems can be combined for biological imaging. A united AR- and OR-PAM system has been implemented by using an optical multimode fiber (MMF) bundle, as shown in Fig. 3.11. A high-repetition-rate laser used for OR-PAM and a high-power laser used for AR-PAM were integrated by a polarizing beam splitter cube (PBS). The two-beam laser is coupled using an optical multimodal fiber to achieve switchable OR-PAM and AR-PAM illumination. A single core (diameter: 2.0 μm) and 10,000 fiber cores were used as the OR-PAM and AR-PAM laser separated beam path, respectively. The OR-PAM beam was focused onto the single core of the (MMF) bundle by aligning the OR-PAM focusing beam at the tip of the fiber bundles, whereas the expanded AR-PAM beam was defocused at the tip to make sure that the beam fills the cores of the bundle. Finally, the fiber bundle was connected to a PA sampler. The integrated laser beam passed through two optical condenser lenses, through a beam combiner composed of an aluminum-coated prism, then focused light propagated into the sample. The acoustic lens collected the generated PA signals, which are aligned with the focal zone of laser beam and then obtained by a high-frequency (50 MHz) ultrasound transducer. In this system, switching and simultaneous modes can be imaged selectively by operators. For switching-mode imaging, after acquiring AR-PAM images, OR-PAM visualized a selected region of interest photoacoustically.

In this system, switching and simultaneous modes can be imaged selectively by operators. For switching mode imaging, after acquiring AR-PAM images, OR-PAM visualized a selected region of interest photoacoustically. Switching mode was used to obtain in vivo AR- and ORPAM images of a mouse ear (Fig. 3.12). First, the large-area PA image (Fig. 3.12b1) was acquired by the AR-PAM system; this image clearly shows the distribution of PA signals of the vasculature. Next, a high-resolution image of a region of interest was imaged using the OR-PAM system; this image shows detailed microvasculature (Fig. 3.12b2) and red blood cells (RBC) (Fig. 3.12b3). For simultaneous mode imaging, both laser beams fire alternately at their own pulse repetition rates. Two PAM images are acquired without any interference by inserting a small time interval between the OR-PAM and the AR-PAM firing times.

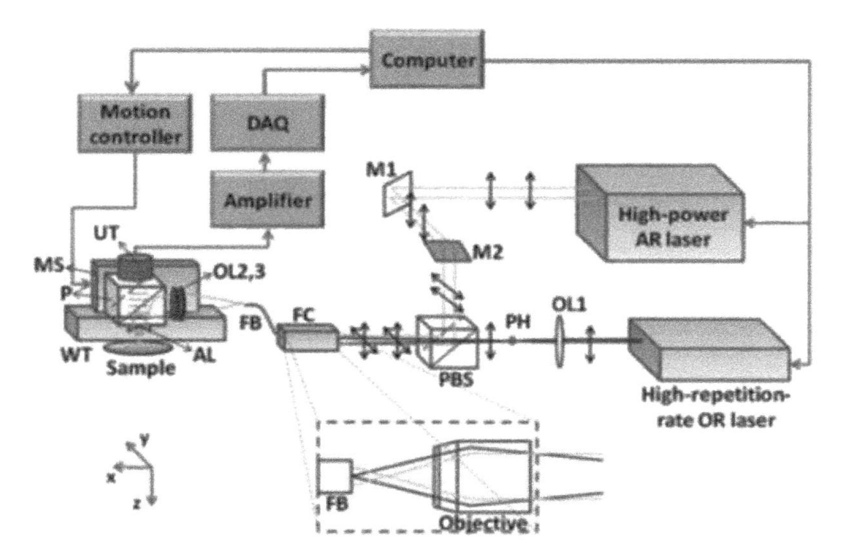

Figure 3.11 Schematic of a united AR- and OR-PAM system using a multimodal fiber bundle. AL, acoustic lens; FB, fiber bundle; FC, fiber coupler; M, mirrors; MS, motor stage; OL, optical lens; P, prism; PBS, polarizing beam splitter; PH, pinhole; UT, ultrasound transducer; WT, water tank. Reproduced with permission from Ref. [11].

This mode was also used to image a small area (Fig. 3.13c1) of a mouse back in vivo. OR-PAM (Fig. 3.13c2) and AR-PAM (Fig. 3.13c3) were acquired simultaneously. The OR-PAM image shows small blood vessels [21] near the surface; the AR-PAM image shows the deep and large blood vessels. A combined image can be obtained by overlaying the images (Fig. 3.13c4).

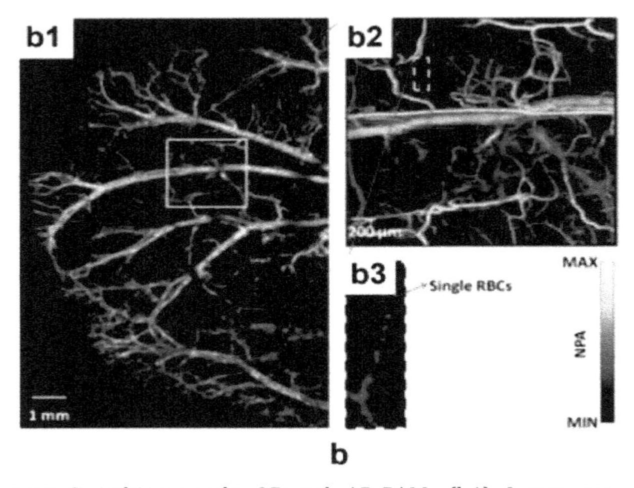

Figure 3.12 Switching mode OR-and AR-PAM. (b1) Large area of the mouse ear of AR-PAM. (b2) Region of interest mapped by OR-PAM. (b3) Enlarged RBC PA image. Switching mode was used to obtain in vivo AR- and OR-PAM images of a mouse ear (Fig. 3.12b). First, the large area PA image (Fig. 3.12b1) was acquired by the AR-PAM system; this image clearly shows the distribution of PA signals of the vasculature. Reproduced with permission from Ref. [11].

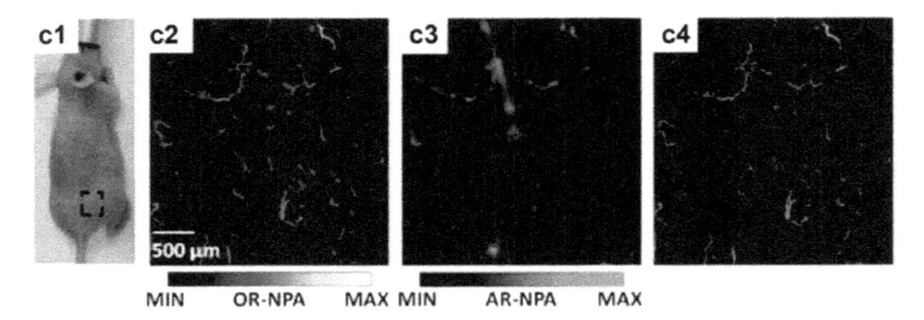

Figure 3.13 Simultaneous mode OR-and AR-PAM. (c1) Photograph of the mouse. Square: area imaged. (c2) OR-PAM MAP image of mouse skin. (c3) AR-PAM MAP image of mouse skin. (c4) Emerged OR-and AR PAM image. Reproduced with permission from Ref. [11].

3.4 Biological Applications of Acoustic-Resolution Photoacoustic Microscopy

As a unique optical absorption microscopy technology and a valuable complement to the existing technologies, OR-PAM has demonstrated broad biomedical applications. Taking advantage of the strong optical absorption of endogenous hemoglobin, OR-PAM enables label-free, noninvasive, volumetric microvascular imaging down to single capillaries, providing both anatomical (such as vessel diameter, connectivity, and tortuosity) and functional (such as SO_2 [22, 23] and blood flow [24]) information. With the aid of exogenous molecular contrast agents [25], OR-PAM is also capable of molecular imaging [26].

In order to verify the in vivo imaging ability of the multi-wavelength OR-PAM system, PA image of the microvasculature in the mouse ear was acquired at 532 nm under normoxia [15], as shown in Fig. 3.14a. Then the mouse inhaled mixed-gases for 10 min, and the two blood vessels across the scan line A were detected at the three wavelengths. The two blood vessels can be easily distinguished in Fig. 3.14b.

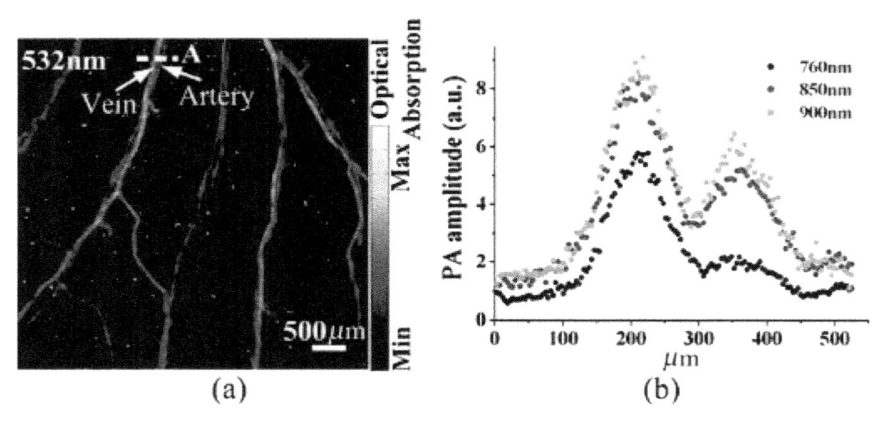

Figure 3.14 (Color online) (a) PAM image of microvasculature in a mouse ear at the wavelength of 532 nm under normoxia. The lateral resolution is ~50 μm; (b) The reconstructed profile of the blood vessel obtained at 760, 850, and 900 nm, respectively, at the position indicated by the dashed line in (a). Reproduced with permission from Ref. [15].

3.4.1 Photoacoustic Identification of Sentinel Lymph Nodes and Lymphatic Systems

Detection of cancers in the sentinel lymph nodes (SLNs) is extremely important to diagnose cancer stage and to guide development of a treatment plan [27–29]. To identify the stage of cancers, including prostate and breast cancers, the general method is to resect lymph nodes (LNs) and then to conduct ex vivo histology on the samples. However, this method can cause lymphedema, sensory nerve injury, and limitation of motion. Therefore, alternative methods to acquire the diagnostic information of SLN noninvasively have been investigated such as ultrasound (US)-guided fine-needle aspiration biopsy (FNAB) [29], fluorescence (FL) and PA imaging [30–35]. US-guided FNAB was less invasive but lymph node identification was not accurate. FL optical microscopy can noninvasively acquire lymph node images but suffers from poor spatial resolution due to its inherent optical scattering issue in deep tissues (>~1 mm). AR-PAM using exogenous contrast agents can provide strong optical contrast and high ultrasound spatial resolution in deep tissues (up to several centimeters). The imaging depth of AR-PAM can be up to a few centimeters, which is much deeper than can be obtained using pure optical imaging. Because the SLNs themselves do not absorb light, exogenous contrast agents with strong optical absorption capacity such as organic dyes and metallic nanostructures are used for in vivo PA imaging of SLNs. Thus, one of the promising biomedical applications in AR-PAM is to noninvasively detect SLNs and lymphatic systems in animals and humans in vivo.

Indocyanine green (ICG) is one of the most popular organic dyes for use as AR-PAM contrast agents. ICG is hydrosoluble and nontoxic and absorbs strongly in the near infrared (NIR) region (λ = ~800 nm). Further, ICG has been approved by the US Food and Drug Administration for diagnosing human hepatic function, cardiology, renal blood flow, and subretinal processes in the eye [1]. Because of its moderate fluorescence quantum yield (i.e., ~10%), ICG can be used for PA imaging. SLNs and lymphatic systems of a rat's axillary region were noninvasively mapped in vivo by PA imaging. The images were post-processed and depicted in 2D using maximum amplitude projection (MAP) method along depth. In the control image (Fig. 3.15a), the surrounding blood vessels were visible, but the lymphatic systems were not. After ICG injection (Fig. 3.15b), the SLN and lymphatic vessels, as well as surrounding vasculatures, were clearly visible. The PA signal of SLN was about seven times higher

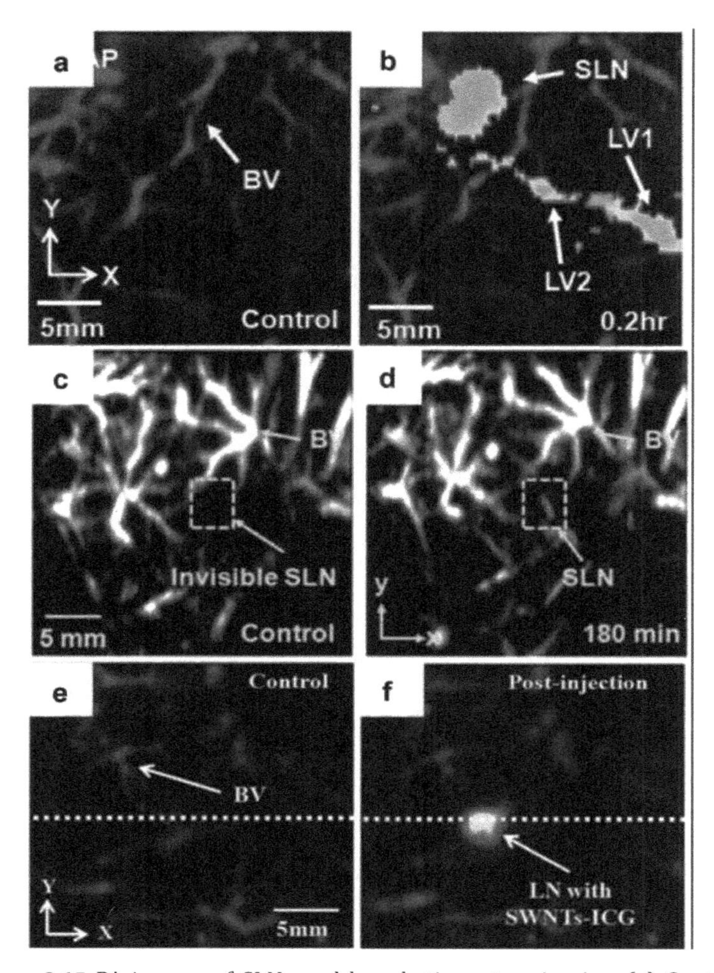

Figure 3.15 PA images of SLNs and lymphatic system in vivo. (a) Control PA MAP image acquired before ICG injection. (b) PA MAP image acquired after ICG injection. (c) Control PA MAP image obtained before $Cu_{2-x}Se$. (d) PA MAP image acquired at 180 min after injection of $Cu_{2-x}Se$. (e) Control PA MAP image acquired before SWNTs-ICG injection. (f) PA MAP image acquired 30 min after injection of SWNTs-ICG. BV, blood vessel; SLN, sentinel lymph node; and LV, lymphatic vessel. Reproduced with permission from Ref. [10].

after injection (0.96 ± 0.038, arbitrary unit) than before injection (0.13 ± 0.13, arbitrary unit). Although use of ICG enables in vivo mapping of SLN and lymphatic vessels, it is cleared rapidly from the circulatory system; this is the main drawback of ICG, and of other organic dyes. To overcome the rapid clearance of organic dyes from the circulatory system, metallic nanostructures have been used as PA contrast agents, because they absorb strongly in the NIR due to localized surface plasmon resonance (LSPR). The metallic nanostructures can generate heat by LSPR under laser irradiation; this is called the photothermal effect. However, the relatively large size of metallic nanostructures causes them to accumulate in organs and lead to potential side effects. Highly doped semiconductor nanocrystals (NCs) such as copper selenide ($Cu_{2-x}Se$) also can generate heat by LSPR. $Cu_{2-x}Se$ is much smaller than metallic nanostructures, so much less of it accumulates in organs and the concerns of potential side effects are thereby relieved. Further, $Cu_{2-x}Se$ can provide strong PA contrast for SLN imaging. To investigate the $Cu_{2-x}Se$ NC as a PA lymphatic tracer, the PA image of a rat's axilla was acquired by injecting 100 μL of a buffer solution containing 0.8 pmol $Cu_{2-x}Se$ NCs intradermally into its left forepaw. In the pre-injection control image, the blood vessel networks of axilla were visible with good contrast but the SLNs and lymphatic vessels were not visible (Fig. 3.15c); 180 min after injection of $Cu_{2-x}Se$ NCs, the blood vessel networks and lymphatic systems were both visible (Fig. 3.15d). The PA amplitude of SLN treated with $Cu_{2-x}Se$ NCs (i.e., 0.12 ± 0.006, arbitrary unit) was about 4.8 times stronger than that of the background (i.e., 0.025 ± 0.02, arbitrary unit). This high sensitivity can provide capability to track disease progression by injecting a minimal dose of contrast agent. As another contrast agent, single-walled carbon nanotubes (SWNTs) have been explored. SWNTs have many merits such as ease of fabrication and bioconjugation, large surface area, and broad optical absorption spectra. Further, PA sensitivity can be significantly enhanced by attaching ICG molecules to the SWNTs. Due to this signal enhancement, the used dose for in vivo imaging can be significantly reduced; so concerns about long-term toxicity can be reduced. The SLNs in the left axilla of rat were visualized utilizing SWNT-ICG conjugate as a PA contrast agent using λ = 820 nm, the LSPR peak of SWNTICG. The control PA image showed the clear vasculatures with good contrast (Fig. 3.15e), but LNs and lymphatic vessels were not able to be visualized. After intradermal injection of SWNTs-ICG, both blood vessels and LNs were obviously identified (Fig. 3.15f). The contrast of SLN (i.e., 30.3 ± 4.3, arbitrary unit) was stronger than that of surrounding blood vessels (i.e., 9.4 ± 1.9, arbitrary unit) because the absorption capacity of SWNT-ICG is higher than that of blood at λ = 820 nm.

3.4.2 Photoacoustic Imaging of Gastrointestinal (GI) Tract

Several gastrointestinal (GI) diseases, including irritable bowel syndrome, inflammatory bowel disease, and constipation, are related to enterokinesia dysfunction. These abnormal intestine conditions influence dangerous diseases such as diabetes, thyroid disorders [68], and Parkinson's disease [69]. Unfortunately, current methods to image the GI tract such as CT, X-ray imaging, magnetic resonance (MRI) and US imaging have limitations such as safety problems, ionization, low spatial resolution, high cost, and low accessibility [70]. Moreover, they have difficulty in imaging intestinal peristalsis and segmentation. Therefore, to improve diagnosis of GI diseases, new GI imaging methods should be developed. AR-PAM is a possible imaging method for this purpose. PA imaging is noninvasive and nonionizing and is therefore appropriate for use in children [29]. By visualizing dynamic intestinal movement, peristalsis and segmentation can be revealed. The AR-PAM system can be fabricated at relatively low cost; it can also be made portable to allow comfortable monitors and accurate access to the GI region of interest. In vivo GI PA imaging used frozen micellar naphthalocyanines (nanonaps) [71]. A dark-field PA imaging system with 5 MHz frequency was used for noninvasive GI PA imaging of the intestine. PA images (Fig. 3.16a) clearly visualized the movement of nanonaps through the intestine during a 6 h period, and individual diverticula were revealed with high spatial resolution (150 μm). Depth-encoding analysis (Fig. 3.6b) of the PA MAP image (Fig. 3.16a at 1 h) provided detailed depth distribution of GI track a resolution of ~5 mm. For dynamic GI track movement monitoring, a commercial Vevo LAZR US/PA system was used; it displays real-time depth-resolved PA and US images simultaneously at 5 frame per second.

The overlain cross-sectional PA (color) and US (gray) image (Fig. 3.16c1) reveals the nanonap distribution in the GI tract. Detailed nanonap movement in the intestine could be monitored in real-time. Moreover, the flow changes (i.e., inflow and outflow) in the intestine were clearly observed (Fig. 3.16c2).

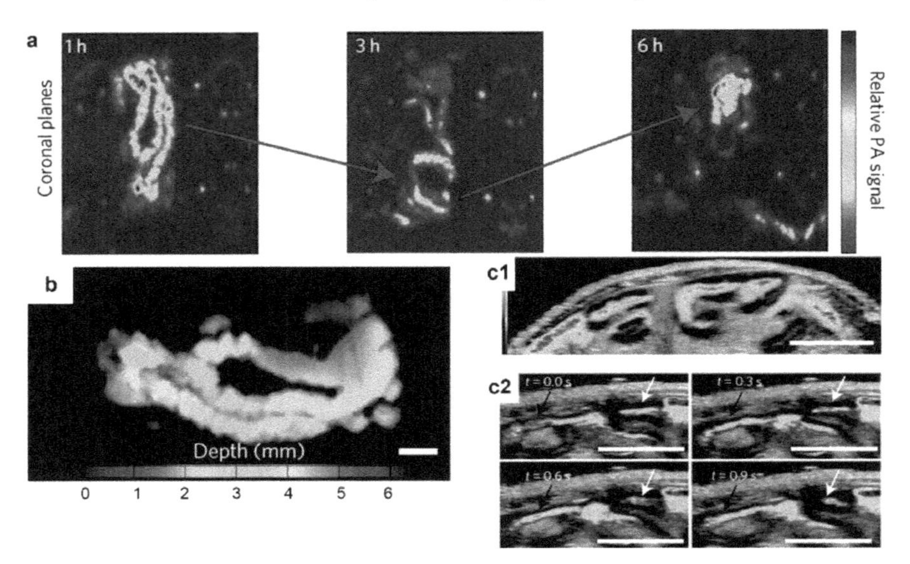

Figure 3.16 (a) PA MAP of nanonap after oral delivery of nanonaps based on a dark-field photoacoustic imaging system. Red arrows indicate nanonap movement in intestine for 6 h. (b) Depth-encoded PA MAP of intestine with nanonaps. (c1) Real-time multimodal cross-sectional mouse PA (Color) and US (gray) simultaneously acquisition after oral delivery of nanonaps. (c2) Nanonap flow movement in the intestine. Black and white arrows indicate inflow and outflow, respectively. Reproduced with permission from Ref. [10].

3.4.3 Whole-Body Photoacoustic Imaging of Small Animals

Whole-body imaging is essential to visualize living animals for biomedical research. The structural and functional information of small animals is useful in the development of new drugs, and in study of cancer physiopathology and brain hemodynamics. To acquire the information of small animals in vivo, US, CT, MRI, positron emission tomography (PET), and optical imaging have been investigated. Each method has its strengths and drawbacks. US imaging can provide real-time image of small animals in vivo, but its speckle artifacts degrade image quality. CT can provide fine spatio-temporal resolution but may cause damage because it uses ionizing radiation; it also has poor sensitivity in soft tissues. MRI can provide excellent spatial resolution and strong soft tissue constant, but it is extremely expensive and slow. PET can provide great sensitivity in molecular activities of biological tissues, but it uses ionizing radioactive materials and is also expensive. Optical imaging methods such as fluorescence imaging and bioluminescence imaging can provide real-time image, are easy to implement, are inexpensive, and have high sensitivity and high specificity, but have the disadvantage of shallow penetration depth. PA imaging can overcome these drawbacks. Due to its acoustic-optical property, PA imaging does not generate speckle artifacts, is completely free from ionizing radiation, is sensitive to soft tissues, is inexpensive, and can penetrate much deeper than pure optical imaging without significant loss of spatial resolution. The field of view of an AR-PAM system was enlarged to encompass the whole body of a BALB/c mouse, then the whole-body PA images were acquired (Fig. 3.17). In vivo PA images of four mice in four different imaging planes (i.e., left sagittal, right sagittal, ventral, and dorsal planes) were taken. The tissues were excited at four different optical wavelength (λ = 532, 700, 850, and 1,064 nm) in all imaging planes. The PA images clearly identify the major blood vessels and internal organs. The imaging views are different, so that visible blood vessels and internal organs varied among the imaging planes. The image of the left sagittal plane (Fig. 3.17a) revealed the blood vessels such as descending aorta, kidney, spleen, intercostal vessels, cranial mesenteric vessels, femoral vessels, cephalic vessels, brachial vessels, transverse marginal vessels, popliteal vessels, and mammalian vessels, and the internal organs such as kidney, spleen, liver, and cecum, transverse marginal vessels, popliteal vessels, and mammalian vessels. The image of the right sagittal plane (Fig. 3.17b) revealed the blood vessels such as descending aorta, intestine, cranial mesenteric vessels, intercostal vessels, femoral vessels, popliteal vessels, transverse marginal vessels, brachial vessels, cephalic vessels, and mammalian vessels and

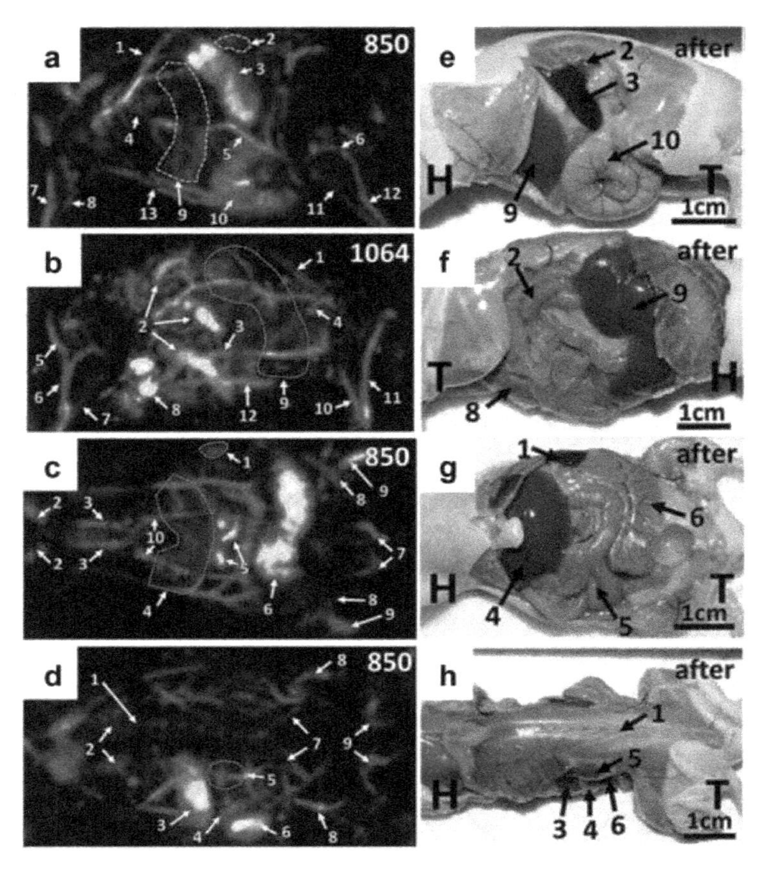

Figure 3.17 Whole-body PA images of BALB/c mice in vivo at (a) left sagittal [1 descending aorta, 2 kidney, 3 spleen, 4 intercostal vessels, 5 cranial mesenteric vessels, 6 femoral vessels, 7 cephalic vessels, 8 brachial vessels, 9 liver, 10 cecum, 11 transverse marginal vessels, 12 popliteal vessels, 13 mammalian vessels], (b) right sagittal [1 descending aorta, 2 intestine, 3 cranial mesenteric vessels, 4 intercostal vessels, 5 femoral vessels, 6 popliteal vessels, 7 transverse marginal vessels, 8 cecum, 9 liver, 10 brachial vessels, 11 cephalic vessels, 12 mammalian vessels], (c) ventral [1 spleen, 2 carotid artery and jugular vein, 3 subclavian vessels, 4 liver, 5 intestine, 6 cecum, 7 caudal vessels, 8 femoral vessels, 9 popliteal vessels, and 10 mammalian vessels], and (d) dorsal plane [1 spinal cord, 2 intercostal vessels, 3 spleen, 4 intestine, 5 kidney, 6 cecum, 7 iliac vessels, 8 femoral vessels, and 9 caudal vessels]. Photographs of skinned mice after in vivo imaging at (e) left sagittal, (f) right sagittal, (g) ventral, and (h) dorsal plane. Reproduced with permission from Ref. [10].

the internal organs such as intestine, cecum, and liver, brachial vessels, cephalic vessels, and mammalian vessels. The image of the ventral plane (Fig. 3.18c) revealed the blood vessels such as spleen, carotid artery, and jugular vein, subclavian vessels, caudal vessels, femoral vessels, popliteal vessels, and mammalian vessels, and the internal organs such as spleen, liver, intestine, and cecum, caudal vessels, femoral vessels, popliteal vessels, and mammalian vessels. The image of the dorsal plane (Fig. 3.17d) revealed the blood vessels such as spinal cord, intercostal vessels, iliac vessels, femoral vessels, and caudal vessels, and the internal organs such as spinal cord, spleen, intestine, kidney, and cecum, iliac vessels, femoral vessels, and caudal vessels. All blood vessels and internal organs were identified by comparing the PA MAP image with commercial mouse anatomy illustration software (Biosphera). After the in vivo images were acquired, the experimental mice are sacrificed, skinned, and photographed (Fig. 3.17e–h). The location and identification of the internal organs in the PA images were confirmed by comparing them with the corresponding photograph.

3.4.4 Conclusion

AR-PAM has several applications in preclinical and clinical biomedical research because of its outstanding imaging capability. AR-PAM can provide high sensitivity and specificity, strong contrast of biological tissues, and relatively deep penetration. This chapter has reviewed the principles and system implementation of AR-PAM, and has presented several applications of AR-PAM to image biological tissues such as sentinel lymph nodes, lymphatic systems, bladders, gastrointestinal tracts, and whole body of small animals. ARPAM can be easily used in other biomedical research, so it is an essential imaging method for this field.

3.5 Single-Wavelength Excited Photoacoustic-Fluorescence Microscopy

The balance between radiative and non-radiative decay is affected by various factors at the excitation state. A simultaneous acquisition of the fluorescence/photoacoustic signals from an appropriate probe provides an efficient and high-resolution means to monitor such a balance in a biological target and thus may render its physiological information. Acidity plays an important role in tissue physiology. Here, we report an integrated photoacoustic-fluorescence microscopy (PA-FLM) for high-resolution (<3 μm) image mapping of interstitial pH by detecting the shift in the signal balance of a pH-sensitive probe. The hypothesis and the technical feasibility are validated with an in vivo tumor model. The results show that the technique can effectively monitor pH changes within the range of biological acidity and are independent of the excitation source fluctuation and local probe concentration. We thus propose that with further research and selection of proper probes, PA-FLM may provide a potential alternative for monitoring tissue physiology.

3.5.1 Introduction

Acidity plays an important role in tissue physiology. It is closely related to the development stage of tumors and commonly used to evaluate therapeutic responses [56–64]. The development of a technique to noninvasively visualize tumor pH at high-resolution could help to predict the potential of cancer metastasis and multidrug resistance as well as to design the most appropriate therapeutic strategy to realize personalized medicine.

There are various methods to detect tumor pH. One invasive method uses microelectrodes inserted into the tumor. However, this approach can only measure the pH at one location at any time and cannot map pH across the tumor bed [65]. Optical imaging is promising for tumor pH mapping because it is noninvasive, sensitive, and affordable and uses no radioactivity. Single-wavelength fluorescence pH imaging has been developed to measure pH in vivo. However, excitation source fluctuations and unpredictable local probe concentrations can affect the signal more than the underlying pH change [66]. This is further compounded by changes in the concentrations of background absorbing/scattering species including water, hemoglobin, and melanin. These can all attenuate the incident excitation energy and emission signal and compromise the accuracy of the measurement.

Dual-wavelength approaches use ratiometric imaging to compensate for these variations. Methods include dual-wavelength fluorescence ratiometric imaging and dual-wavelength photoacoustic ratiometric imaging method—both have been used to measure pH [67, 68]. With these techniques, a ratiometric probe is designed with a pH sensitive indicator molecule and a pH insensitive reference molecule. The pH measurement uses the ratio of the intensity of two fluorescence peaks or photoacoustic peaks rather than that of a single peak. The ratio is less sensitive to local probe concentration. However, this method does require switching between excitation wavelengths, and this effectively doubles the time needed to generate a tumor map. In addition, the different wavelengths will be differentially absorbed by the tissue. This could potentially compromise the detection accuracy.

Previously [62, 69, 70], a pH-sensitive NIR fluorescence probe (DiIRB-S) was developed and shown to have emission intensity that changed as a function of pH. DiIRB-S is made up of two heptamethine cyanine derivatives: IR783 groups conjugated via an acid liable hydrazine bond. In neutral environments, the bound IR783s with a hydrophobic core create a face-to-face aggregate, and their radiative relaxation (fluorescence) is suppressed via intermolecular nonradioactive decay. At acidic pH, hydrazone bonds are cleaved with a concomitant decrease in the self-quenching effects. This activates fluorescence. It is well known that stimulated emission from an excited fluorescence probe has two paths: radiative decay via fluorescence and non-radiative decay via thermal dissipation [71]. Non-radiative decay was not usually measured/monitored due to technical difficulty. If the excitation light is short pulse (e.g., nanosecond), then the heat generated during non-radiative decay produces an ultrasonic wave via thermoelastic expansion—this is the photoacoustic effect [72, 73]. Thus, this change in the ratio between fluorescence and photoacoustic signal can indicate a change in the underlying radiative and non-radiative relaxation pathways. In turn, this signal can be a sensitive tool to measure pH.

In this chapter, we describe a single-wavelength excited photoacoustic-fluorescence microscope (PA-FLM) that was developed to detect shifts in the fluorescence/photoacoustic signal balance of a single excitation process. This was then used for high-resolution mapping of the tumor interstitial pH in vivo. Lateral resolution of this PA-FLM system was tested. We verified the ratio of photoacoustic and fluorescence signal intensities as a function of pH. Furthermore, we demonstrate that this pH measurement technique is independent of excitation source fluctuations and local probe concentration and can provide high imaging contrast to measure relatively small changes in pH. Finally, the proposed technique was validated with a mouse superficial tumor model to demonstrate that, the interstitial pH could be mapped in high-resolution at different stages of the tumor development. As the fluorescence and photoacoustic waves propagate differently in biological tissues, the PA-FLM will likely be depth-limited due to light scattering, similar to that in conventional fluorescence imaging.

3.5.2 Photoacoustic-Fluorescence Microscopy Imaging System

First, we developed an integrated PA-FLM to synchronize detection of probes' fluorescence and photoacoustic signals in a single excitation process (Fig. 3.18a). A 2.5 kHz tunable wavelength laser system (pulse width is 10 ns; spectral tuning range 670–2600 nm, NT252-2.5K, EKSPLA) was used as the light source for both excitation photoacoustic and fluorescence signal. The laser was tuned to 685 nm and was coupled to a 1 m fiber via a port coupler (PAF-X-7-B, Thorlabs). The collimating lens on the other end of the fiber was mounted to the upward side of an objective lens (a field flattening lens with 0.3 NA) and a home-made ring ultrasound transducer (18 MHz center frequency, −6 dB bandwidth of 9 MHz, 15.8 mm focal length, 2 mm inner diameter, and 2.6 mm outer diameter). This transducer was mounted to the bottom of objective lens (Fig. 3.18a insert). The light passing through collimating lens was then focused by a second laser in the coupling cup and then projected onto the test sample surface. The photoacoustic signal was collected by the transducer beneath the objective lens. Simultaneously, the resulting fluorescence was collected by the objective lens and detected with a photomultiplier tube (PMM02, Thorlabs) through a dichroic lens (FF731-Di01-25x36, Semrock) and filters (FF01-769/41–32, Semrock). A silicon photodiode (ET 2000, Electro-Optics Technology, Inc., Traverse City, USA) monitors and calibrates the intensity of the excitation light. The photoacoustic and fluorescence data were synchronously recorded via a dual-channel data acquisition card. The sampling rate was 200 M samples per second. A computer-controlled ultrasonic motor moved the integrated collimating lens, and an objective lens with two-dimensional imaging and analysis collected the photoacoustic and fluorescence signal generated at each scan point in real-time. The fluorescence images, photoacoustic images, and the ratio of fluorescence and photoacoustic images within the region of interest can be obtained simultaneously.

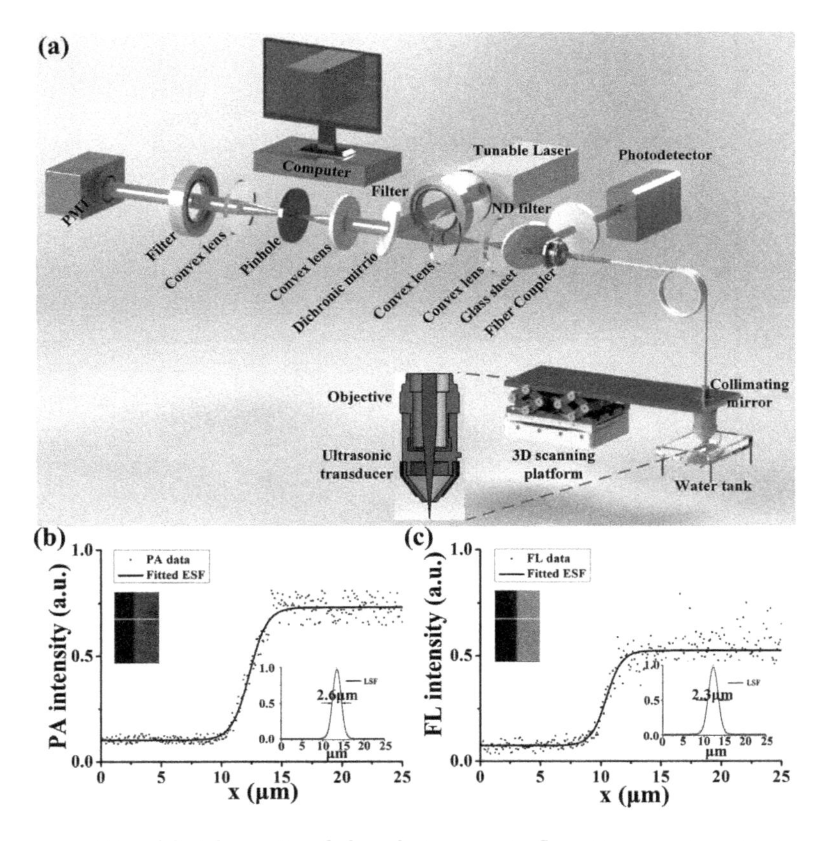

Figure 3.18 (a) Schematic of the photoacoustic-fluorescence microscopy system (PA-FLM). Panels (b) and (c) show the lateral resolution of PA-FLM. Edge spread function (ESF) and line spread function (LSF) of photoacoustic channel (b) and fluorescence channel in PA-FLM (c). The LSF in the inserts is extracted from the fitted ESF. PA, photoacoustic; FL, fluorescence; PMT, photomultiplier tube; filter; ND, neutral density filter; and 3D, three-dimensional. Reproduced with permission from Ref. [74].

A surgical blade was imaged to measure the lateral resolution of the PA-FLM system. In this experiment, the scanning step is 0.2 μm, and the rate is 200 Hz. The maximum amplitude projection (MAP) images of the blade and the edge spread function (ESF) acquired by the photoacoustic and fluorescence channels are shown in Fig. 3.18b,c, respectively. The derivative of the ESF yielded the line spread function (LSF), which is shown by a black line in the insets of Fig. 3.18b,c, respectively. The lateral resolution of photoacoustic and florescence channels was defined by the (FWHM) of the LSF. These values were 2.6 μm and 2.3 μm, respectively.

3.5.3 Single-Wavelength Excited Photoacoustic-Fluorescence Microscopy for in vivo pH Mapping

The pH-sensing mechanism is shown in Fig. 3.19a. A pH-sensitive NI R fluorescence probe (DiIRB-S) is made up of two IR783 that were conjugated via an acid liable hydrazine bond. Hydrazone bond is an acid liable bond that usually keeps stable under physiologically neutral condition (pH 7.4) but can be cleaved rapidly under acidic microenvironment (pH 5.5–6.5). The cleavage of hydrazone bond is inreversible. In our previous work, the results confirmed that most hydrazone bonds were broken at 2 h post incubation in acidic buffered solution (pH 5.5 and 6.5). However, the hydrolytic rate of hydrazone bond at neutral pH was very low and only trace amounts of hydrolytic products were produced [62, 69]. In neutral environments, the DiIRB-S has low fluorescence and high photoacoustic signal after excitation. In acidic pH, the hydrazone bonds are cleaved resulting in the activation of fluorescence that weakens the photoacoustic signal. The DiIRB-S exhibited two absorption bands at 685 and 756 nm in an aqueous solution—these were assigned to the H-type aggregate and free monomer of IR783. While absorbance at 756 nm increased consistently at acidic pH, the absorbance at 685 nm decreased gradually. This indicated the conversion of the aggregate to a monomer in an acidic environment (Fig. 3.19b). Fluorospectroscopy studies indicated remarkable fluorescence enhancements of DiIRB-S at acidic pHs (Fig. 3.19c) confirming cleavage of the hydrazone bonds. PA-FLM simultaneously detected the fluorescence and photoacoustic signals of DiIRB-S

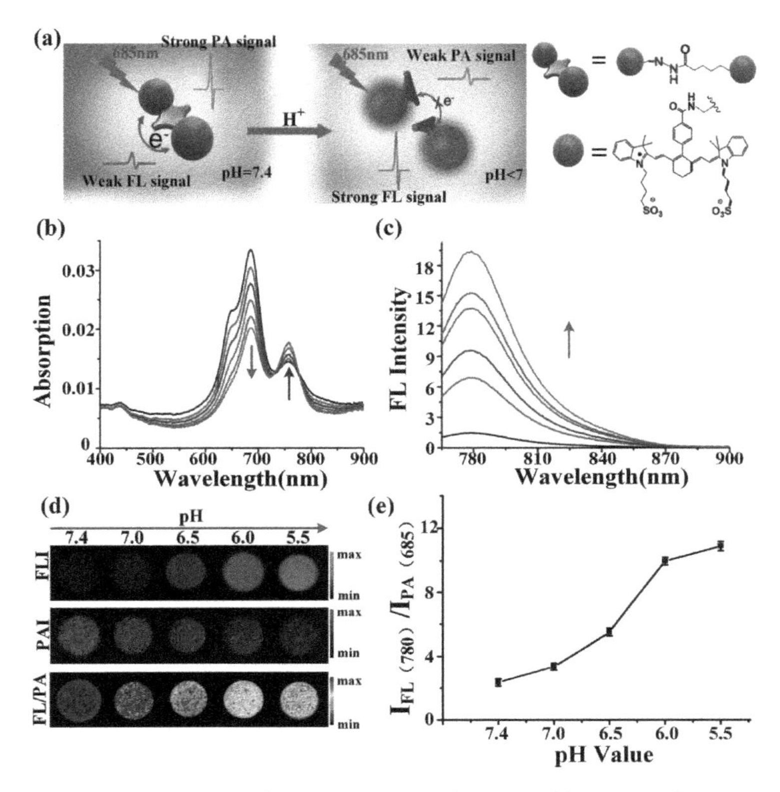

Figure 3.19 (a) Proposed pH-sensing mechanism. (b) UV-vis absorption spectra and (c) fluorescence spectra of DiIRB-S (1 μM) in phosphate buffered saline (PBS) responding for acid environment. (d) Fluorescence imaging, photoacoustic imaging and the ratio of fluorescence/photoacoustic data for DiIRB-S (2 μM, after 2 h incubation) in PBS at pH = 7.4, 6.8, 6.5, 6.0, and 5.5. Photoacoustic and fluorescence signal of DiIRB-S in a single excitation process are synchronously detected by PA-FLM. (e) Quantification of the ratio of fluorescence and photoacoustic signal intensities at different pH values. The error bars represent standard deviations of three separate measurements. PA, photoacoustic; FL, fluorescence; and FLI, fluorescence imaging; PAI, photoacoustic imaging. Reproduced with permission from Ref. [74].

in solution with different pH values (7.4, 7.0, 6.5, 6.0, and 5.5). The ratio of fluorescence/photoacoustic signal intensities at pH = 6.0 is 9.99 ± 0.24, pH = 6.5 is 5.52 ± 0.22, pH = 7.0 is 3.36 ± 0.21 and pH = 7.4 is 2.37 ± 0.23. The results showed that the DiIRB-S photoacoustic signal reduced with decreasing solution pH. The fluorescence at 780 nm increased (Fig. 3.19d). The ratio of fluorescence and photoacoustic signals at 685 nm excitation changed sensitively with pH (Fig. 3.19d). Furthermore, the ratio of fluorescence and photoacoustic signals showed pH dependence from 5.5–7.0 (Fig. 3.19e)—these are clinically relevant values in the tumor microenvironment.

Next, we show that the photoacoustic and fluorescence signal intensities are not affected by fluctuations in excitation source intensity and local probe concentration. After excitation, the emitted fluorescence intensity equals [74]

$$F_1 = AF\mu_a\eta, \tag{3.1}$$

where A is a constant, F is optical fluence (J/cm^2), and μ_a is the absorption coefficient (cm^{-1}). These terms are dependent on the fluorophore's concentration and its molecular absorption cross section; is quantum yield of the probe. If the excitation light is a short pulse, then the heat generated during non-radiative decay produces an ultrasonic wave via thermoelastic expansion. The photoacoustic amplitude is [74]

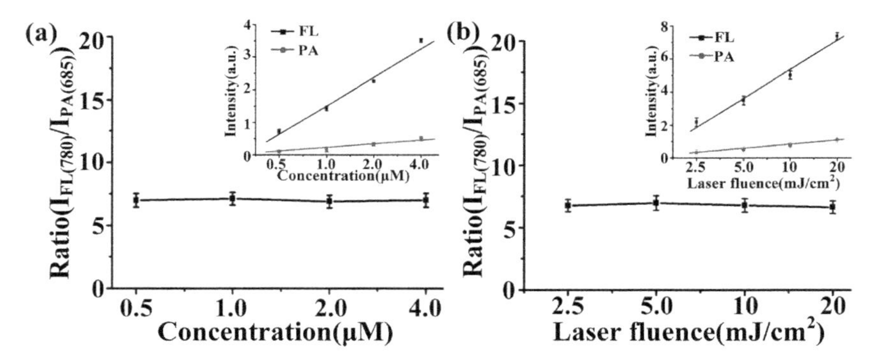

Figure 3.20 (a) The ratio of fluorescence/photoacoustic signal intensities at various concentrations (0.5 μM–4 μM). Laser energy fluence is 5 mJ/cm^2. Inset: The fluorescence and photoacoustic signal intensities at various concentrations. (b) The ratio of fluorescence/photoacoustic signal intensities at different excitation laser fluences. The concentration of the probe is 2 μM. Inset: The fluorescence and photoacoustic signal intensities at different excitation laser fluences. The error bars represent standard deviations of three separate measurements.

$$P = BF\mu_a(1 - \eta)\Gamma. \tag{3.2}$$

In Eq. (3.2), B is a constant, and Γ is the Grüneisen coefficient (nearly constant at room temperature). The ratio of photoacoustic and fluorescence signal intensities in this probe indicates pH. Thus, pH is proportional to the ratio of fluorescence and photoacoustic signals in a single excitation:

$$PH \sim \frac{F_1}{P} = \frac{A}{B\Gamma} \cdot \frac{\eta}{1-\eta} \tag{3.3}$$

Thus, pH is dependent on the quantum yield η of DiIRB-S. If the probe concentration is in a certain range (from 0.5 μM to 4 μM in our tests) and the saturation and self-quenching effect is negligible, then the ratio of fluorescence/photoacoustic signal intensities is not affected by excitation source fluctuations (Fig. 3.20a) or local probe concentration (Fig. 3.20b).

Next, this system was used to map the pH in murine 4T1 tumors with the same genetic background at different stages (4, 8, and 12 days, n = 3 mice per group) after 2 h intratumoral injection of DiIRB-S (20 μL, 20 μM). The tumor from the surface is ~100 μm. For comparison, the same amount of DiIRB-S was injected into muscle on the contralateral side. In vivo photoacoustic and fluorescence signal was then simultaneously collected via PA-FLM (Imaging duration is 15 min) (Fig. 3.21a). Both fluorescence and photoacoustic signals from the same focus region after excitation were acquired. The photoacoustic images were reconstructed via maximum projection algorithm with signals from the focal region. The pH distribution images were reconstructed by mapping the ratio of fluorescence/photoacoustic signal intensities. The pH was calculated based on the relationship in (Fig. 3.19e). There is a non-uniform pH distribution in the tumor; partial areas show the special low pH [the third line in Fig. 3.21a]. The high-resolution PA-FLM data show heterogeneous patterns of pH in the tumor. The averaging intratumoral pH was lower than in muscle—7.3 on average. The pH values in the tumors were 6.8, 6.5, and 6.2 at days 4, 8, and 12 post-tumor inoculation, respectively. The tumors (n = 3 mice per group) that underwent scanning in the PA-FLM system were subsequently measured by pH/mV-Meter (9 pH measurements were performed per mouse). Mice were sacrificed after pH/mV-Meter measurements over. Animal procedures were in agreement with the guidelines of the Institutional Animal Care and Use Committee.

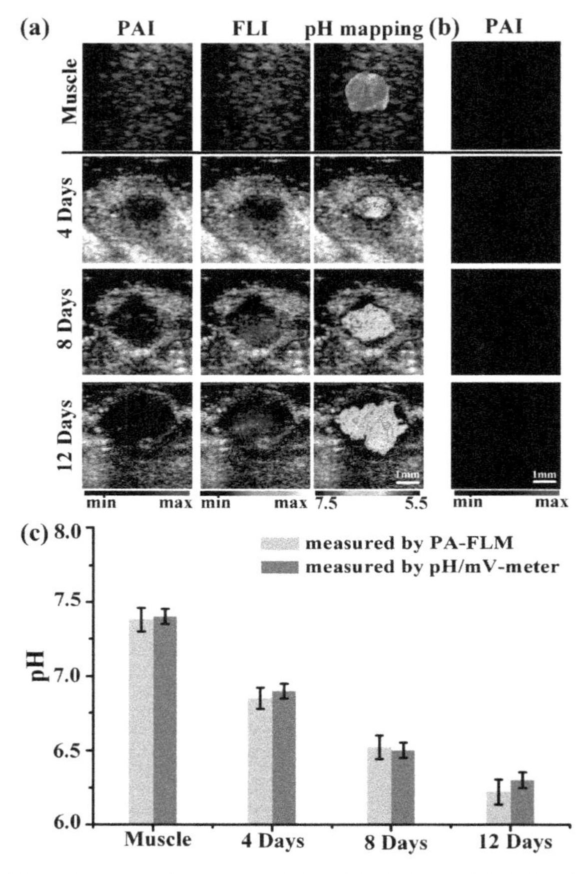

Figure 3.21 In vivo mapping tumor interstitial pH by PA-FLM. Two hours after intratumoral injection of DiIRB-S, regions of interest were scanned with PA-FLM. (a) Representative ultrasound (gray), photoacoustic (the first column, red), fluorescence (the second column, green) and pH distribution (the third column) sectional 2D images in horizontal slice (6 × 6 mm) through the tumors at different stages (4, 8, 12 days). The ultrasound images show the skin and tumor boundaries. (b) Representative in vivo photoacoustic images of the vasculature in the normal tissue and the margin of tumor (exited at 532 nm) and this is the different region than (a). (c) The comparison of the average intratumoral pH obtained by PA-FLM and pH/mV-Meter. The error bars represent standard deviations of three separate measurements. Reproduced with permission from Ref. [74].

The comparison of the average intratumoral pH at different stages obtained by PA-FLM and pH/mV-Meter is shown in Fig. 3.21c. The low pH is expected because as the tumor grows larger, the tumor microenvironment becomes hypoxic and more acidic due to the production of lactate via the anaerobic glycolytic pathway.

These results show that PA-FLM measured pH with high resolution via shifts in the fluorescence/photoacoustic signal balance. To the best of our knowledge, this is the first use of ratiometric fluorescence and photoacoustic signal to measure pH. This technique measures pH in a single scan based on change in the pH-sensitive probes' quantum yield (Eq. (3.3)). It is not affected by the excitation source fluctuations and local probe concentration, and it provides high imaging contrast for distinguishing pH changes. The method provides high-resolution tumor pH mapping via a single-scan technique that use less than half of the time of dual-wavelength ratiometric imaging.

However, the PA-FLM technique still has some challenges. Here, PA-FLM imaging speed is limited by the repetition frequency of the laser system (2.5 K). Higher repetition laser system (MHz) applied in PA-FLM will enhance the speed for mapping of the tumor pH. However, considering that the ultrasound speed in tissue is ~1.5 mm/μs, the ideal maximum imaging depth is ~1.5 mm if the repetition rate of laser system is 1 MHz. To maintain the stability of the data acquisition, the maximum imaging depth may be below 1 mm for photoacoustic imaging. With the repetition rate increase (> 1 MHz), the photoacoustic imaging depth will further decrease.

3.5.4 Conclusion

We showed that PA-FLM is a high-resolution tool to map tumor pH in vivo. We show for the first time that pH can be accurately measured by the ratio of photoacoustic and fluorescence signal intensities with no impact from excitation source fluctuations or local probe concentration. The capacity of PA-FLM to map tumor pH was successfully demonstrated in mice bearing tumors at different stages. PA-FLM is a promising tool to predict tumor invasive/metastatic potential and evaluate the therapeutic response non-invasively by dynamically monitoring intra-tumoral acidosis [58–62]. We thus propose that with further research and selection of proper probes, PA-FLM may provide a potential alternative for monitoring tissue physiology.

References

1. Zhang, H. F., Maslov, K., Stoica, G., and Wang, L. V. (2006). Functional photoacoustic microscopy for high-resolution and noninvasive in vivo imaging. *Nat. Biotechnol.,* **7**, pp. 848–851.
2. Wang, L. V. (2008). Photoacoustic microscopy and computed tomography. *Nat. Photonics,* **9**, pp. 503–509.
3. Zhang, C., Maslov, K., and Wang, L. V. (2010). Subwavelength-resolution label-free photoacoustic microscopy of optical absorption in vivo. *Opt. Lett.,* **19**, pp. 3195–3197.
4. Oh, J. T., Li, M. L., Zhang, H. F., Maslov, K., Stoica, G., and Wang, L. V. (2006). Three-dimensional imaging of skin melanoma in vivo by dual-wavelength photoacoustic microscopy. *J. Biomed. Opt.,* **3**, pp. 034032–034032.
5. Maslov, K., Zhang, H. F., Hu, S., and Wang, L. V. (2008). Optical-resolution photoacoustic microscopy for in vivo imaging of single capillaries. *Opt. Lett.,* **9**, p. 929.
6. Xie, Z., Jiao, S., Zhang, H. F., and Puliafito, C. A. (2009). Laser-scanning optical-resolution photoacoustic microscopy. *Opt. Lett.,* **12**, pp. 1771–1773.
7. Hu, S., Maslov, K., and Wang, L. V. (2011). Second-generation optical-resolution photoacoustic microscopy with improved sensitivity and speed. *Opt. Lett.,* **7**, p. 1134.
8. Wang, L., Maslov, K., Yao, J., Rao, B., and Wang, L. V. (2011). Fast voice-coil scanning optical-resolution photoacoustic microscopy. *Opt. Lett.,* **2**, pp. 139–141.
9. Hu, S., Maslov, K., and Wang, L. V. (2009). In vivo functional chronic imaging of a small animal model using optical-resolution photoacoustic microscopy. *Med. Phys.,* **6**, pp. 2320–2323.
10. Park, S., Lee, C., Kim, J., and Kim, C. (2014). Acoustic resolution photoacoustic microscopy. *Biomed. Eng. Lett.,* **3**, pp. 213–222.
11. Xing, W., Wang, L., Maslov, K., and Wang, L. V. (2013). Integrated optical-and acoustic-resolution photoacoustic microscopy based on an optical fiber bundle. *Opt. Lett.,* **1**, pp. 52–54.

12. Maslov, K., Zhang, H. F., Hu, S., and Wang, L. V. (2008). Optical-resolution photoacoustic microscopy for in vivo imaging of single capillaries. *Opt. Lett.*, **9**, p. 929.

13. Schmitt, J. M. (1999). Optical coherence tomography (OCT): A review. *IEEE J. Sel. Top. Quant.*, **4**, pp. 1205–1215.

14. Webb, R. H. (1996). Confocal optical microscopy. *Rep. Prog. Phys.*, **3**, p. 427.

15. Chen, Z., Yang, S., and Xing, D. (2012). In vivo detection of hemoglobin oxygen saturation and carboxyhemoglobin saturation with multiwavelength photoacoustic microscopy. *Opt. Lett.*, **16**, pp. 3414–3416.

16. Li, B., Qin, H., Yang, S., and Xing, D. (2014). In vivo fast variable focus photoacoustic microscopy using an electrically tunable lens. *Opt. Express*, **17**, pp. 20130–20137.

17. Xing, W., Wang, L., Maslov, K., and Wang, L. V. (2013). Integrated optical-and acoustic-resolution photoacoustic microscopy based on an optical fiber bundle. *Opt. Lett.*, **1**, p. 52–54.

18. Jeon, M., Kim, J., and Kim, C. (2016). Multiplane spectroscopic whole-body photoacoustic imaging of small animals in vivo. *Med. Biol. Eng. Comput.*, **2–3**, pp. 283–294.

19. Maslov, K., Stoica, G., and Wang, L. V. (2005). In vivo dark-field reflection-mode photoacoustic microscopy. *Opt. Lett.*, **6**, pp. 625–627.

20. Ogawa, S., and Lee, T. M. (1990). Magnetic resonance imaging of blood vessels at high fields: In vivo and in vitro measurements and image simulation. *Magn. Reson. Med.*, **1**, pp. 9–18.

21. Veeranjaneyulu, K., N'soukpoé-Kossi, C. N., and Leblanc, R. M. (1991). SO_2 effect on photosynthetic activities of intact sugar maple leaves as detected by photoacoustic spectroscopy. *Plant Physiol.*, **1**, pp. 50–54.

22. Petkovska, L. T., Radak, B. B., Miljanić, Š. S., Bailey, R. T., Cruickshank, F. R., and Pugh, D. (1991, March). SO_2 absorption of the CO_2-laser emission measured by the photoacoustic technique. In *Proceedings of the Indian Academy of Sciences-Chemical Sciences* (vol. **103**, no. 3, pp. 401–404). Springer India.

23. Yao, J., Maslov, K. I., Shi, Y., Taber, L. A., and Wang, L. V. (2010). In vivo photoacoustic imaging of transverse blood flow by using Doppler broadening of bandwidth. *Opt. Lett.*, **9**, pp. 1419–1421.

24. Luke, G. P., Yeager, D., and Emelianov, S. Y. (2012). Biomedical applications of photoacoustic imaging with exogenous contrast agents. *Ann. Biomed. Eng.*, **2**, pp. 422–437.

25. Cai, X. Li, L., Krumholz, A., Guo, Z., Erpelding, T. N., Zhang, C., Zhang, Y., Xia, Y., Wang, L. V. (2012). Multi-scale molecular photoacoustic tomography of gene expression. *Plos One*, **8**, p. e43999.

26. Kell, M. R., and Kerin, M. J. (2004). Sentinel lymph node biopsy. *New Engl. J. Med.*, **7452**, pp. 1330–1331.

27. Shahla, M. M. D., Veronesi, U., Paganelli, G., Galimberti, V., Viale, G., and Zurrida, S., et al. (1997). Sentinel-node biopsy to avoid axillary dissection in breast cancer with clinically negative lymph-nodes. *Lancet*, **9069**, p. 1864.

28. Khosla, R., Rohatgi, P. K., Seam, N. (2009). Ultrasound-guided fine needle aspiration biopsy of pleural-based intrathoracic lesions. *J. Bronchol. Interv. Pulmonol.*, **16**(2), 87–90.

29. Kim, S., Yong, T. L., Soltesz, E. G., Grand, A. M. D., Lee, J., and Nakayama, A., et al. (2004). Near-infrared fluorescent type ii quantum dots for sentinel lymph node mapping. *Nat. Biotechnol.*, **1**, pp. 93–97.

30. Kitai, T., Inomoto, T., Miwa, M., and Shikayama, T. (2005). Fluorescence navigation with indocyanine green for detecting sentinel lymph nodes in breast cancer. *Breast Cancer*, **3**, pp. 211–215.

31. Tagaya, N., Yamazaki, R. A., Abe, A., Hamada, K., Kubota, K., and Oyama, T. (2008). Intraoperative identification of sentinel lymph nodes by near-infrared fluorescence imaging in patients with breast cancer. *Am. J. Surgery*, **6**, pp. 850–853.

32. Erpelding, T. N., Kim, C., Pramanik, M., Jankovic, L., Maslov, K., and Guo, Z., et al. (2010). Sentinel lymph nodes in the rat: Noninvasive photoacoustic and us imaging with a clinical us system. *Radiology*, **1**, pp. 102–110.

33. Song, K. H., Stein, E. W., Margenthaler, J. A., and Wang, L. V. (2008). Noninvasive photoacoustic identification of sentinel lymph nodes containing methylene blue in vivo in a rat model. *J. Biomed. Opt.*, **5**, p. 054033.

34. Kim, C., Song, K. H., Gao, F., Wang, L. V. (2010). Sentinel lymph nodes and lymphatic vessels: Noninvasive dual-modality in vivo mapping by using indocyanine green in rats-volumetric spectroscopic photoacoustic imaging and planar fluorescence imaging. *Radiology*, **2**, pp. 442–450.

35. Kim, G., Huang, S. W., Day, K. C., O'Donnell, M., Agayan, R. R., and Day, M. A., et al. (2007). Indocyanine-green-embedded pebbles as a contrast agent for photoacoustic imaging. *J. Biomed. Opt.*, **4**, p. 044020.

36. Wang, L. V., and Wu, H. I. (2012). *Biomedical Optics: Principles and Imaging*. John Wiley & Sons.

37. Scardapane, A., Pagliarulo, V., Ianora, A. A. S., Pagliarulo, A., and Angelelli, G. (2008). Contrast-enhanced multislice pneumo-ct-cystography in the evaluation of urinary bladder neoplasms. *Eur. J. Radiol.,* **2**, pp. 246–252.

38. Rothwell, R. I., Ash, D. V., and Jones, W. G. (1983). Radiation treatment planning for bladder cancer: A comparison of cystogram localisation with computed tomography. *Clin. Radiol.,* **1**, pp. 103–11.

39. Browne, R. F. J., Murphy, S. M., Grainger, R., and Hamilton, S. (2005). Ct cystography and virtual cystoscopy in the assessment of new and recurrent bladder neoplasms. *Eur. J. Radiol.*, **1**, pp. 147–153.

40. Lim, R. (2009). Vesicoureteral reflux and urinary tract infection: Evolving practices and current controversies in pediatric imaging. *Am. J. Roentgenol.*, **192**, pp. 1197–1208.

41. Brown, M. C., Sutherst, J. R., Murray, A., and Richmond, D. H. (1985). Potential use of ultrasound in place of x-ray fluoroscopy in urodynamics. *Brit. J. Urol.*, **1**, pp. 88–90.

42. Vining, D. J., Zagoria, R. J., Liu, K., and Stelts, D. (1996). Ct cystoscopy: An innovation in bladder imaging. *Ajr Am. J. Roentgenol.*, **2**, pp. 409–410.

43. Kim, C., Cho, E. C., Chen, J., Song, K. H., Au, L., Favazza, C., Zhang, Q., Cobley, C. M., Gao, F., Xia, Y., Wang, L. V. (2010). In vivo molecular photoacoustic tomography of melanomas targeted by bioconjugated gold nanocages. *ACS Nano*, **8**, pp. 4559–4564.

44. Jeon, M., Jenkins, S., Oh, J., Kim, J., Peterson, T., and Chen, J., et al. (2014). Nonionizing photoacoustic cystography with near-infrared absorbing gold nanostructures as optical-opaque tracers. *Nanomedicine,* **9**, pp. 1377–1388.

45. Srivatsan, A., Jenkins, S. V., Jeon, M., Wu, Z., Kim, C., and Chen, J., et al. (2014). Gold nanocage-photosensitizer conjugates for dual-modal image-guided enhanced photodynamic therapy. *Theranostics,* **2**, pp. 163–174.

46. Soffer, E. E. (2000). Small bowel motility: Ready for prime time?. *Curr. Gastroenterol. Rep.*, **5**, pp. 364–369.

47. Ohama, T., Hori, M., and Ozaki, H. (2007). Mechanism of abnormal intestinal motility in inflammatory bowel disease: How smooth muscle contraction is reduced?. *J. Smooth Muscle Res.*, **2**, p. 43.

48. Abrahamsson, H. (1995). Gastrointestinal motility disorders in patients with diabetes mellitus. *J. Intern. Med.*, **4**, pp. 403–409.

49. Shafer, R. B., Prentiss, R. A., and Bond, J. H. (1984). Gastrointestinal transit in thyroid disease. *Gastroenterology*, **1**, pp. 852–855.

50. Jost, W. H. (1997). Gastrointestinal motility problems in patients with Parkinson's disease. *Drugs Aging*, **4**, pp. 249–258.

51. Dye, C. E., Gaffney, R. R., Dykes, T. M., and Moyer, M. T. (2012). Endoscopic and radiographic evaluation of the small bowel in 2012. *Am. J. Med.*, **12**, pp. 1228-e1.

52. Zhang, Y., Jeon, M., Rich, L. J., Hong, H., Geng, J., Zhang, Y., Seshadri, M. (2014). Non-invasive multimodal functional imaging of the intestine with frozen micellar naphthalocyanines. *Nat. Nanotechnol.*, **8**, pp. 631–638.

53. Kim, C., Favazza, C., Wang, L. V. (2010). In vivo photoacoustic tomography of chemicals: High-resolution functional and molecular optical imaging at new depths. *Chem. Rev.*, **5**, pp. 2756–2782.

54. Kagadis, G. C., Loudos, G., Katsanos, K., Langer, S. G., Nikiforidis, G. C. (2010). In vivo small animal imaging: Current status and futureprospects. *Med. Phys.*, **12**, pp. 6421–6442.

55. Xu, L., and Fidler, I. J. (2000). Acidic ph-induced elevation in interleukin 8 expression by human ovarian carcinoma cells. *Cancer Res.*, **16**, pp. 4610–4616.

56. Fais, S., De, M. A., You, H., Qin, W. (2007). Targeting vacuolar h^+-ATPases as a new strategy against cancer. *Cancer Res.*, **22**, pp. 10627–10630.

57. Rofstad, E. K., Mathiesen, B., Kindem, K., and Galappathi, K. (2006). Acidic extracellular ph promotes experimental metastasis of human melanoma cells in athymic nude mice. *Cancer Res.*, **66**, pp. 6699–6707.

58. Webb, B. A., Chimenti, M., Jacobson, M. P., and Barber, D. L. (2011). Dysregulated pH: A perfect storm for cancer progression. *Nat. Rev. Cancer*, **9**, pp. 671–677.

59. Estrella, V., Chen, T., Lloyd, M., Wojtkowiak, J., Cornnell, H. H., and Ibrahimhashim, A., et al. (2013). Acidity generated by the tumor microenvironment drives local invasion. *Cancer Res.*, **5**, pp. 1524–1535.

60. Robey, I. F., Baggett, B. K., Kirkpatrick, N. D., Roe, D. J., Dosescu, J., Sloane, B. F., Gillies, R. J. (2009). Bicarbonate increases tumor pH and inhibits spontaneous metastases. *Cancer Res.*, **6**, pp. 2260–2268.

61. Wang, L., Fan, Z., Zhang, J., Changyi, Y., Huang, C., and Gu, Y., et al. (2015). Evaluating tumor metastatic potential by imaging

intratumoral acidosis via ph-activatable near-infrared fluorescent probe. *In. J. Cancer,* **4**, pp. 107–116.

62. Helmlinger, G., Yuan, F., Dellian, M., and Jain, R. K. (1997). Interstitial ph and po2 gradients in solid tumors in vivo: High-resolution measurements reveal a lack of correlation. *Nat. Med.*, **3**, pp. 177–182.
63. Tannock, I. F., and Rotin, D. (1989). Acid ph in tumors and its potential for therapeutic exploitation. *Cancer Res.*, **16**, pp. 4373–4384.
64. Antonsson, J. B., Boyle, C. C. D., Kruithoff, K. L., Wang, H. L., Sacristan, E., Rothschild, H. R., and Fink, M. P. (1990). Validation of tonometric measurement of gut intramural pH during endotoxemia and mesenteric occlusion in pigs. *Am. J. Physiol. Gastr. L.*, **4**, pp. G519–G523.
65. Wang, L., and Li, C. (2011). pH responsive fluorescence nanoprobe imaging of tumors by sensing the acidic microenvironment. *J. Mater. Chem.*, **40**, pp. 15862–15871.
66. Martin, G. R., and Jain, R. K. (1994). Noninvasive measurement of interstitial pH profiles in normal and neoplastic tissue using fluorescence ratio imaging microscopy. *Cancer Res.*, **21**, pp. 5670–5674.
67. Chen, Q., Liu, X., Chen, J., Zeng, J., Cheng, Z., and Liu, Z. (2015). A self-assembled albumin-based nanoprobe for in vivo ratiometric photoacoustic pH imaging. *Adv. Mater.*, **43**, pp. 6820–6827.
68. Wang, L., Zhu, X., Xie, C., Ding, N., Weng, X., and Lu, W., et al. (2012). Imaging acidosis in tumors using a ph-activated near-infrared fluorescence probe. *Chem. Commun.*, **95**, pp. 11677–11679.
69. Huang, C., Qin, H., Qian, J., Zhang, J., Zhao, S., and Changyi, Y., et al. (2015). Multi-parametric imaging of the invasiveness-permissive acidic microenvironment in human glioma xenografts. *RSC Adv.*, **69**, pp. 55669–55677.
70. Valeur, B., and Berberan-Santos, M. N. (2012). *Molecular Fluorescence: Principles and Applications.* John Wiley & Sons.
71. Xu, M., and Wang, L. V. (2006). Photoacoustic imaging in biomedicine. *Rev. Sci. Instrum.*, **44**, p. 041101.
72. Ntziachristos, V., Ripoll, J., Wang, L. V., and Weissleder, R. (2005). Looking and listening to light: The evolution of whole-body photonic imaging. *Nat. Biotechnol.*, **3**, pp. 313–320.
73. Gao, L., Zhang, C., Li, C., and Wang, L. V. (2013). Intracellular temperature mapping with fluorescence-assisted photoacoustic-thermometry. *Appl. Phys. Lett.*, **19**, p. 193705.

74. Yan, B., Qin, H., Huang, C., Li, C., Chen, Q., and Xing, D. (2017). Single-wavelength excited photoacoustic-fluorescence microscopy for in vivo pH mapping. *Optics Lett.*, **42**(7), 1253–1256.

Chapter 4

Photoacoustic Endoscopy and Its Biomedical Applications

4.1 Introduction

Photoacoustic tomography has great potential for in vivo medical applications, because it is safe and has a high ratio of imaging depth to resolution [1–11]. It provides image information by detecting ultrasonic waves generated by absorbed energy impulses, typically laser pulses. Its deep imaging capability is attributed to its energy delivery mechanism, which uses diffused light that can penetrate up to several centimeters into soft tissue. Unfortunately, at large depths, imaging resolution can be inadequate. In such cases, the photoacoustic probe must be positioned close to the area of interest by means of endoscopy. Other optical imaging modalities, such as ultrasound imaging, optical coherence tomography (OCT) and confocal microscopy, have already shown their endoscopic potential to demonstrate the structural and functional imaging of the cavity body. Consequently, much interest has developed in the endoscopic embodiment of photoacoustic imaging.

A photoacoustic endoscopic system has to deliver light pulses, detect ultrasonic waves, and perform area or line scanning at the tip of a small probe. A flexible-shaft-based mechanical scanning

Biomedical Photoacoustics
Sihua Yang and Da Xing

ISBN 978-981-4774-58-1 (Hardcover), 978-0-203-70365-6 (eBook)
www.jennystanford.com

mechanism can potentially be used for the scanning method, as is done in endoscopic ultrasonography and OCT. Because the photoacoustic imaging depth and resolution can be reach 1–5 mm with 10–300 μm, the applications of photoacoustic endoscopy are related to imaging of internal organs, such as the wall of the stomach, esophagus, intestines, and coronary artery. In fact, endoscopic PA imaging can obtain the internal organ image; therefore, it is necessary to develop endoscopic PA imaging for detecting lesions of internal organs.

4.2 Reconstruction Algorithm of Photoacoustic Endoscopy

In general, signal acquisition strategies are closely related to the image reconstruction algorithms and the type of detector. For example, if a single focused detector is used, its signal acquisition mode is one-dimension with line-by-line scanning or plane-by-point scanning, and its two-dimensional image reconstruction algorithm is limited to a linear projection algorithm or an amplitude maximum imaging algorithm (amplitude maximum imaging method generally corresponds to the maximum value of the amplitude of each set of photoacoustic signals is taken as the pixel value of the corresponding point in the reconstructed image). If a single non-focused ultrasonic detector is used, the general use of rotating detectors or samples of the way to collect photoacoustic signals, the image reconstruction algorithm is generally used back projection class algorithm; if the multidetector acquires photoacoustic signals, with different kinds of detector and signal acquisition methods, the signal acquisition methods are different, and the corresponding image reconstruction algorithm is not the same. Endoscopic photoacoustic imaging acquisition mode is fixed in the probe to the rotation or helical scan, the following will introduce the conventional endoscopic photoacoustic tomography algorithm and multi-wavelength excitation endoscopic photoacoustic component imaging algorithm.

4.2.1 Endoscopic Photoacoustic Tomography Algorithm

The imaging mode of the endoscopic tomography is achieved by inserting a miniature photoacoustic scanning probe into the body cavity. The scanning probe acquires the tomographic image of the visceral organ wall by 360° scan and finds the lesion according to the information on the image. Therefore, the endoscopic tomography algorithm reflects the projection of light absorption distribution of the sample with the commonly used inverse of the filtered back-projection method. As shown in Fig. 4.1, the relationship between the sample and the point source photoacoustic signal is the intermediate divergence relation.

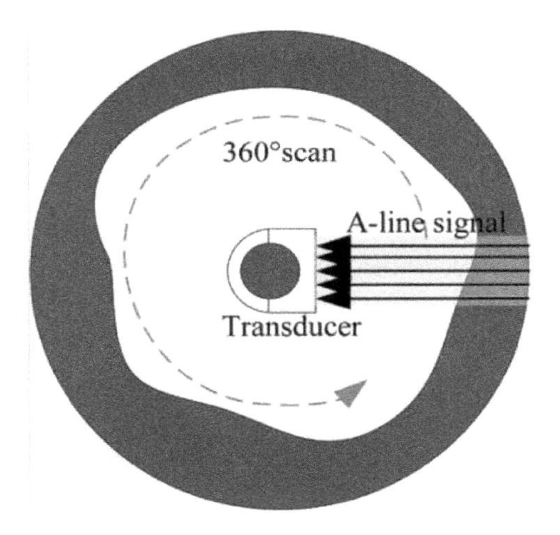

Figure 4.1 Diagram of endoscopic photoacoustic tomography signal acquisition.

The position of the detector is the origin of the coordinates. In the spherical coordinate system, the time is defined as zero when the laser begins to be irradiated, so we get

$$p(t)=\frac{\beta}{4\pi C_{\mathrm{p}}}\int\left(\frac{1}{t'}\iint A(ct',\theta,\phi)(ct')^2\sin\theta d\theta d\phi\right)I'(t-t')dt'. \tag{4.1}$$

Equation (4.1) can be written as

$$p(t)=\left(\frac{1}{t}\iint A(ct,\theta,\phi)(ct)^2\sin\theta d\theta d\phi\right)\times I'(t) \tag{4.2}$$

Assume that a point absorber produces a sound pressure $P_{\mathrm{point}}(t)$,

$$p_{\text{point}}(t) = k\frac{1}{r_0}I'\left(t - \frac{r_0}{c}\right), \tag{4.3}$$

where r_0 is the distance from the point source to the field point and k is the coefficient determined by the point source absorption and the incident laser parameters. Make

$$p_0(t) = p_{\text{point}}\left(t + \frac{r_0}{c}\right) = k\frac{1}{r_0}I'(t). \tag{4.4}$$

Then Eq. (4.2) can be written as

$$p(t) = \left(\frac{r_0}{kt}\iint A(ct,\theta,\phi)(ct)^2 \sin\theta d\theta d\phi\right) \times p_0(t). \tag{4.5}$$

In the case where the time at which the laser irradiation is started is defined as the time zero point, $r = ct$, $\iint A(ct, \theta, \phi)$ $(ct)^2 \sin\theta d\theta\, d\phi$ can be written as $\iint A(r)dS|_{|r|=ct}$, Where $\iint A(r)ds|_{|r|=ct}$ is the integral of the absorption coefficient on the spherical surface with radius of $r = ct$, and P is the center point. For two-dimensional problems, $\iint A(r)ds|_{|r|=ct}$ can be transformed into $\iint A(r)dl|_{|r|=ct}$, which can be considered as the integral of the absorption coefficient on the arc. In image reconstruction, the integral along a certain direction is called the projection in this direction.

The meaning of $\iint A(ct,\theta,\phi)(ct)^2 \sin\theta d\theta d\phi$ can be demonstrated in Fig. 4.1. P represents the position of the transducer, $\iint A(ct,\theta,\phi)(ct)^2 \sin\theta d\theta d\phi$ is the projection of the absorption distribution $A(ct, \theta, \phi)$ on the spherical surface with radius of ct. In order to remove the pulse response influence of the transducer, we assume that the pulse response of the transducer is $h(t)$. Assuming that the detected PA signal is $p_d(t)$, we can get

$$\begin{aligned} p_d(t) &= p_d(t) \times h(t) \\ &= \left(\frac{r_0}{kt}\iint A(ct,\theta,\phi)(ct)^2 \sin\theta d\theta d\phi\right) \times p_0(t) \times h(t). \end{aligned} \tag{4.6}$$

If the same transducer is used to detect PA signal from a laser spot, the obtained PA signal thus can be considered as $p_0(t) \times h(t)$. Assume that

$$p_{d0}(t) = p_0(t) \times h(t). \quad (4.7)$$

So we obtain

$$p_d(t) = \left(\frac{r_0}{kt} \iint A(ct,\theta,\phi)(ct)^2 \sin\theta d\theta d\phi \right) \times p_{d0}(t). \quad (4.8)$$

Equation (4.8) shows that the detected PA signal of samples equals the convolution of the signal from a laser point and the optical absorption distribution by a same transducer. It also indicates that in view that the transducer is a linear time invariant system, an absorber can be regarded as a set of absorbing points, and thus the PA signal of the samples can be considered as a linear superposition of signals from these absorbing points.

From Eq. (4.8), we can see that the projection of optical absorption $\iint A(r)ds\Big|_{|r|=ct}$ can be calculated by deconvolution of the PA signal of the sample and the PA signal of a point source, rather than knowing the pulse response of the transducer. Yet, we need to obtain the PA signal of a point source $p_{d0}(t)$. After obtaining $p_{d0}(t)$ by from a point source by laser-focusing on an absorber, $p_{d0}(t)$ can be obtained through time-shift according to the distance r_0 between the point source and the transducer. After deconvolution, the optical absorption projection then can be obtained. Meanwhile, $\iint A(r)ds\Big|_{|r|=ct}$ can also be acquired by inverse Fourier transform.

$$\oiint_{|\mathbf{r}|=v_0 t} A(r)dr = IFFT\left(\frac{P_d(\omega)W(\omega)}{P_0(\omega)} \right) \frac{tk}{r_0} = \iint A(ct,\theta,\phi)(ct)^2 \sin\theta d\theta d\phi \quad (4.9)$$

IFFT represents the fast inverse Fourier transform. $P_d(\omega)$ and $P_0(\omega)$ are the Fourier transform of $p_d(t)$ and $p_{d0}(t)$, respectively. $W(\omega)$ is a window function, which slows the truncation to zero to avoid Gibbs oscillations. Here, we use Hanning function:

$$W(\omega)=\frac{1}{2}\left(1+\cos\left(\frac{2\pi n}{N}\right)\right),\quad n=0,+1,+2,N \tag{4.10}$$

By limiting the projection angle, each A-sweep signal is projected on the projection plane in the form of an arc, and the time information and intensity information of the signal are projected into a small sector area. The angle of the projection must satisfy the receiving angle of the detector. After the completion of the 360° projection, it is possible to reconstruct the optical absorption image of the lumen section, as shown in Fig. 4.2.

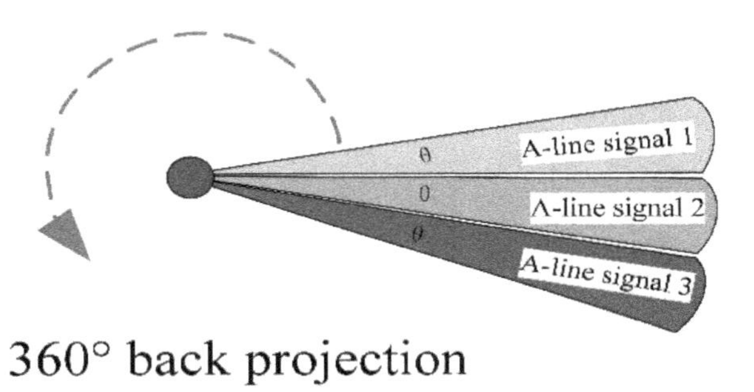

Figure 4.2 Projection diagram of light absorption distribution for intravascular photoacoustic tomography.

4.2.2 Multi-Wavelength Excitation Endoscopic Photoacoustic Component Imaging Algorithm

Hybrid system refers to two or more substances in the same system that can maintain stable on their properties. They usually have several forms, including gas (air), liquid (oil), solid (ore) and mixed state (such as biological tissue). In the industrial and agricultural production, in many cases especially in the field of medicine and health, quantitative measurements are needed to analyze the distribution of elements in the mixed system, and their corresponding concentrations. However, the material light absorption capacity is related to the concentration of the substance. It can be imagined that, the higher the concentration leads to higher light absorption. So, even though the material is same, PA images are severely affected by the concentrations of the absorbers. Thus, conventional PA imaging cannot better reflect the traditional mixture of single elements concentration distribution information.

We can solve the distribution of each mixture in the photoacoustic imaging using multi wavelength PA technique. The principle as follows: In the frequency domain, multi wavelength PA equation in a non-uniform acoustic medium can be expressed as [12]

$$\nabla^2 P(r,\omega,\lambda) + k_0^2(1+o)P(r,\omega,\lambda) = ik_0 \frac{v_0 \beta \mu_{\mathrm{a}}(r,\lambda)\phi(r,\lambda)}{C_{\mathrm{p}}}. \tag{4.11}$$

Here, P is PA pressure, $k_0 = \omega/v_0$ is the wave number represented by the angular frequency ω, v_0 is the velocity of sound wave in the reference medium or the coupling fluid, β is the thermal expansion coefficient, C_{P} is heat capacity, μ_{a} is the absorption coefficient, ϕ is the photo density, λ is the wavelength of the incident light. According to Bill's law, the absorption coefficient of the tissue can be expressed as a function of wavelengths as

$$\mu_{\mathrm{a}}(\lambda) = \sum_{i=1} \varepsilon_i(\lambda)c_i. \tag{4.12}$$

Here, c_i is concentration, $\varepsilon_i(\lambda)$ is the extinction coefficient of i-th at the wavelength λ. We assume that $A = \mu_{\mathrm{a}} = \varepsilon_{\mathrm{c}}\varphi$, thus Eq. (4.11) becomes

$$\nabla^2 P(r,\omega,\lambda) + k_0^2(1+o)P(r,\omega,\lambda) = ik_0 \frac{v_0 \beta A(r,\lambda)}{C_{\mathrm{p}}}. \tag{4.13}$$

Based on Eq. (4.13), when irradiated by single wavelength, there exist multiple solutions, which can be expressed as

$$(\tau)P(r,\omega,\lambda) = \frac{v_0 \beta \sum_{i=1} \varepsilon_i(\lambda)c_i\phi(\lambda)}{C_{\mathrm{P}}} = \frac{v_0 \beta \sum_{i=1} A_i(\lambda)}{C_{\mathrm{P}}}. \tag{4.14}$$

It is required that Eq. (4.14) applies for all incident wavelengths. Thus, Eq. (4.14) only makes sense when m wavelengths and n absorbers are applied.

$$\begin{bmatrix} A(\lambda_1) \\ \vdots \\ A(\lambda_m) \end{bmatrix} = \begin{bmatrix} \varepsilon_1(\lambda_1)\cdots\varepsilon_n(\lambda_n) \\ \vdots \quad \vdots \quad \vdots \\ \varepsilon_1(\lambda_m)\cdots\varepsilon_n(\lambda_m) \end{bmatrix} \begin{bmatrix} c_1 \\ \vdots \\ c_n \end{bmatrix} \tag{4.15}$$

By transforming Eq. (4.15), the relation between the concentration, the extinction coefficient ε and the final absorption A is obtained.

$$c = (\varepsilon^T \varepsilon)^{-1} \varepsilon^T A \tag{4.16}$$

According to Eq. (4.16), the PA signal intensity PA has linear correlation with the optical absorption, and the PA signal intensity can reflect the absorption, and the final results reflect the distribution of various components. Therefore, the inversion of the concentration distribution of absorbing matter in the cavity can be obtained by the above calculation.

4.3 Photoacoustic Endoscopy

Endoscopy is an important technique in practical medicine to diagnose internal organs, such as respiratory and gastrointestinal (GI) tracts. The most common technique is video endoscopy, which can provide clear real-time video images of organs' surface to the operator. However, these video images lack depth information and are insufficient to diagnose diseased tissues that develop in endothelial tissues.

Endoscopic ultrasonography (EUS) is an advanced technique to overcome this obstacle. An intravascular ultrasound (IVUS) probe and gastrointestinal ultrasound probe are representative tools for cardiovascular imaging and GI tract imaging, respectively. They can image very deep areas of organs, up to several centimeters, based on ultrasonographic techniques. Even up to several centimeters, however, this kind of ultrasound probe has some limitations for the early diagnosis of tissue abnormalities, such as tumors, because its intrinsic contrast is based on the tissue mechanical properties. Therefore, novel optical endoscopic techniques have been developed. Representative techniques are endoscopic OCT and confocal endoscopy. These techniques can produce high-resolution 3D images based on optical contrast, which is more sensitive than current techniques for detecting tissue abnormalities. However, their limitation is poor penetration depth because they utilize ballistic or quasi-ballistic photons. So, the photoacoustic endoscopy technique is expected to be superior to other endoscopies due to the hybrid imaging modality of optics and ultrasound, which combines the strong points of optical and ultrasound imaging so that it can image deep areas of organs with high resolution and optical absorption contrast.

We divide the photoacoustic endoscopy into two categories according to the biomedical applications as follows: (1) the photoacoustic endoscopy for respiratory and gastrointestinal (GI) tracts and (2) the intravascular photoacoustic (IVPA) endoscopy for plaque detection.

4.3.1 Photoacoustic Endoscopy System

4.3.1.1 The first photoacoustic endoscope

The first photoacoustic endoscope is developed by integrating a light-guiding optical fiber, an ultrasonic sensor, and a mechanical scanning unit to enable circumferential sector scanning of internal organs [13, 14]. In the probe, as shown in Fig. 4.3, a light-guiding optical fiber (0.22 NA, 365 μm core diameter), a single element ultrasonic transducer ($LiNbO_3$, 43 MHz, 2.0 mm aperture, unfocused), and a mechanical micromotor are placed into a stainless steel tube. Laser pulses from a diode-pumped, solid-state, Nd:YLF pumped dye laser are guided by the optical fiber and emitted through a central hole (0.5 mm diameter) in the transducer. Circumferential sector scanning is accomplished by rotating a mirror (3.0 mm diameter, protected aluminum on glass substrate, with the reflection surface at 45° to the probe's axis). The mirror, driven by a 1.5 mm diameter, 12.0 mm-long geared micromotor, steers both the light beam from the optical fiber to the tissue and the acoustic wave from the tissue to the transducer. The optical fiber, the transducer's signal wires, and the micromotor's wires are encapsulated in the flexible endoscope body. The implemented probe size for the distal end is 4.2 mm in diameter and 48 mm in length. The field of view for the photoacoustic probe is a ring partially blocked by the stainless-steel housing bridge. For each circular B-scan, 254 time-resolved photoacoustic signals (A-lines) are recorded, leading to an angular step size of 360°/254 = 1.42°. However, 76 A-lines, corresponding to the 110° stainless-steel sector, are blocked.

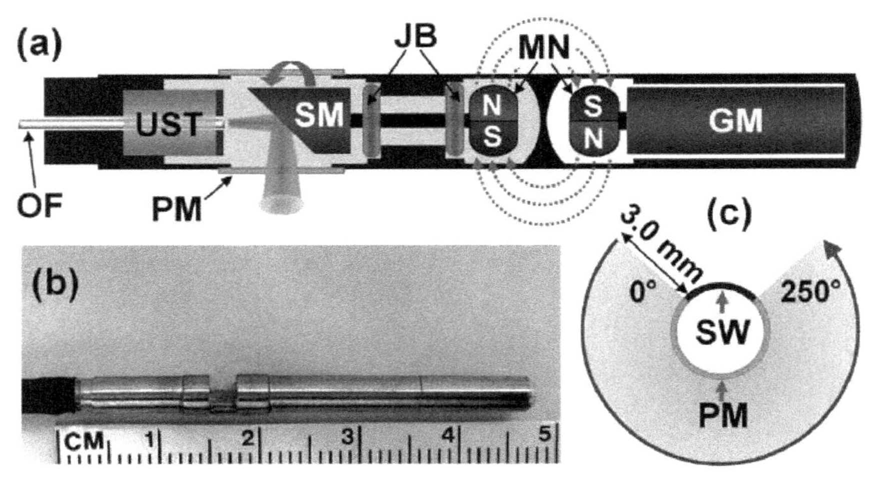

Figure 4.3 (a) Schematic of the photoacoustic endoscopic probe: GM, geared micromotor; JB, jewel bearings; MN, magnets; OF, optical fiber; PM, plastic membrane (imaging window); SM, scanning mirror; UST, ultrasonic transducer. (b) Photograph of the probe. (c) Field of view: SW, stainless-steel wall (blocked zone, 110°); PM, plastic membrane (imaging zone, 250°). Reproduced with permission from Ref. [13].

Based on this probe, the concept and a scanning-mirror-based system for photoacoustic endoscopy demonstrated the feasibility of ex vivo imaging experiments for the first time.

4.3.1.2 A 2.5 mm diameter photoacoustic endoscope

The diameter of above photoacoustic endoscope is too large to fit in the instrument channel (typically ~2.8 mm or ~3.7 mm diameter) of a standard video endoscope, and thus small probes should be further miniaturized. A small-diameter focused ultrasonic transducer provides adequate signal sensitivity and enables the miniaturization of the probe. Therefore, a smaller photoacoustic probe with diameter of 2.5 mm and a ~35 mm rigid length (distal end of the endoscope) is developed [15].

A photograph and a schematic diagram of the probe's rigid distal end are presented in Figs. 4.4a,b, respectively. The distal section has a streamlined structure with a dome-shaped end for smooth intracavitary advancement and is sheathed with biocompatible stainless steel and PET plastic tubes. The endoscopic probe comprises three key units: an optical fiber and US transducer unit (Fig. 4.4c), a scanning mirror unit (Fig. 4.4d), and a micromotor unit (Fig. 4.4e). A single strand of multimode optical fiber (0.22 NA, 365 μm core diameter) delivers laser pulses for PA imaging, and a ring-shaped focused US transducer detects both PA and US pulse-echo signals. The optical fiber and US transducer are coaxially aligned so that the optical illumination and acoustic detection overlap to optimize the sensitivity. A mechanically rotated, dielectric-coated fused silica mirror with a 45°-deflected reflection surface serves as the principle component of the scanning mechanism (referred to as a scanning mirror). The mirror reflects both the laser light and acoustic waves and performs rotational scanning. A liquid medium (de-ionized water) fills the sealed inner cavity of the endoscope and provides acoustic coupling between the probe's imaging window and the transducer. However, to provide an in-air working environment, the micromotor is physically isolated from the liquid medium. The torque required to rotate the mirror is transferred through magnetic coupling of the mirror and the micromotor. All metal frames that secure the three key units are constructed from stainless steel or brass. However, the imaging window is formed from an optically and acoustically transparent plastic tube with a wall thickness of ~70 μm. The rigid stainless steel housing is sheathed with another PET tube (~25 μm thickness) to fix the micromotor's four electric wires (~200 μm thickness for each wire). The endoscope's angular field of view is ~310° partially blocked by the electric wires. Another PET tube encloses all of the flexible section of the wires (optical fiber, US transducer's signal wire, and the motor wires) over the 2 m-long, flexible section.

This new mini-probe can be inserted into the instrument channel of a standard video endoscope and be used under the guidance of the video endoscope. Moreover, it also permits simultaneous PAE and EUS imaging that provides complementary contrast.

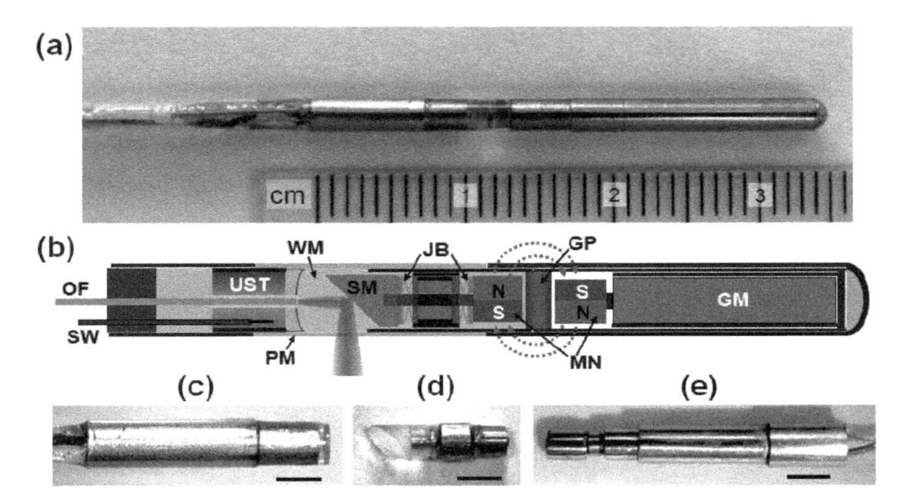

Figure 4.4 (a) Photo of the integrated PAE-EUS mini-probe's rigid distal end (2.5 mm diameter and ~35 mm length). (b) Schematic of the mini-probe. GM, geared micromotor; GP, glass partition; JB, jewel bearings; MN, magnets; OF, optical fiber; PM, plastic membrane (imaging window); SM, scanning mirror; SW, signal wire; UST, ultrasonic transducer; WM, water medium. (c) Optical fiber and US transducer unit. (d) Scanning mirror unit. (e) Micromotor unit. In (c–d), the scale bars represent 2 mm. Reproduced with permission from Ref. [15].

4.3.1.3 Catheter-based photoacoustic endoscope

In order to image the human esophagus, it is essential to use a video endoscope for image guidance to approach the target organ safely. Thus, the PAE probe must be narrow and flexible to pass through the instrument channel of the video endoscope freely. The probe must also be fully encapsulated to avoid direct contact of the scanning tip with the target tissue during mechanical scanning. To meet these technical requirements, a new catheter-based PAE system was developed in 2014 [16]. This endoscopic system is based on mechanical scanning of a single-element US transducer. Importantly, by using a flexible shaft instead of a micromotor, the probe is able to achieve a rigid distal section length of only 16 mm, two times shorter than the above-mentioned micromotor-based endoscope and to enable the catheter section to pass freely through the commercial video endoscope. The flexible catheter section, 3.2 mm in outer diameter and 2.5 m in length, can be inserted into a 3.7 mm diameter instrument channel as shown in Fig. 4.5.

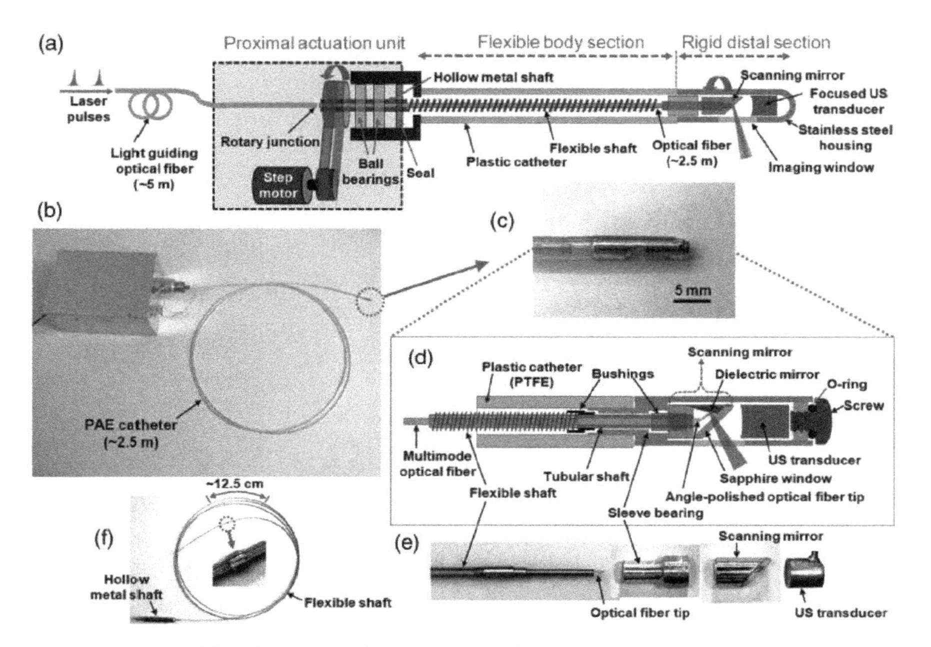

Figure 4.5 (a) Schematic of the catheter-based photoacoustic endoscopy (PAE) system. (b) Photo of the entire endoscopic system. (c) Photo of the rigid distal section. (d) Schematic showing the detailed structure of the rigid distal section presented in (c). (e) Photos of the flexible shaft and optical fiber tip, the bronze sleeve bearing, the scanning mirror, and the focused US transducer. (f) Photo of the 2.5 m long flexible shaft connected to the hollow metal shaft depicted in (a). Reproduced with permission from Ref. [16].

The endoscopic system comprises a proximal actuation unit, a ~2.5 m-long flexible body section sheathed with a plastic catheter, and a rigid distal section that includes a scanning mirror. A step motor installed in the proximal actuation unit provides torque. The torque is transferred to a hollow metal shaft supported by two ball bearings through a timing belt and pulleys and further transmitted to the scanning mirror through the 2.5 m-long flexible shaft. Finally, the scanning mirror receives the torque and performs side-view rotational scanning. For PA imaging, a pulsed laser beam from a laser source is guided by a multimode optical fiber and is transferred via a rotary junction (Fig. 4.5a) to the endoscope's optical fiber located inside the flexible shaft, sent to the target tissue by the optics located in the scanning mirror to generate PA waves. The generated PA waves that propagate to the scanning mirror are reflected to a focused US transducer and converted into electrical signals. Figures 4.5b,c show the entire endoscopic system and the distal section of the probe.

A more detailed structure of the rigid distal section and photos of several key components are showed in Figs. 4.5c–e. The rigid distal section is ~16 mm long, and its housing was fabricated from a 3.05 mm diameter stainless steel tube (wall thickness: 0.33 mm). Inside the housing, the scanning mirror (2.1 mm outer diameter) and the focused US transducer (2.1 mm outer diameter) are encapsulated. To provide a smooth rotation to the scanning mirror, a bronze sleeve bearing and placed bushings at the sleeve bearing's ends are utilized to eliminate longitudinal movement of the scanning mirror. The optical fiber's tip is polished at a ~30° angle, and the scanning mirror's inner space is filled with air, so that the laser beam that impinges on the polished fiber surface experiences an optical total internal reflection, and it exits the fiber with an oblique angle of ~43°. The laser beam is then reflected by a dielectric-coated borofloat mirror and finally sent to the target tissue after passing through an optically transparent sapphire window and an optically clear polyethylene terephthalate (PET) plastic membrane. The PET membrane, a medical grade plastic, forms an imaging window by sealing the inner cavity of the stainless steel housing. To provide acoustic matching from the imaging window to the US transducer, the inner cavity of the endoscope was filled with deionized water. Figure 4.5e shows the components of the rigid distal section, including the flexible shaft and optical fiber tip, the bronze sleeve bearing, the scanning mirror, and the focused US transducer. Figure 4.5f shows the entire flexible shaft. Its proximal end is connected to the hollow metal shaft, and its distal end, which corresponds to the optical fiber tip.

The development of such a catheter endoscope has been an important challenge to realize the technique's benefits in clinical settings. A prototype PAE system was implemented that has a 3.2 mm diameter and 2.5 m-long catheter section. As the instrument's flexible shaft and scanning tip are fully encapsulated in a plastic catheter, it easily fits within the 3.7 mm diameter instrument channel of a clinical video endoscope.

4.3.1.4 Optical-resolution photoacoustic endoscope

As shown in Fig. 4.6, an optical-resolution photoacoustic endomicroscopy (OR-PAEM) system [17] has been developed, which enables internal organ imaging with a much finer resolution than conventional acoustic resolution PAE systems.

For this endomicroscope, the confocal optical illumination and acoustic detection methods are applied for superior signal sensitivity. As shown in Fig. 4.6f, a focused ultrasonic (US) ring transducer, fabricated through the press-focusing technique, is placed inside a 3.4 mm diameter stainless steel housing and then coaxially configured a 1.2 mm outer diameter gradient-index (GRIN) lens unit (Fig. 4.6g) inside the transducer. The unit has an optical working distance of 6.5 mm in the water medium that filled the imaging probe. The working distance is chosen by considering the large path length (3.6 mm) between the transducer and the plastic membrane (i.e., imaging window) at the given probe outer diameter of 3.8 mm, and the optical focal spot was confocally placed within the acoustic focal zone of the transducer, at ~0.8 mm outside the membrane surface. To accommodate the GRIN lens unit inside the transducer while maximizing the effective area of the piezo-electric element, the transducer is fabricated with stepped center hole diameters: a 1.3 mm hole for the installation of the GRIN lens and a 0.9 mm hole for the exit of the laser beam. As shown in Fig. 4.6g, the GRIN lens unit is enclosed in a stainless steel tube and completely sealed with epoxy at both ends to avoid water permeation. Inside the stainless steel tube, the GRIN lens was coupled with a single-mode optical fiber, which guided the laser beam from a light source. The distance between the optical fiber's tip and the beam entrance surface of the GRIN lens was set at 0.8 mm. This distance and the pitch of the GRIN lens were determined from a ZEMAX simulation. After completing the GRIN lens unit, we the output beam profile is analyzed using a beam profiler and measured the FWHM-based beam diameter to be ~9.2 μm (Fig. 4.6i).

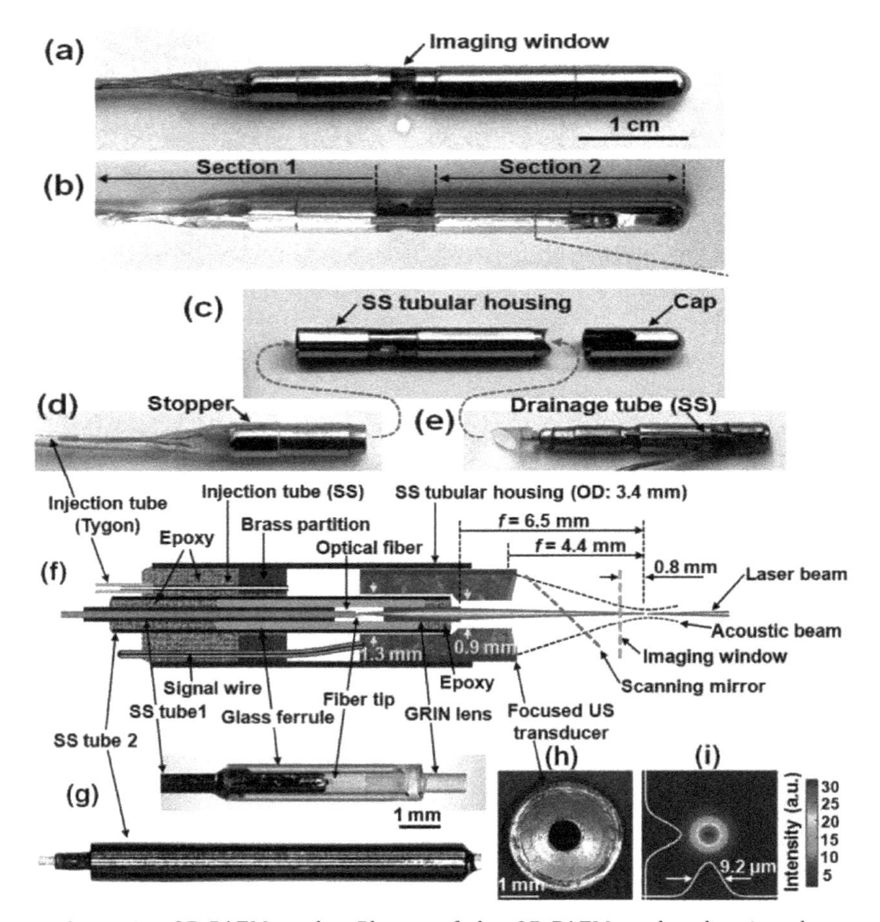

Figure 4.6 OR-PAEM probe. Photos of the OR-PAEM probe showing the imaging window side (a) and bridge side (b). (c) Photo of the SS tubular housing and distal cap. SS, stainless steel. (d) Photo of the optical illumination and acoustic detection unit. (e) Photo of the scanning mirror and micromotor unit. (f) Schematic of (d). The virtual locations of the scanning mirror and the plastic membrane (imaging window) are indicated by two colored dashed lines. (g) Photos of the GRIN lens unit before (upper) and after (lower) being enclosed by a SS tube 2. (h) Front view of the US transducer. (i) Laser beam intensity profile at the focal distance. Reproduced with permission from Ref. [17].

The implemented OR-PAEM system with a full intravital microscopy imaging capability and achieved a transverse resolution as low as 10 μm, the finest among reported optical resolution endoscopic PA images, and an SNR of 29 dB for a 3 μm diameter microsphere illuminated by a 500 nJ pulse energy in a beam of 9.2 μm in diameter. As the system can be utilized in small animals and also can potentially accommodate many other multi-functional molecular probes, it could be a useful tool in many biological experiments, such as tumor and metabolic disease studies.

4.3.1.5 Ring transducer array–based photoacoustic endoscope

The above systems all need to rotate the detector or mirror to receive PA signals in a 2π view. To curtail the time of signal acquisition, a fast preclinical photoacoustic imaging endoscope (PAIE) based on a ring transducer array is developed [18]. The sound, light, and ring transducer array are coaxial, the structure of which makes the PAIE receive the PA signals in a 2π view without rotating the ultrasound transducer. The ring transducer array is connected with the parallel acquisition system, and PA signals from all elements are collected simultaneously. The parallel acquisition system can collect data quickly, and the acquisition time of a set of data is only 50 ms.

Figure 4.7a shows the schematic of the PAIE. It includes a Plexiglas tube (polymethyl methacrylate), a ring transducer array with 64 elements, a taper reflector, an optical fiber (core diameter of 0.6 mm), and an ultrasonic coupling medium with water and glycerin. The length and the diameter of the Plexiglas tube are 95 mm and 30 mm, respectively. The vertex angle of the taper reflector shown in Fig. 4.7b is ~123°, the length and diameter of which are 15 mm and 10 mm, respectively. The ring transducer array shown in Fig. 4.7c is hollow, and the optical fiber is inserted in the hole as the excitation light source. The resonance frequency of the ring transducer array is 6 MHz with a -6 dB bandwidth of 60%. Its inner and outer diameters are 5 mm and 10 mm, respectively. The optical fiber and the ultrasonic coupling medium are separated by the transparent plastic film to prevent contact between each other. In the PAIE system, the laser reflected by the taper reflector is formed into a ring beam to irradiate the sample, and the PA signal is generated due to thermal expansion. The PA signal is detected by the ring transducer array and then collected by the parallel acquisition system. The data are stored by a personal computer. The components of the parallel acquisition system consist of an analog signal processing board and a digital signal processing board. The 64-element ring transducer array is connected with the parallel acquisition system. The 64-channel PA signals are sent to the analog signal processing board and processed by amplification and two antialiasing filters, and then the analog signals are sent to the digital signal processing board for analog-to-digital conversion. The PAIE system can collect the PA signals in a 2π view, simultaneously, without rotating the ultrasound transducer, which provide the potential for detecting colorectal cancer tissue in vivo based on the different optical absorption coefficient between normal colorectal tissue and colorectal cancer tissue.

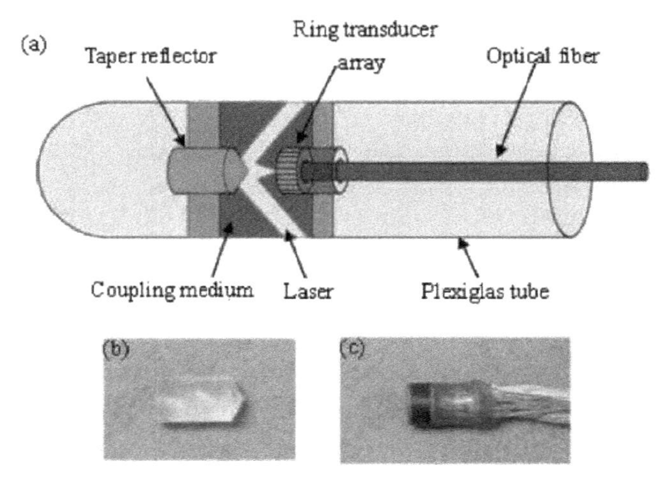

Figure 4.7 (a) Schematic of the PAIE. (b) Photograph of the taper reflector. (c) Photograph of the hollow ring transducer array. Reproduced with permission from Ref. [18].

4.3.1.6 Multi-modalities photoacoustic endoscope

The first prototype system that integrates OCT, US, and PAI modalities for endoscopy applications was developed in 2011 [19]. In this report, an integrated OCT, ultrasound (US) and photoacoustic imaging (PAI) prototype endoscopy system is present for ovarian tissue characterization. The overall diameter of the prototype endoscope is 5 mm, which is suitable for insertion through a standard 5–12.5 mm endoscopic laparoscopic port during minimally invasive surgery. It consists of a ball-lensed OCT sample arm probe, a multimode fiber having the output end polished at 45° angle so as to deliver the light perpendicularly for PAI, and a high-frequency ultrasound transducer with 35 MHz center frequency.

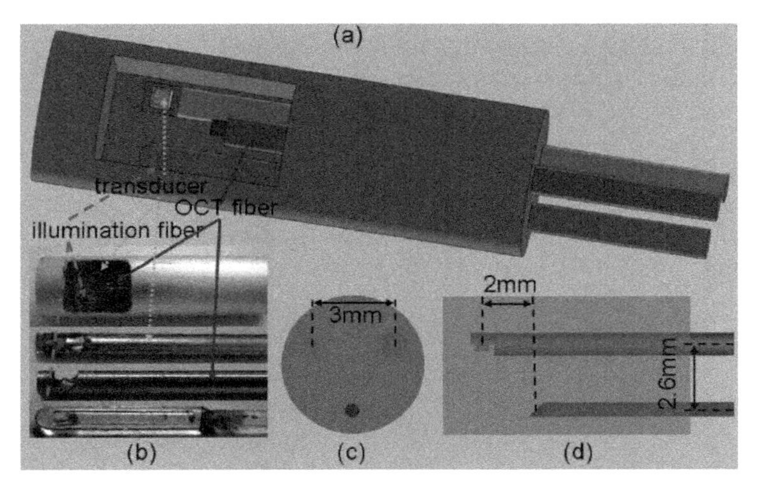

Figure 4.8 Integrated OCT-US-PAI tri-modality endoscopic probe. (a) sketch of the tri-modality probe; (b) photographs of probe and components (transducer, OCT fiber and light illumination fiber); (c) left view of the probe configuration; (d) side view of the probe configuration. Reproduced with permission from Ref. [19].

Figure 4.8a depicts the combined three-modality endoscopic probe which consists of a ball-lensed OCT fiber, a multimode fiber having the distal end polished at 45° angle for delivering the laser beam for PAI, and a high-frequency unfocused ultrasound transducer. The photograph of the probe is shown in Fig. 4.8b. The diameters of the ball-lensed OCT fiber, PAI-light-delivering fiber, and the transducer are 0.5 mm, 0.9 mm, and 0.9 mm, respectively. The square aperture of the ultrasound transducer is 0.5 mm × 0.5 mm. The 35 MHz center frequency was chosen as a compromise between the axial resolution and penetration depth. The three components (OCT fiber, illumination fiber for PAI, and ultrasound transducer) are fixed inside a homemade structure. The tip of the illumination fiber and the transducer element are aligned side-by-side with a 3 mm center-to-center separation as shown in Fig. 4.8c. The OCT fiber is about 2.6 mm under the other two components and indents 2 mm towards the proximal end as shown in Fig. 4.8d. The overall diameter of the endoscopic probe is 5 mm. The light exiting the illumination fiber is directed towards the imaging medium about 4 mm away from the center of the probe.

The optical absorption contrast provided by PAI, the high-resolution subsurface morphology provided by OCT, and the deeper tissue structure imaged by US demonstrate the synergy of the combined endoscopy and the superior performance of this hybrid device over each modality alone tissue characterization. Therefore, combining OCT, US and PAI, the multi-modalities photoacoustic endoscope would further provide complementary tissue optical absorption, scattering information, and deep tissue structures.

4.3.2 Biomedical Applications of Photoacoustic Endoscopy

4.3.2.1 Photoacoustic endoscope of melanoma tumor in rat model

Photoacoustic endoscopic imaging study of melanoma tumor growth in a nude rat in vivo is showed in Fig. 4.9 [20]. After inducing the tumor at the colorectal wall of the animal, the tumor development in situ by using a photoacoustic endoscopic system.

The serial PA radial-maximum amplitude projection (RMAP) images acquired from the rat using a 578 nm laser wavelength. To show the image reproducibility, two sets of image data are present in parallel for each experiment day. As shown in the images, the tumor region was clearly visualized in PA images because the melanoma tumor tissue includes a lot of highly light-absorbing melanin. Also, one can see the increase of the tumor size and adjacent vasculature change in the colorectal wall clearly between the first and second experiment days. The results demonstrated PAE could provide the melanoma tumor region and its adjacent blood vasculature clearly without using any contrast agent.

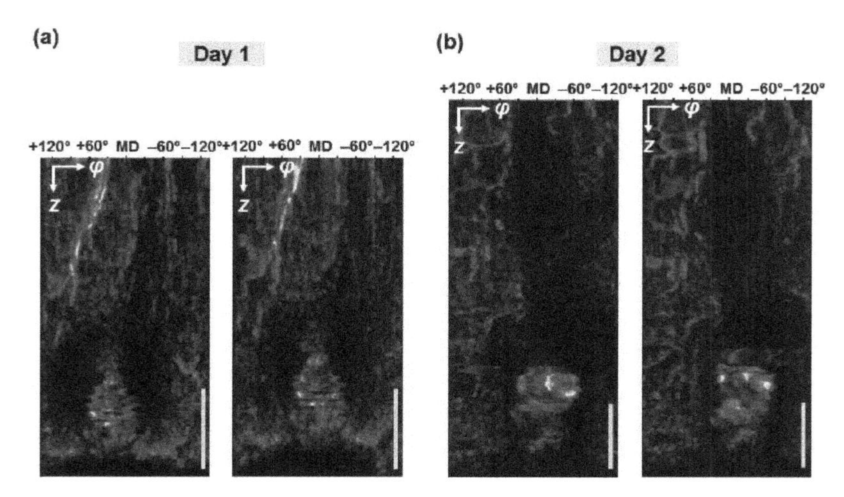

Figure 4.9 Endoscopic PA images from the rat colorectum with a melanoma tumor (views from the inside of the colon). In each image, the horizontal φ-axis corresponds to the angular FOV covering 270°, and the vertical z-axis corresponds to the pullback length of 2–3 cm. The approximate mid-dorsal (MD) position and angular measures from the MD are marked along the horizontal φ-axis, where the positive and negative values correspond to the right and left sides of the animal. The scale bars represent 5 mm. Reproduced with permission from Ref. [20].

4.3.2.2 Photoacoustic endoscope of internal organs in animal model in vivo

A section of colorectum of a New Zealand white rabbit demonstrated the endoscope's in vivo imaging ability [21, 22]. Figure 4.9a shows a 3D-rendered, co-registered PA and US image from the rabbit colorectum acquired in vivo over a scanning time of ~4.5 min. The image was processed from a volumetric data set acquired from a 27 mm diameter and 53 mm-long image volume. In this experiment, only 7 mm imaging depth data are recorded because the transverse resolutions are degraded in the distant region from the focal point. The working distance is 3.5 mm from the probe surface. As shown in PA and US radial maximum amplitude projection (RMAP) images (Figs. 4.9b,c), which were produced from the volumetric image, the PA signals acquired at the 532 nm wavelength show vasculature distributed in the colorectum and adjacent organs, while the US signals reveal a spine structure very clearly because of the high acoustic reflection caused by tissue density variation near the cavity wall. In Figs. 4.9d,e, two combined PA and US B-scan images were selected from the marked positions in Fig. 4.9b. PA imaging shows the cross sections of blood vessels distributed in the intestine wall, which was estimated to be ~1 mm thick, and detects PA signals from a depth of more than 6 mm.

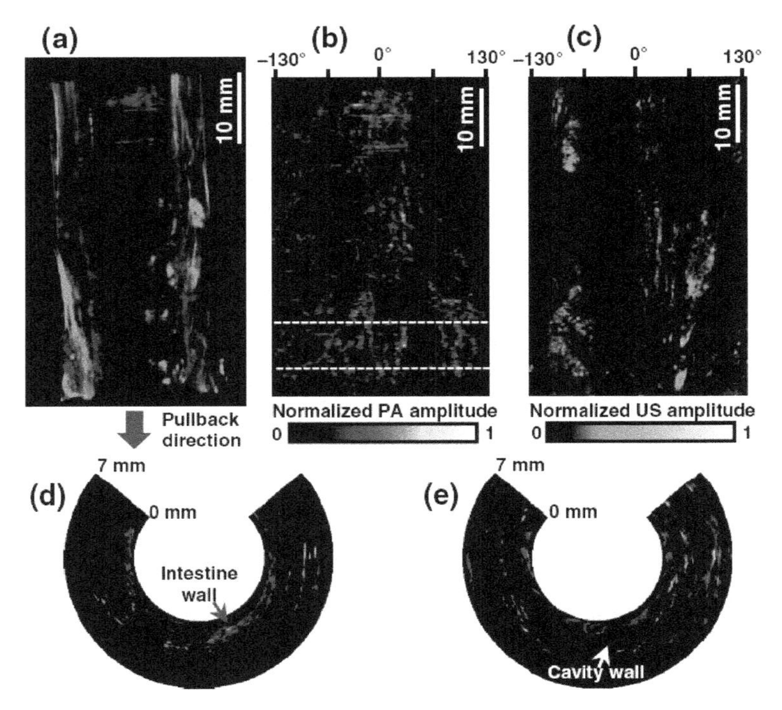

Figure 4.10 (a) Co-registered PA and US volumetric image from a rabbit colorectum acquired in vivo over a 53 mm range with a 27 mm image diameter. The lower portion of the image corresponds to the anus. (b), (c) PA- and US-RMAP images of (a). (d), (e) Representative combined PA and US B-scan images near the location indicated by the dashed line in (b). Reproduced with permission from Ref. [21].

Figures 4.11a–d are the OCT, US, PAI superimposed on US, and corresponding H&E-stained histology images, respectively. The OCT image shows the shallow tissue features and the well-defined boundary indicating there is a big cyst or follicle underneath. The bright spots indicated by the pink arrows in the OCT image represent the collagen bundles, which are marked in the H&E histology (Fig. 4.11d) as well. In Fig. 4.11b, the US image shows a big follicle whose shape and bottom structures are clearly identified. However, these structures were too deep and OCT could not adequately image them. PAI reveals a very high optical absorption at the surface which corresponds to significant amount of red blood cells (see from H&E stains) resulting from endometriosis. This example demonstrates the exquisite sensitivity of the PAI, however, it also suggests that multiple wavelengths are needed to distinguish between old and fresh hemoglobin content for increasing the specificity of the PAI.

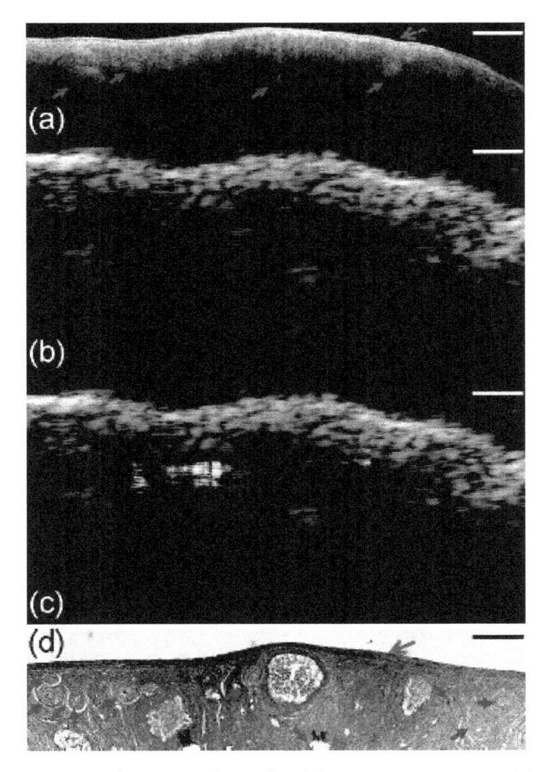

Figure 4.11 One set of images from healthy porcine ovary. (a) OCT image (10 × 2.5 mm); (b) US image (10 × 5 mm); (c) superimposed PAI and US image (10 × 5 mm); (d) corresponding histology (10 × 2.5 mm). Blue arrow, primordial follicle; green arrow, surface epithelium; red stealth arrow, blood vessel; scale bar, 1 mm. Reproduced with permission from Ref. [19].

So far, the international photoacoustic research group is still committed to the development of photoacoustic endoscope with more flexible, more stable and it can match the use of existing clinical equipment. In addition to rectal examination, the photoacoustic endoscopy technology continues to explore new clinic applications, such as the digestive tract, uterus, ovary and other tumor detection. It is to be believed that photoacoustic endoscopy can realize clinical applications in the near future.

4.4 Intravascular Photoacoustic Endoscopy

Despite widespread efforts, cardiovascular disease still accounts for the majority of deaths in industrialized countries nowadays [23]. Histological studies demonstrate that the primary cause of acute cardiovascular events is the rupture of atherosclerotic plaques [24]. Therefore, reliable detection and diagnosis of vulnerable plaque is of paramount importance. The typical vulnerable atherosclerotic plaque consists of a large necrotic lipid core and a thin fibrous cap.

At present, intravascular imaging techniques such as IVUS, OCT, and near-infrared spectroscopy (NIRS) have provided morphological information with high imaging depth or high resolution, or lipid content information for clinical assessment of plaque [25–27]. Nevertheless, these techniques have inherent limitations that do not allow a complete and thorough assessment of the lipid within atherosclerotic plaques. Consequently, there is an ever-increasing demand for a new imaging method that can comprehensively provide detailed information about atherosclerotic lipid for assessing plaque vulnerability.

Photoacoustic tomography (PAT) is a hybrid imaging technique. In PAT, the tissues are irradiated with nanosecond pulsed laser. The deeply penetrable diffused laser energy was absorbed by tissues, which subsequently undergo the rapid thermoelastic expansion and further generate high-frequency ultrasonic signals. The optical absorption within the tissues can be imaged by reconstructing the detected ultrasonic signal. 10 Therefore, PAT provides the volumetric image of tissues with high optical contrast and high ultrasonic spatial resolution at sufficient imaging depth (up to centimeters). Since 2010, many groups have been studied the intravascular photoacoustic imaging of atherosclerosis plaque [28–58]. In the following, we introduce the intravascular photoacoustic imaging system and medical applications.

4.4.1 Intravascular Photoacoustic Probe

The feasibility of performing IVPA imaging using a clinical IVUS imaging catheter was reported for the first time [28]. The setup for intravascular photoacoustic imaging is presented in Fig. 4.12. The imaging system comprises two major components. First, an optical excitation mechanism was used for photoacoustic wave generation in the sample. Second, the photoacoustic waves were detected using an IVUS imaging catheter. A Q-switched pulsed Nd:YAG laser operating at a second harmonic wavelength (λ = 532 nm) delivered short laser pulses at a repetition rate of 20 Hz. The laser source has a pulse width of 5 ns and a maximum energy of 24 mJ. Laser light is delivered to the sample from outside using a prism. Diffused optical illumination is achieved by delivering light through a ground-glass, optical diffuser. The spot size of the optical beam is increased to 20 mm by using diffused light. Consequently, the total energy fluence per optical pulse incident on the sample is reduced and maintained at approximately 1 mJ/cm^2 during the entire experiment.

The clinical realization of an intravascular photoacoustic imaging requires an integrated intravascular imaging catheter. The designs of IVUS/IVPA imaging catheters—side fire fiber-based and mirror-based catheters—are reported [30]. As shown in Fig. 4.13 (left), a commercially available IVUS imaging catheter was utilized for both pulse-echo ultrasound imaging and detection of photoacoustic transients. Laser pulses were delivered by custom-designed fiber-based optical systems. The optical fiber and IVUS imaging catheter were combined into a single device. As shown in Fig. 4.13 (right), the distal end of the optical fiber was polished flat and perpendicular to the optical axis of the fiber, and a mirror was used to redirect the light. The photograph and the diagram of mirror-based IVUS/IVPA imaging catheter are shown in Figs. 4.13. A mirror with a laser damage threshold of 170 mJ/cm^2 for 5 ns pulses was fabricated by thermal evaporation of silver powder on 1 mm-thick glass. The mirror was attached to the fiber using a custom-made brass fixture.

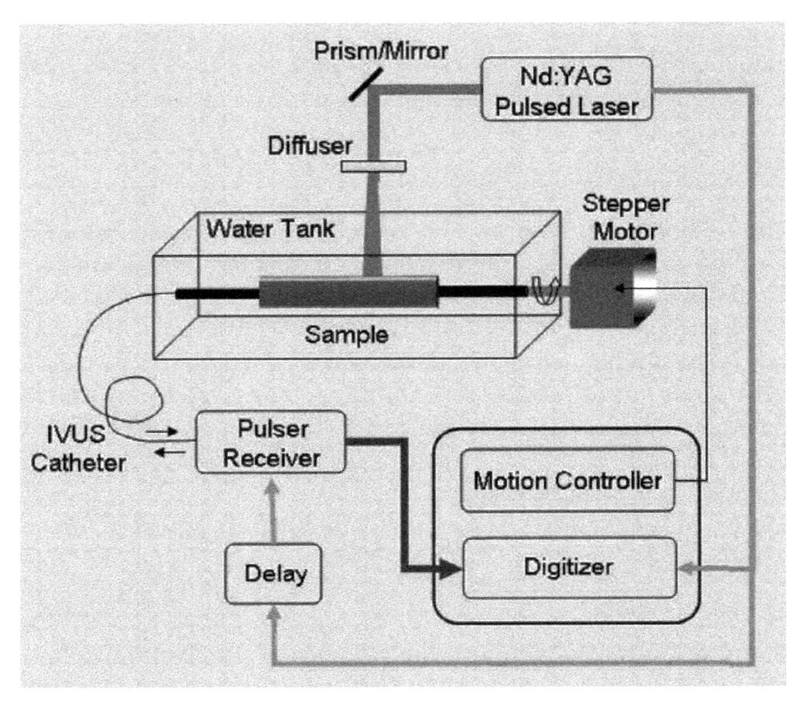

Figure 4.12 Block diagram of the experimental setup for IVPA and IVUS imaging. Reproduced with permission from Ref. [43].

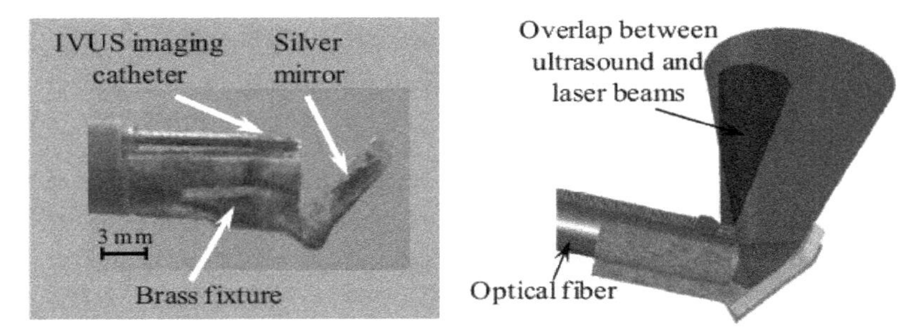

Figure 4.13 Left: Photograph of distal end of the combined IVUS/IVPA mirror-based imaging catheter. Right: a diagram of the combined IVUS/IVPA imaging catheter showing an alignment of the ultrasound and light beams. Reproduced with permission from Ref. [30].

The angle between optical fiber's axis and the mirror was chosen to be 52° for better overlap between light and ultrasound beams. The IVUS imaging transducer was fixed facing away from fiber in the position resulting in the maximum overlap of the ultrasound and light beams and also avoiding a direct interaction of light with the ultrasound transducer. Since reflection of the light from a mirror does not rely on the refractive indices of fiber's core and the surrounding medium, there is no need to keep air near distal end of the light delivery system as in the first design. However, the light divergence of this imaging catheter depends on both the fiber's NA and the relationship between refractive coefficients of fiber's core and the medium.

A 1.25 mm diameter intravascular imaging catheter was built, comprising an angle-polished optical fiber adjacent to a 30 MHz ultrasound transducer [31]. As shown in Fig. 4.14, the hybrid IVPA/IVUS catheter was built, consisting of a 400 μm diameter core optical fiber and a lead–zirconium–titanate (PZT) ultrasound transducer with a diameter of 1.0 mm. The tip of the fiber was polished at an angle of 34° and covered with a glued-on quartz cap to preserve an air–glass interface deflecting the beam by total reflection. The transducer had a center frequency of 30 MHz and a 6 dB fractional bandwidth of 65%. The fiber and the transducer were mounted in an assembly with an outer diameter of 1.25 mm (see Fig. 4.14b). The fiber tip and transducer center were separated by approximately 1 mm. The angle between the optical and acoustical beams is 22°, and the beams overlap between 0.5 and 4.5 mm from the transducer. Cross-sectional scans at sites of interest were made by rotation of the catheter in 1° steps using a motorized rotary stage. At every position, IVPA and IVUS image lines were acquired, ensuring image coregistration. Pulse echo imaging was performed using an arbitrary waveform generator transmitting a Gaussian-modulated cosine wave (30 MHz, 10 V peak to peak, 100% bandwidth). A tunable laser provided the excitation light (pulse width 5 ns, repetition rate 10 Hz, pulse energy 1.2 mJ at catheter tip) for photoacoustic imaging. Images were acquired at several excitation wavelengths between 715 and 1800 nm.

Normally, the probe is designed by mounting the polished fiber tip and US transducer either side by side or mounting the transducer in front of the fiber tip. Based on this geometry, the overlap portion of the laser beam and ultrasound beam is determined by the face angle of the polished fiber and the distance between the fiber tip and the transducer. In this case, the working distance of the probe is finite. However, if the vessel size is larger than this working distance, parts of the vessel would not be covered in one circumferential B-scan. The uncovered area would increase when the probe is not positioned at the center of the vessel lumen. Additionally, in order to get highly efficient acoustic signals, the multiple components need to be precisely aligned. Therefore, a coaxial integrated probe design was presented [33], which can ease the requirement of precise alignment and potentially improve the probe's working distance. Figure 4.15a shows the schematic of our integrated US and PA probe. The 532 nm pulse laser beams were delivered by a 200 μm core multimode optical fiber and emitted through the central hollow of the ring-shaped US transducer. The design steered both the coaxial laser beams and the US beams into the sample by a customized microrod mirror (platinum coating, 2.0 mm diameter, 4 mm length, with the reflective surface angled at 45° to the probe's axis). Also, the US echoes and the excited PA waves from the sample were deflected by the mirror and detected by the US transducer. The microrod mirror shared by the coaxial laser and acoustic beam makes alignment of the US transducer and optical probe more precise and robust. Moreover, this design guaranteed that the laser beam and acoustic beam overlapped along the whole transmitting path. Snell's law governs the path of reflected and refracted waves when the acoustic wave encounters an interface of two media. The ratio of sound-propagation speeds (1.5/5.1, longitudinal wave; 1.5/3.3, shear wave) in water and glass was large enough so that the total internal reflection occurred at the interface of water and glass. In other words, there was no additional propagation loss on the transmitting path of the ultrasonic wave. The mirror, optical fiber, and US transducer were fixed and packaged in a polyimide tube in which a window was made to allow the laser beam and US beam to pass through. Figure 4.15b shows the ring-shaped US transducer with 2.2 mm outer diameter and 0.6 mm inner diameter. Figure 4.15c is the whole integrated probe that was packaged in a polyimide tube with a final packaged diameter of 2.3 mm.

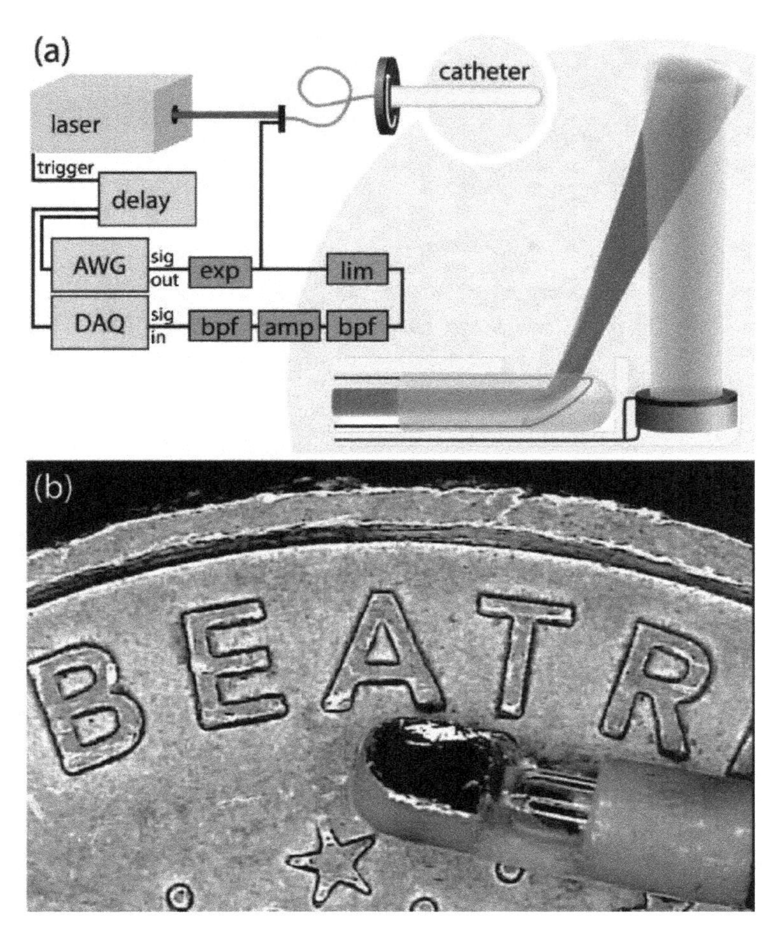

Figure 4.14 (a) Diagram of the experimental setup, including a detailed schematic of the catheter tip, showing the beam layout. AWG, arbitrary wave generator; DAQ, data acquisition; exp, expander; lim, limiter; bpf, bandpass filter; amp, amplifier. (b) Photograph of the catheter tip on the edge of a 10 eurocent coin. Reproduced with permission from Ref. [31].

A first-of-its-kind intravascular optical-resolution photoacoustic tomography (OR-PAT) system with a 1.1 mm diameter catheter is developed, offering optical-diffraction limited transverse resolution as fine as 19.6 mm, 10-fold finer than that of conventional intravascular photoacoustic and ultrasonic imaging.

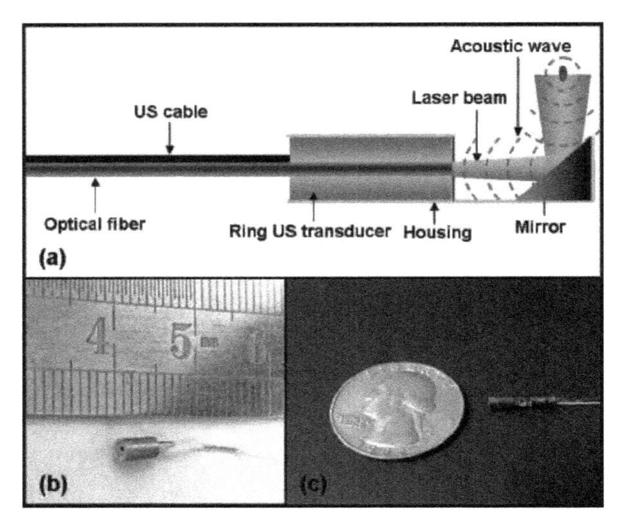

Figure 4.15 (a) Schematic of the combined US and PA probe. (b) Photo of the hand-crafted ring-shaped US transducer. (c) Photo of the combined US and PA probe after packaging. Reproduced with permission from Ref. [33].

Figure 4.16a is a schematic of the distal end of the miniature OR-PAT catheter, consisting of multiple optical and acoustic components assembled in a stainless steel tube housing. The outer diameter of the catheter (housing) was 1.1 mm, comparable to the size of IVUS catheters used in the clinic, making it highly translatable for in vivo intravascular imaging. The single mode fiber (SMF) was capsuled by a 0.5 mm outer diameter flexible stainless steel coil to ensure effective transmission of rotational torque from the rotary joint to the probe distal end. A custom-designed gradient-index (GRIN) lens with a diameter of 0.5 mm and a working distance of 3 mm in air was used to focus the laser from the SMF tip to a 0.5 mm micro-prism, which then reflected the focused laser beam to the inner surface of the vessel wall. The distal end of the flexible stainless steel tube, the GRIN lens, and the micro-prism were fixed and sealed in a thin-wall polyimide tube (PT1) with an inner diameter of 0.7 mm and a wall thickness of 25 mm. An aperture was opened at the tip of the tube to allow the delivery of light for photoacoustic excitation.

A custom-made ultrasonic transducer with a dimension of 0.6 mm 60.5 mm 60.2 mm, a center frequency of 40 MHz, and a fractional bandwidth of 60% was sequentially aligned with the optical components and mounted on a metal base behind the micro-prism. The center distance between the ultrasonic transducer and the microprism was finely adjusted to be 1 mm. The ultrasonic transducer is purposely tilted with an angle of 20° to increase the detection sensitivity to the excited photoacoustic waves. Finally, the rest of the flexible stainless steel coil out of PT1 and the ultrasonic transducer cable were sealed in another thin-wall polyimide tube (PT2) with an inner diameter of 0.74 mm and a wall thickness of 30 mm. All the optical and ultrasonic components at the distal end of the catheter were ultimately housed in the stainless steel tube housing with an outer diameter of 1.1 mm and an inner diameter of 0.9 mm, with an aperture opened at its tip to allow the free traveling of both light and ultrasound.

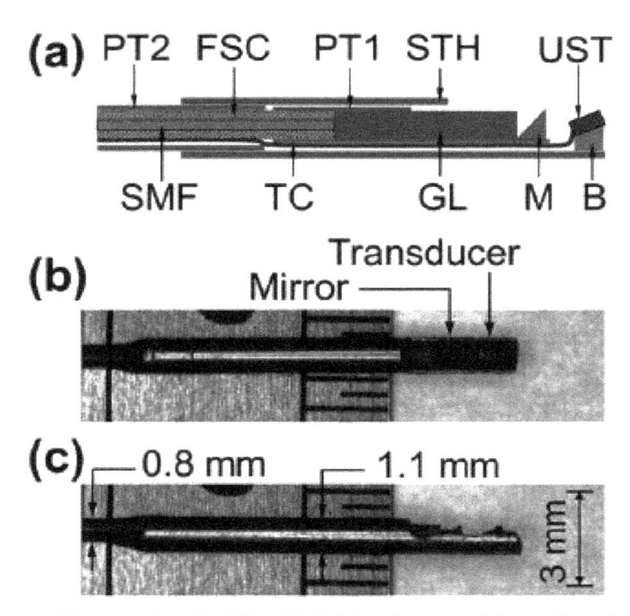

Figure 4.16 The distal end of the OR-PAT catheter. (a) Overall architecture, (b) side and (c) top view of the catheter under a conventional optical microscope. FSC, flexible stainless steel coil; PT1, polyimide tube 1 with an inner diameter of 0.7 mm; PT2, polyimide tube 2 with an inner diameter of 0.74 mm; STH, stainless steel tube housing; UST, ultrasonic transducer; SMF, single-mode fiber; TC, transducer cable; GL, gradient-index lens; M, micro-prism; B, metal base. Reproduced with permission from Ref. [34].

An intravascular confocal photoacoustic (PA) endoscope with symmetrically aligned dual-element ultrasonic transducers was developed [34]. By combining focused laser excitation and focused acoustic collection, the intravascular confocal PA endoscope is capable of realizing resolution-enhanced intravascular PA imaging with improved signal-to-noise ratio (SNR) to ameliorate the resolution reduction caused by laser scattering with increasing tissue depth. This intravascular confocal PA endoscope with an outer diameter of 1.2 mm supports potential for clinical applications in intravascular plaque imaging and subsequent diagnosis.

Figure 4.17a shows a schematic of the intravascular confocal PA endoscope, which consists of a nickel tube, an SMF, a customized gradient index (Grin) lens, a prism, and a dual-element transducer. The length and outer diameter of the nickel tube are 20 mm and 1.2 mm, respectively. A custom-made Grin lens with a diameter of 0.5 mm and a focus distance of 5 mm is used to focus the laser from the SMF to the prism. The inclined plane of the prism is purposely tilted at an angle of 60° to reflect the focused laser beam to top of the transducers. Following the widely used method to evaluate the surface laser fluence in optical-resolution PA imaging, the output laser energy from the endoscope is limited to ~400 nJ per pulse, corresponding to a fluence of ~12 mJ/cm^2 per pulse at the tissue surface (assuring the optical focus is 1 mm below the vessel), which is below the 20 mJ/cm^2 safety standard of the American National Standard Institute (ANSI) safety standard. A window is opened at the side of the tube to allow the delivery of the laser beam for PA excitation and acoustic reception. The transducer elements with dimensions of 1.1 mm × 0.3 mm are symmetrically placed on both sides of the reflected laser to form an acoustic focus with focal length of ~2.5 mm and an angle of about 148°, in accordance with the length from the center to the vessel wall. Each element of the transducer has a coaxial cable with an outer diameter of 200 μm, the end of which is combined with bayonet nut connectors (BNC). The laser beam is focused into the acoustic focus area to improve the detection sensitivity. Figure 4.17b shows a photograph of the intravascular confocal PA endoscope with an outer diameter of 1.2 mm.

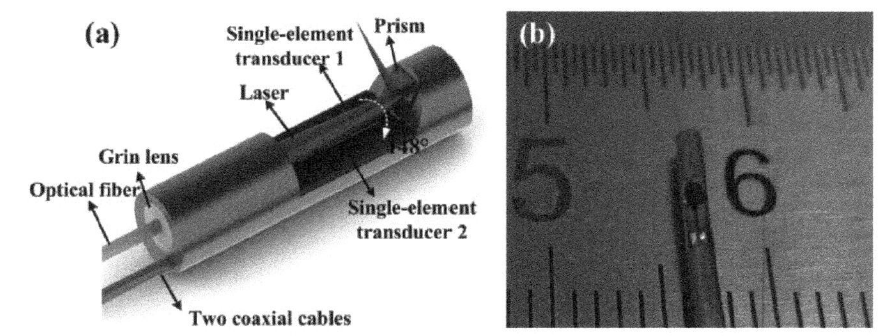

Figure 4.17 (a) Architecture of the endoscopic probe. Transducer elements with an angle of about 148° are symmetrically placed on both sides of the reflected laser. (b) Photograph of the intravascular confocal PA endoscope. Reproduced with permission from Ref. [37].

4.4.2 High-Speed Intravascular Photoacoustic System

A high-speed IVPA system capable of multiwavelength spectroscopic photoacoustic imaging was developed [38]. Figure 4.18 illustrates the overall architecture of the high-speed IVPA system, which can perform both IVPA and IVUS imaging. A ns-pulsed OPO laser (EKSPLA, NT242) with a repetition rate of 1 kHz, operating at a tunable wavelength ranging from 1185 to 1235 nm (a lipid absorption band), was used for photoacoustic excitation. The pump source in NT242 is a diode-pumped solid–state Q-switched laser operating at 1064 nm. Upon third harmonic generation, the ultimate pump energy is about 3.5 mJ at 355 nm (used as the pump wavelength for the entire wavelength tunable band). The efficiency of the OPO crystal is about 4% at 1210 nm. The output laser beam was reshaped by an iris, attenuated by a neutral density filter, and then focused by a condenser lens into the multimode fiber of the catheter. A photodiode (PD) was used to record the fluctuation of the laser pulse energy for fluence compensation during postexperiment data processing. The laser pulse energy out of the catheter was maintained to be ~30 μJ throughout the study in this study. A custom-made miniature single-element ultrasonic transducer was used to detect the photoacoustic waves from the sample as well as to perform pulse-echo ultrasound imaging (a short delay of ~5 μs was set to separate each photoacoustic and ultrasonic

A-line). Both the photoacoustic and ultrasound signals were amplified (39 dB) using an ultrasound pulser receiver, and then digitized with a data acquisition card (12 bit digitization; 250 MHz sampling rate) in a personal computer. To form an A-line image, the acquired (photoacoustic and ultrasonic) signals were band-pass filtered (1–50 MHz for photoacoustic signal and 10–50 MHz for ultrasound signal, with a 128th-order zero-phase forward and reverse finite impulse response filter), followed by a Hilbert transform for envelope detection. Finally, the image was converted into a polar coordinate for display. To acquire cross-sectional images (B-scans), the catheter was rotated at a speed as fast as 300 RPM via an optical-electric rotary joint. Furthermore, three-dimensional (3-D) images can be obtained via a pull-back stage driven by a step motor. For imaging at each wavelength, a B-scan rate up to 5 Hz (200 A-lines/B-scan) was achieved, about two orders of magnitude faster than conventional IVPA systems operating at a similar wavelength tuning range.

Thus far, IVPA systems suffered from slow imaging speed due to the lack of a suitable laser source for high-speed excitation. An improvement in IVPA imaging speed by two orders of magnitude, to 1.0 s per frame, enabled by a custom-built, 2 kHz master oscillator power amplifier (MOPA)-pumped, barium nitrite [$Ba(NO_3)_2$] Raman laser. This advancement narrows the gap in translating the IVPA technology to the clinical setting.

Herein, a $Ba(NO_3)_2$-based Raman shifter with a major Raman mode (V) at 1047 cm was applied to convert the 1064 nm pump to a 1197 nm output (Fig. 4.1A). A 2 kHz MOPA system was employed as a pumping source and obtained an output of 2.0 mJ pulse energy at 1197 nm at 2 kHz repetition rate from the Raman shifter [39]. The conversion efficiency of our Raman laser is 32%, which is, one order of magnitude higher that the commercially available OPO system. This high-repetition-rate laser system enabled the IVPA imaging of lipid-laden plaque with 1 Hz frame rate, which is nearly two orders of magnitude faster than the reported systems. Owing to the utilization of a polarization-maintaining single-mode fiber in fiber amplifier before the free space amplifiers, the beam quality of 1064 nm output from the MOPA was measured to be M2 5 1.6. This high-beam quality ensured the high conversion efficiency of the Raman shifter. Moreover, the pulse duration of the MOPA system is tunable, and it is known that the bandwidth of the photoacoustic response is dictated by the optical excitation pulse duration and the relaxation response of the sample. Although the general paradigm for photoacoustic imaging is that the optimal pulse duration is around 5 to 10 ns, the optimal pulse duration for IVPA imaging using a high-frequency transducer depends on many factors, including the penetration depth. This pulse-duration-tunable MOPA system can potentially provide optimization of the signal for different sample conditions. By design and development of a kHz-repetition-rate Raman laser, the intravascular photoacoustic imaging speed was improved by two orders of magnitude.

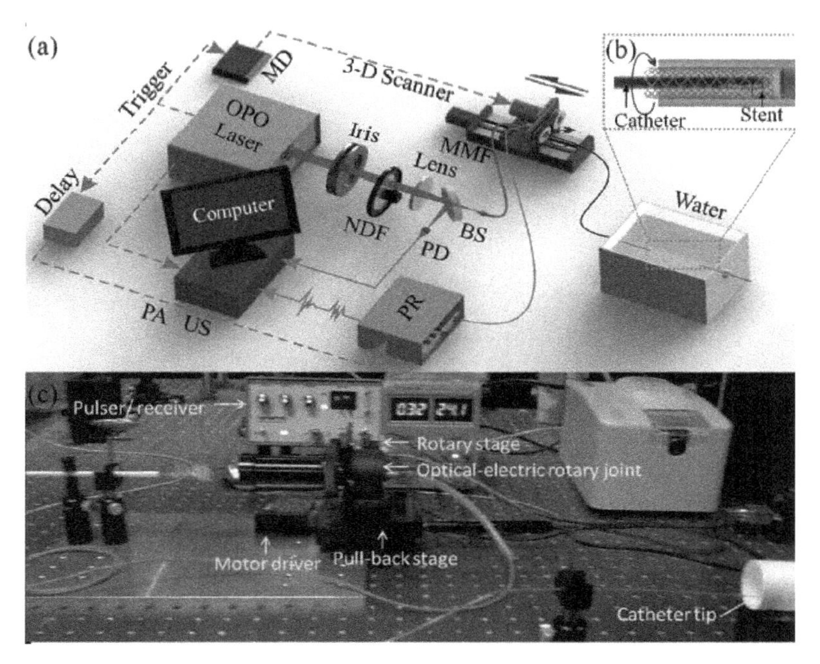

Figure 4.18 Illustration of the high-speed intravascular photoacoustic (IVPA) system. (a) Overall architecture of the system. (b) Enlarged view of the dashed box in (a). (c) Photo of the high-speed IVPA system. Optical parametric oscillator (OPO); neutral density filter (NDF); beam splitter (BM); photodiode (PD); ultrasonics (US); photoacoustics (PA); multimode fiber (MMF); 3-D Scanner, consisting of an optical-electric rotary joint, a motorized rotary stage and a motorized pull-back stage; PR, ultrasound pulser/receiver; MD, motor driver. Reproduced with permission from Ref. [38].

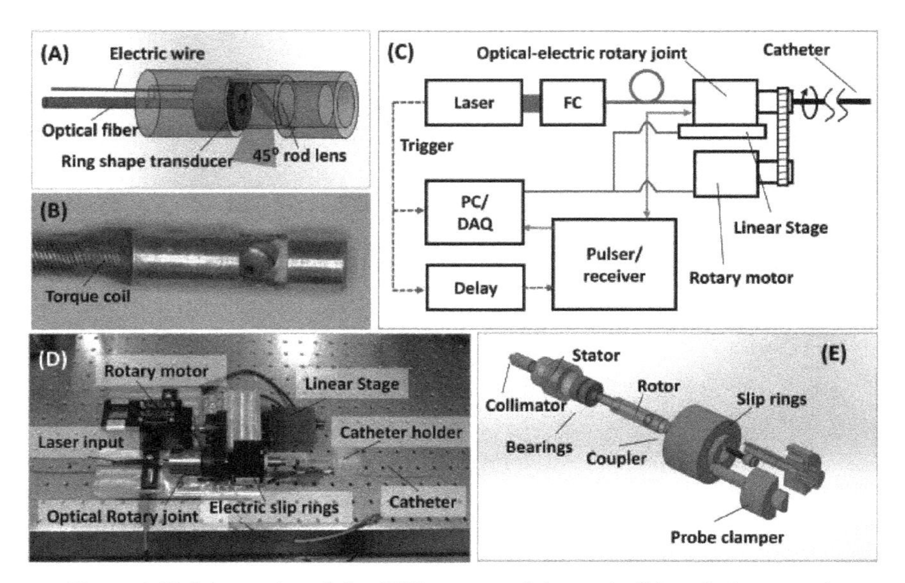

Figure 4.19 Schematics of the IVPA system: Schematic (A) and photograph (B) of the IVPA probe. (C) Block diagram showing the data acquisition system. FC: fiber coupler; PC/DAQ: personal computer, data acquisition. (D) Photograph of the scanning assembly. (E) Schematic of the assembly of optical rotary joint and electrical slip rings. Reproduced with permission from Ref. [39].

4.4.3 Ex vivo Characterization of Atherosclerosis with Intravascular Photoacoustic Tomography

Representative IVUS, IVPA, and histological cross-sectional images of the aorta excised from cholesterol fed rabbit are shown in Fig. 4.20 [43]. The 6.75 mm diameter grayscale IVUS image in Fig. 4.20a is presented using 40 dB display dynamic range. The IVUS image suggests the presence of a diffused plaque all along the inner layer of the aorta (marked by the arrow). The contrast in the IVUS B-Scan is not sufficient to clearly demarcate the lesion. However, the thickening of the intimal layer combined with a definite presence of hypoechoic ultrasonic signals may suggest the presence of a lipid filled plaque. The mean thickness of the vessel wall in the IVUS image is 1.2 mm and the diameter of the lumen measures about 3 mm. The attenuation-compensated IVPA image in Fig. 4.20b represents the photoacoustic response from the atherosclerotic rabbit aorta.

This color image is displayed using 37 dB dynamic range. An important feature in the IVPA image is the dark region in deeper regions of the plaque. The low magnitude of IVPA response could be attributed to the presence of lipids—a weak light absorber at 532 nm. Furthermore, the IVPA image also indicates relatively strong photoacoustic response from the media, medial-adventitial boundary and the inner layer of the intimal plaque as marked by arrows in Fig. 4.20b. The combined IVUS/IVPA image in Fig. 4.20c allows the analysis of the photoacoustic response from tissue within the structural content of the vessel wall provided by IVUS image. The IVUS/IVPA image shows ultrasound echo and photoacoustic signal correspondence in the fibrous regions of the aorta. The low-amplitude photoacoustic signal matches the plaque-media boundary containing hypoechoic IVUS signals. The concentric rings in the center of image in Figs. 4.20b,c are the strong photoacoustic response generated on the surface of the transducer element. The contribution of the plaque components to the photoacoustic response was evaluated by analyzing the tissue histology. The high-resolution photographs of the cross sections stained with H&E, RAM-11 and Picrosirius red are presented in Figs. 4.20d–f. The H&E-stained image in Fig. 4.20d indicated the general morphology of the aorta and plaque. The increase in the thickness of the intima is clearly seen in the image. The migration of the smooth-muscle cells from the media into the intima in addition to the accumulation of lipids cause an increase in the size of the intima, whereas a normal intimal layer is usually a thin endothelial layer lining the arterial wall. The result of progressive cellular events involving monocyte migration into the intima, accumulation of macrophages and subsequent encapsulation of lipids into the macrophage cells is seen as dense clusters of macrophage foam cells in Fig. 4.20e. These macrophages filled with lipids are stained brown by RAM-11 and are accumulated all over the plaque. Furthermore, the smooth muscle cells migrating into the plaque is responsible for the synthesis of fibrous collagen. The result of this process is manifested as thick type I collagen fibers. The polarization photomicrograph (Fig. 4.20f) of the tissue section stained with Picrosirius red suggests the presence of normal and thin Type III collagen all along the vessel and focally dense deposits of the thick Type I collagen in the plaque. The histological report analysis also suggests that the plaque appears to be more cellular (lipid filled macrophage cells) than fibrous (collagen).

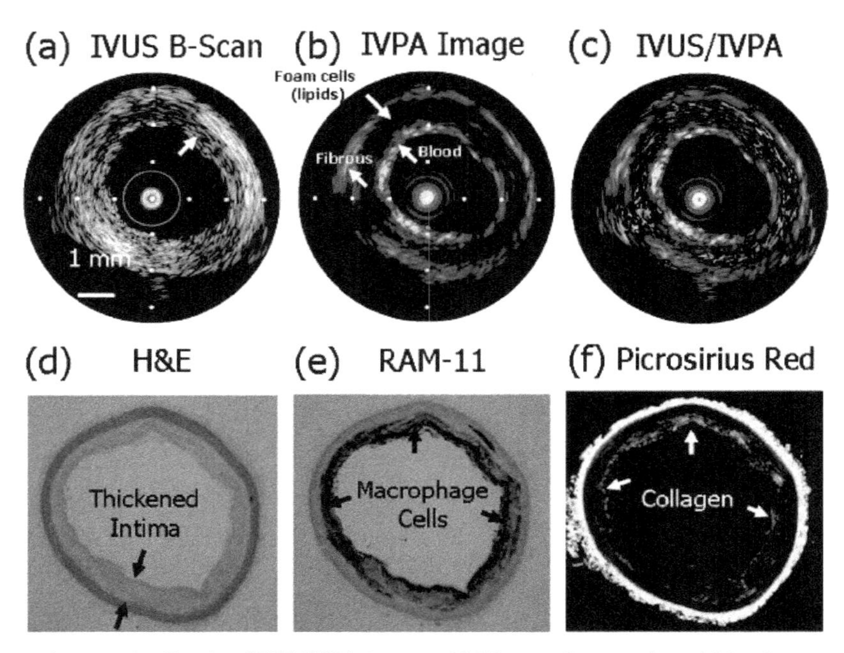

Figure 4.20 Ex vivo IVUS/IVPA images (6.75 mm diameter) and histology of an advanced atherosclerotic lesion. (a) The IVUS image of the arterial cross section shows a hypoechoic concentric plaque with significant structural thickening of the intima. (b) The IVPA image at 532 nm optical excitation with lipids in the plaque indicated by low photoacoustic signals. The higher photoacoustic response from the rest of the plaque corresponds to the presence of the fibrous collagenous cap infiltrated with macrophages. (c) Coregistered IVUS and IVPA image. (d) H&E-stained histology image. (e) RAM11-stained image showing the highly expressed RAM11 antigen with intense macrophages embedded in the plaque. (f) Polarized image of the Picrosirius red-stained cross section showing the presence of focally dense collagen in the plaque. Reproduced with permission from Ref. [43].

Intravascular photoacoustic (IVPA) imaging, a minimally invasive imaging modality, can spatially resolve the optical absorption property of arterial tissue. Based on the distinct optical absorption spectrum of fat in the near infrared wavelength range, spectroscopic IVPA imaging may distinguish lipid from other water-based tissue types in the atherosclerotic artery.

A bench-top spectroscopic IVPA imaging system was used to ex vivo image both atherosclerotic and normal rabbit aortas [45]. By combing the spectroscopic IVPA image with the IVUS image, lipid regions in the aorta were identified. The results demonstrated that IVUS-guided spectroscopic IVPA imaging is a promising tool to differentiate lipid in atherosclerosis. Given the distinct optical absorption spectra of various tissue types (Fig. 4.21), spectroscopic IVPA imaging may be used to assess tissue composition.

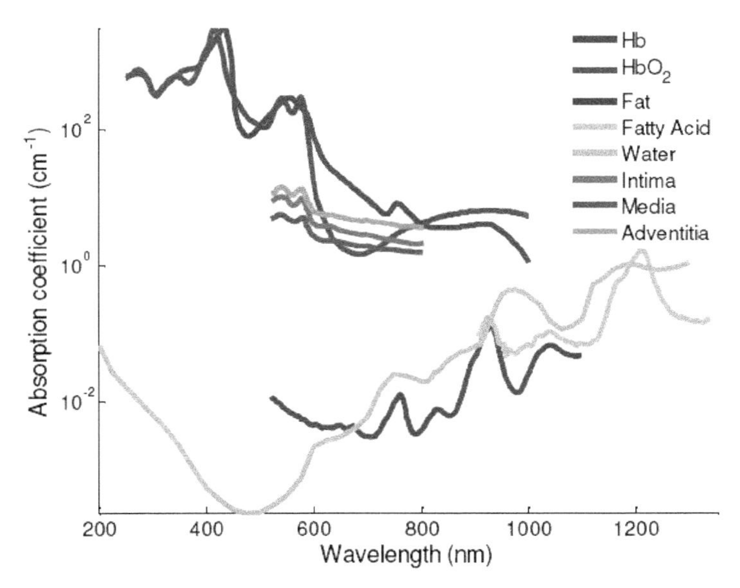

Figure 4.21 Optical absorption spectra of various tissue types. Reproduced with permission from Ref. [45].

Given the differences in tissue type-dependent spectral behavior of photoacoustic signals, analysis of multi-wavelength IVPA images was performed to identify regions with enhanced content of lipid. The lipid-rich regions, i.e., the regions with anticipated spectroscopic behavior of IVPA signals, were color-coded and plotted over co-registered IVUS images (Figs. 4.22a,d). Clearly, lipid-rich regions were detected in the thickened intima of the diseased aorta, but no significant lipid-rich regions were present in normal aorta. The location of lipid-rich areas was further confirmed by examination of tissue cross sections adjacent to the imaged location. Both the Oil red O stain for lipid (Fig. 4.22b) and H&E stain (Fig. 4.22c) show that the plaques in the diseased aorta are lipid-rich, whereas no lipid was present in the intima of the normal aorta (Figs. 4.22e,f). Both normal and diseased aortas show some lipid in the peri-adventia, but this is unrelated to atherosclerosis. In conclusion, a method to differentiate lipid-rich regions in atherosclerotic vessels using spectroscopic IVPA imaging together with IVUS imaging was introduced. Ex vivo tissue studies demonstrated that the spectroscopic IVPA imaging in the 100–1230 nm wavelength range can successfully identify lipid-rich regions in the atherosclerotic rabbit aorta. Generally, spectroscopic IVPA imaging has the potential to identify tissue composition based on intrinsic optical absorption contrast between various types of tissues.

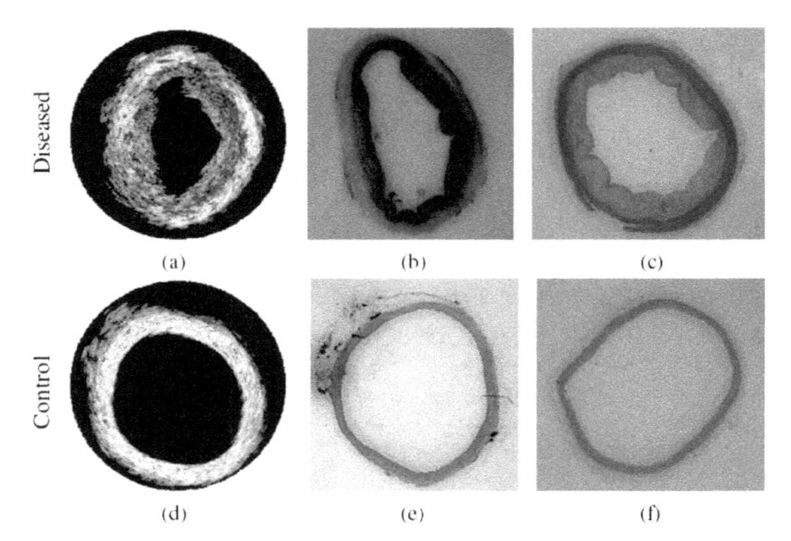

Figure 4.22 Combined IVUS and spectroscopic IVPA images and corresponding histological slices of the diseased atherosclerotic aorta (a–c) and normal (i.e., control) aorta (d–e). Lipid-rich regions (orange color) were identified from multi-wavelength photoacoustic imaging and displayed over the IVUS images. Lipid-rich regions were detected in the thickened intima layer of the diseased aorta (a) confirmed by Oil red O stain for lipid (b) and H&E stain (c). In contrast, spectroscopic IVPA imaging (d) and tissue histology (e and f) show insignificant lipid-rich regions in normal rabbit aorta. Both normal and diseased aortas show some insignificant deposits of lipid in the peri-adventitia. Reproduced with permission from Ref. [45].

Lipid detection in atherosclerotic human coronaries by spectroscopic intravascular photoacoustic imaging was demonstrated [48]. Co-registered cross-sectional images were acquired at several wavelengths near 1200 nm and a lipid-specific absorption band. Correlating the photoacoustic spectra at 6 or 3 wavelengths from 1185 to 1235 nm with the absorption spectrum of cholesterol and peri-adventitial tissue, we could detect and differentiate the lipids in the atherosclerotic plaque and peri-adventitial lipids, respectively. With two wavelengths, both plaque and peri-adventitial lipids were detected but could not be distinguished.

IVPA/IVUS was performed on a human coronary artery specimen exhibiting early-stage atherosclerosis (left anterior descending artery, female aged 41, imaged < 24 h post mortem). Figures 4.23a,b show the co-registered combined IVPA/IVUS images at 1205 nm (high lipid absorption) and 1235 nm (low lipid absorption), respectively. In the IVUS image, the luminal border and the external elastic lamina are clearly visible. The images show two thickened intimal regions that exhibit a brighter signal in the 1205 nm IVPA image than in the 1235 nm IVPA image. A brighter signal is also observed in the 1205 nm IVPA image in the peri-adventitial regions all around the vessel. The corresponding lipid stain, shown in Fig. 4.23c, confirms the presence of lipids in the regions with enhanced 1205 nm IVPA signal. The three different lipid detection methods were applied to the co-registered IVPA data. The findings are comparable to the lipid detection results in the phantom: Fig. 4.23d clearly depicts the lipids in the fatty streak at the bottom right while suppressing successfully the peri-adventitial lipids. However, it fails to distinctly show the lipids present in the smaller fatty streak at the top of the vessel. In contrast, Fig. 4.23e depicts the peri-adventitial lipids while suppressing the lipids present within the thickened intima. The plaque and peri-adventitial lipid maps created using the three-wavelength (1185, 1205 and 1235 nm) correlation with the associated reference spectra, overlaid on the IVUS image, are shown in Figs. 4.23g,h, respectively. The three-wavelength correlation with cholesterol seems to be working equally well as the six-wavelength correlation, whereas the three-wavelength peri-adventitial lipid correlation degrades slightly. In summary, sIVPA data to detect atherosclerotic lesion lipid content were demonstrated on human coronary arteries ex vivo and have the potential to be used in an in vivo setting.

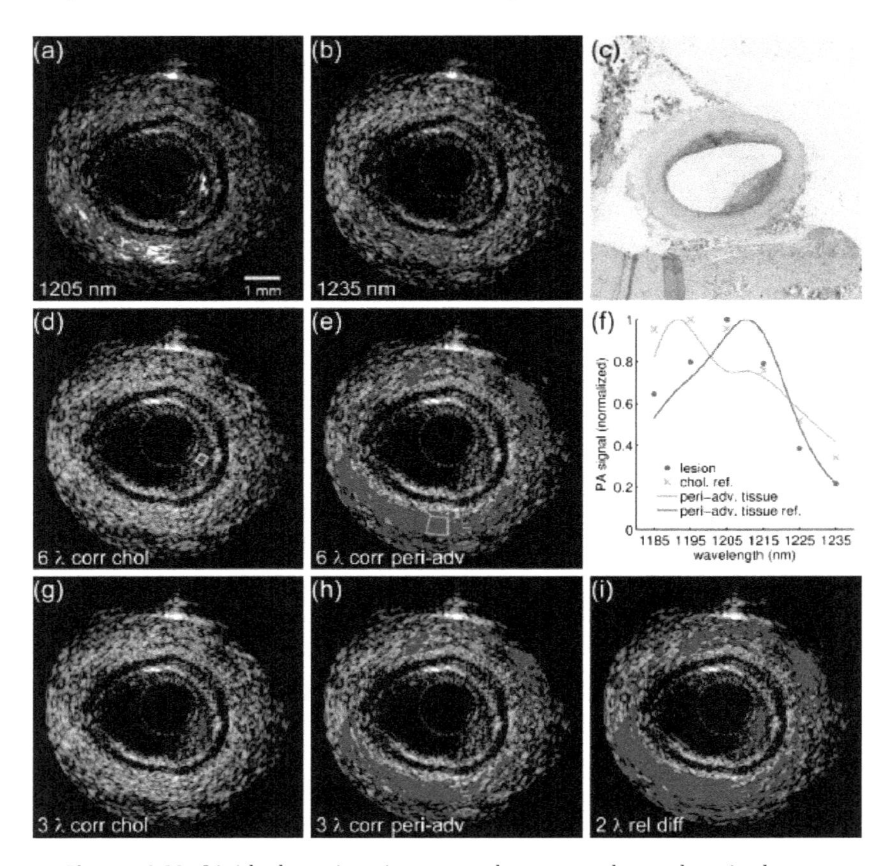

Figure 4.23 Lipid detection in an early-stage atherosclerotic human coronary artery. (a) 1205 nm and (b) 1235 nm combined IVPA/IVUS images (IVPA 35 dB, IVUS 40 dB). (c) Lipid histology stain (ORO); lipids are stained red. Lipid map based on six-wavelength correlation with the cholesterol (d) and peri-adventitial reference spectrum (e). (f) Average IVPA signal strength in the lesion and peri-adventitial areas indicated in (d) and (e), respectively, shows high correlation with the respective reference spectra. Lipid map based on three-wavelength correlation with the cholesterol (g) and peri-adventitial reference spectrum (h). (i) Lipid map based on two-wavelength relative difference. All lipid maps are shown overlaid on the corresponding IVUS image. Reproduced with permission from Ref. [48].

4.4.4 In vivo Characterization of Atherosclerosis with Intravascular Photoacoustic Tomography

Pilot studies of in vivo combined IVUS and IVPA imaging are reported [53]. A recently introduced prototype of an integrated IVUS/IVPA imaging catheter consisting of a single-element ultrasound transducer and a light delivery system based on a single optical fiber was adapted and used for in vivo imaging of a coronary stent deployed in a rabbit's thoracic aorta in the presence of luminal blood.

Two sets of in vivo IVUS, IVPA, and combined IVUS/IVPA images of the coronary stent deployed inside the rabbit's thoracic aorta in the presence of luminal blood are shown in Fig. 4.24. The images in Fig. 4.24 (a–c) and in (d–f) were obtained at two different locations within the rabbit's stented artery. The anatomy of the aorta and connective tissues are clearly visible in both IVUS images shown in Figs. 4.24a,d using 30 dB display dynamic range. Speckles inside the lumen indicate the presence of blood, while the black circle in the center indicates the location of the integrated IVUS/IVPA imaging catheter. However, the IVUS contrast between the metallic struts and soft tissues is poor. IVPA images of the same cross sections of the artery are shown in Figs. 4.24b,e using a linear display dynamic range. The struts of the stent are clearly depicted, while both blood and arterial tissues do not provide noticeable photoacoustic transients due to the high optical absorption coefficient of metal compared to that of soft tissues and blood at 1064 nm wavelength. Such high optical contrast indicates that the laser energy can be potentially decreased to decrease laser fluence without sacrificing the quality of IVPA images. The combined IVUS/IVPA images shown in Figs. 4.24c,f indicate that highly absorbing areas (struts of the stent) are located at the inner surface of the lumen. The obtained results are in agreement with our previously reported ex vivo studies.

In the reported pilot studies, a coronary stent deployed into a rabbit's thoracic aorta was imaged without removing the luminal blood. Both the resolution and the imaging depth of the in vivo IVPA imaging are comparable to that of IVUS imaging. Overall, the results indicate that the concept of integrated IVUS/IVPA imaging catheter is suitable for in vivo imaging of metallic implants and vascular tissues.

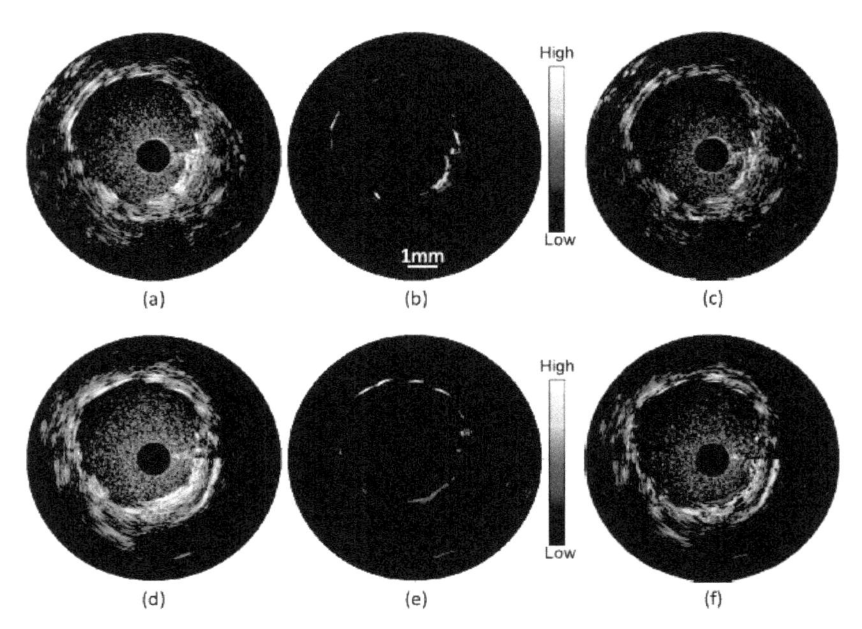

Figure 4.24 In vivo IVUS (a and d), IVPA (b and e), and combined IVUS/IVPA (c and f) images of the rabbit thoracic artery with a deployed stent. The images (a)–(c) and (d)–(f) correspond to two different cross sections of the rabbit's artery. Reproduced with permission from Ref. [53].

The theoretical advantage of IVPAT can be derived from the limitations of other imaging modalities, specifically IVUS, sound wave based) and near-infrared spectroscopy (light wave based). Grayscale IVUS has a high imaging depth, which enables a full-field evaluation of plaques with large lipid cores. And virtual histology IVUS can distinguish plaque components. However, detailed assessment of lipid content is limited. Near-infrared spectroscopy provides several valuable indexes for evaluating lipid content. However, near-infrared spectroscopy is an en-face imaging mode that only enables the evaluation of lipid from a two-dimensional image plane (endothelial surface), without depth information. As a hybrid imaging technique, IVPAT integrates the advantages of optical contrast and ultrasonic resolution, which can simultaneously demonstrate the spatial distribution and relative concentration of lipid content in atherosclerotic plaques along entire vessel segments. Furthermore, the unique combination of the optical and ultrasonic components suggests prospective multiple-imaging modalities with IVUS or optical coherence tomography for comprehensively evaluating plaque structures (e.g., calcification, arterial wall remodeling, and fibrous cap).

Lipid imaging by in vivo IVPAT was demonstrated in ref. [54]. Magnetic resonance imaging visualizes plaque morphology (Figs. 4.25A,D,G), but it was insufficient to specify lipid content within plaques. IVPAT obtained high-contrast LRC maps at the same imaging planes (Figs. 4.25B,E,H), which simultaneously showed content and distribution of lipid within the plaques. The whole boundary of the lipid-rich intima could be distinguished, which matched well with the histology (Figs. 4.25C,F,I). Further measurements demonstrated that all three intima had >95% maximum LRC and >40% mean LRC. Although the blood flow was kept during the data acquisition, IVPAT probe could be located by x-ray (Fig. 4.25J). The relative error between IVPAT and histology was approximately 10% (Fig. 4.25K).

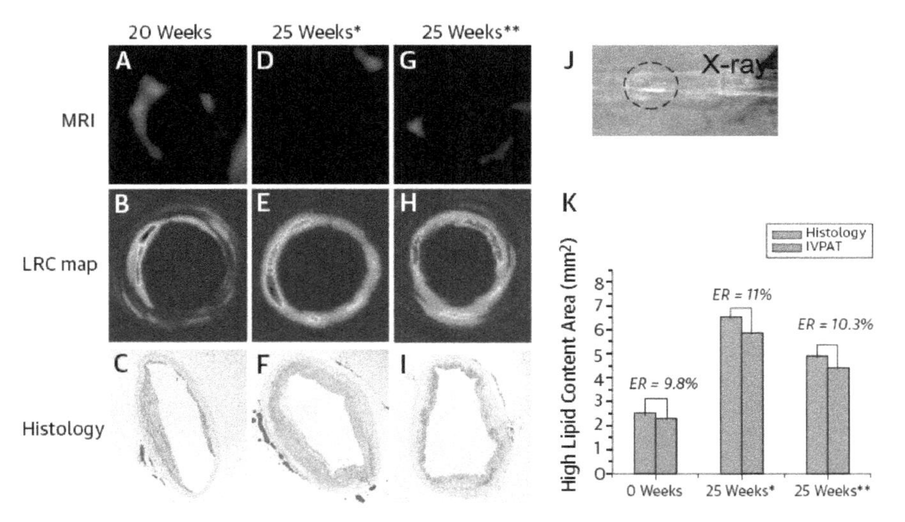

Figure 4.25 In vivo IVPAT Imaging of Lipid in the Atherosclerotic Plaques Magnetic resonance imaging (MRI) results, LRC map, and histological image of three rabbits in vivo examined with in vivo IVPAT after 20 weeks (A to C) or 25 weeks (D to F, G to I) HFC diet feeding. (J) X-ray image showed the location of IVPAT probe within the rabbit aorta. (K) Comparison of the high lipid content area measured by IVPAT and histology. Reproduced with permission from Ref. [54].

4.4.5 Inflammation-Targeted Intravascular Photoacoustic Tomography

To detect macrophages in atherosclerotic plaques, plasmonic gold nanoparticles are introduced as a contrast agent for intravascular photoacoustic imaging [56]. The ex vivo tissue studies show that the individual spherical nanoparticles, resonant at 530 nm wavelength, produce a weak photoacoustic signal at 680 nm wavelength while photoacoustic signal from nanoparticles internalized by macrophages is very strong due to the plasmon resonance coupling effect.

To demonstrate that IVPA imaging with Au NPs can be used for imaging of macrophages in atherosclerotic plaques, ex vivo tissue experiments were performed on a diseased rabbit aorta. A section of the aorta was extracted from a rabbit that was on a high cholesterol diet for 3 months. Macrophages loaded with Au NPs (approximately 4×10^4 nanoparticles per cell) were mixed with gelatin (2×10^7 cells per ml) and injected into the outer and inner boundary of the aorta. The sample was imaged by the same IVPA/IVUS imaging system used in phantom experiments. The hypoechoic regions in the IVUS image, denoted by green arrows in Figs. 4.26a,b,e, indicate the areas of the injection of macrophages loaded with Au NPs. IVPA images of this cross section were taken at 700, 750, and 800 nm wavelengths. As shown in Fig. 4.26, Au NPs produce the strongest photoacoustic signal when irradiated with laser pulses at 700 nm wavelength. At a laser fluence of 15 mJ/cm^2, the particles produce readily detectable photoacoustic signal from the laser light traveling through more than 1 mm of the arterial tissue. As the wavelength increased, the photoacoustic signal strength from Au NPs decreased; the photoacoustic response is hardly visible at 800 nm wavelength. This decreasing trend of photoacoustic signal magnitude closely corresponds to the absorbance spectra of aggregated Au NPs. Therefore, the ex vivo tissue experiment demonstrates the ability of IVPA imaging to detect macrophages loaded with Au NPs.

These results suggest that intravascular photoacoustic imaging can assess the macrophage mediated aggregation of nanoparticles and therefore identify the presence and the location of nanoparticles associated with macrophage-rich atherosclerotic plaques.

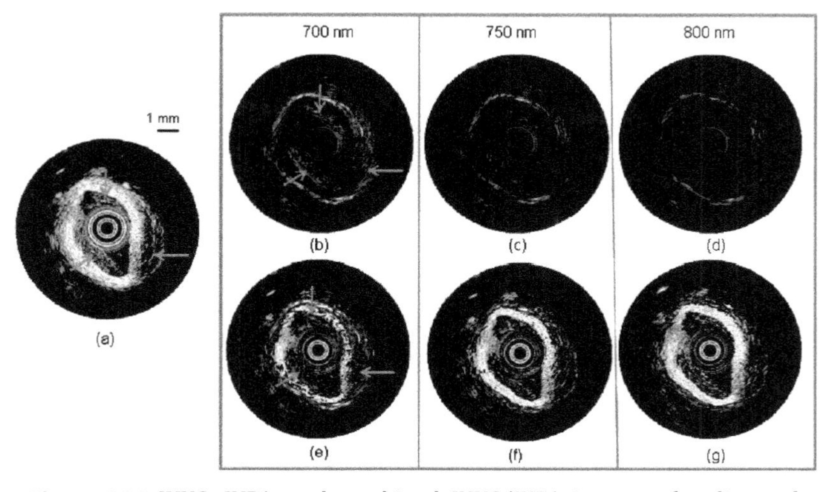

Figure 4.26 IVUS, IVPA, and combined IVUS/IVPA images of a diseased rabbit aorta injected with macrophages loaded with Au NPs. The IVUS image (panel a) is displayed using 50 dB dynamic range. The injected macrophages in the outer and inner regions of aorta are denoted in the images (a, b, e) with green arrows. The normalized IVPA images (panels b–d) and combined IVUS/IVPA images (panels e–g) obtained using 700, 750, and 800 nm wavelength are displayed using 20 dB display dynamic range. The IVPA and combined IVUS/IVPA images taken at 700 nm wavelength (panels b and e) showed high photoacoustic signal at the injected regions denoted by arrows. Reproduced with permission from Ref. [55].

Elevated expression of matrix metalloproteinase-2 (MMP2) is one of the clinical features of high-risk atherosclerotic plaques. Many methods such as fluorescence imaging methods have been used to detect MMP2 for evaluating plaque vulnerability. However, so far no imaging method has been able to resolve the expression of MMP2 with high fidelity and resolution beyond microscopic depths. In this result, gold nanorods conjugated with MMP2 antibody (AuNRs-Abs) were developed as a highly efficient photoacoustic imaging (PAI) probe for mapping MMP2 in atherosclerotic plaques. AuNRs-Abs could specifically target MMP2 as demonstrated by scanning electron microscope and immunofluorescence imaging. After labeling with AuNRs-Abs, area of distribution of MMP2 from the surface to the depths of the atherosclerotic plaques was revealed using intravascular PAI. AuNRs-Abs has the ability to enable quantitative detection of MMP2 in atherosclerotic plaques [58].

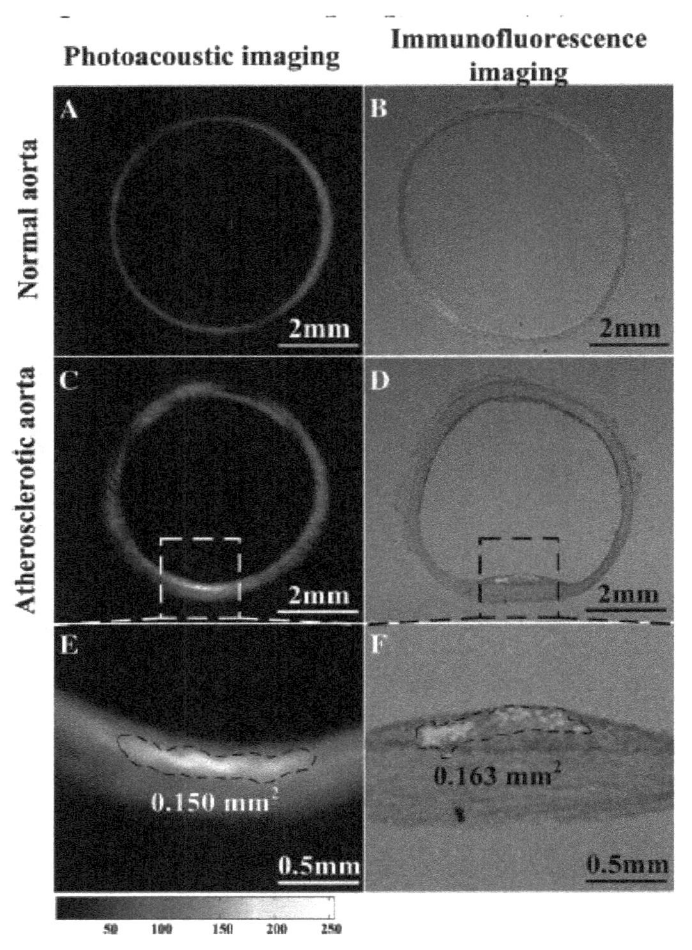

Figure 4.27 After the injection of AuNRs-Abs through the rabbit ear vein, the intravascular PAI of (A) the normal aorta and (C) the aorta containing atherosclerotic plaque were taken in situ, in which the yellow area indicates the area of expression of MMP2. The contrast/background ratio in the photoacoustic image is about 1.47. The overlay results of microscopic image and immunofluorescence result of MMP2 (B) in normal aorta and (D) in the atherosclerotic plaque were obtained soon after the PAI experiment. Images (E) and (F) are the enlarged images of the dotted rectangular areas in (C) and (D). In PAI, the area of distribution of MMP2 is 0.150 mm^2. Correspondingly, the immunofluorescence area (indicating the area of MMP2) is 0.163 mm^2. Reproduced with permission from Ref. [58].

The function of AuNRs-Abs in the intravascular photoacoustic imaging for in situ mapping MMP2 was demonstrated. Experiments were performed in a MMP2-rich plaque model with normal aortas used as controls. High-resolution photoacoustic images of the arterial structure were obtained throughout the intact tissue (Figs. 4.27A,C), which corresponded well with the histology (Figs. 4.27B,D. Intravascular PAI visualized plaque morphology clearly, as the high photoacoustic signal intensity region indicated the areas of MMP2 (yellow area). The white rectangle area in Fig. 4.27C was analyzed and the result was presented in Fig. 4.27E. The MMP2 distribution area in this cross section was 0.150 mm^2. The black rectangle area in Fig. 4.27D was analyzed and the result was presented in Fig. 4.27F. The immunofluorescence imaging of histologic section in the same location demonstrated the closed characteristics (0.163 mm^2, green area). Thus, with AuNRs-Abs, intravascular photoacoustic imaging can provide high-resolution morphological structure and quantitative information of the plaque inflammation.

High-resolution imaging and quantitative detection of MMP2 are critical to differentiate low- and high-risk plaques. The study data demonstrate that AuNRs-Abs are a promising probe for intravascular photoacoustic imaging quantitative detection of distribution of MMP2. These results could encourage further development of AuNRs-based photoacoustic probes for basic research in science and clinical diagnosis of atherosclerosis.

References

1. Wang, X., Pang, Y., Ku, G., Xie, X., Stoica, G., and Wang, L. V. (2003). Noninvasive laser-induced photoacoustic tomography for structural and functional in vivo imaging of the brain, *Nat. Biotechnol.*, **21**, pp. 803–806.
2. Yin, B. Z., Xing, D., Wang, Y., Zeng, Y. G., Tan, Y., and Chen, Q. (2004). Fast photoacoustic imaging system based on 320-element linear transducer array, *Phys. Med. Boil.*, **49**, pp. 1339–1346.
3. Zeng, Y. G., Xing, D., Wang, Y., Yin, B. Z., and Chen, Q. (2004). Photoacoustic and ultrasonic coimage with a linear transducer array, *Opt. Lett.*, **29**, pp. 1670–1672.

4. Zhang, H. F., Maslov, K., Stoica, G., and Wang, L. V. (2006). Functional photoacoustic microscopy for high-resolution and noninvasive in vivo imaging, *Nat. Biotechnol.*, **24**, pp. 848–851.
5. Yang, S. H., Xing, D., Zhou, Q., Xiang, L. Z., and Lao, Y. Q. (2007). Functional imaging of cerebrovascular activities in small animals using high-resolution photoacoustic tomography, *Med. Phys.*, **34**, pp. 3294–3301.
6. Lao, Y. Q., Xing, D., Yang, S. H., and Xiang, L. Z. (2008). Noninvasive photoacoustic imaging of the developing vasculature during early tumor growth, *Phys. Med. Boil.*, **53**, pp. 4203–4212.
7. Ntziachristos, V. (2010). Going deeper than microscopy: The optical imaging frontier in biology, *Nat. Methods*, **7**, pp. 603–614.
8. Wang, H. W., Chai, N., Wang, P., Hu, S., and Dou, W. (2011). Label-free bond-selective imaging by listening to vibrationally excited molecules, *Phys. Rev. Lett.*, **106**, p. 238106.
9. Yang, J. M., Favazza, C., Chen, R., Yao, J., Cai, X., Maslov, K., Zhou, Q., Shung, K. K., and Wang, L. V. (2012). Simultaneous functional photoacoustic and ultrasonic endoscopy of internal organs in vivo, *Nat. Med.*, **18**, pp. 1297–1302.
10. Wang, L. V., and Hu, S. (2012). Photoacoustic tomography: In vivo imaging from organelles to organs, *Science*, **335**, pp. 1458–1462.
11. Wang, L. V., and Yao, J. (2016). A practical guide to photoacoustic tomography in the life sciences, *Nat. Methods*, **13**, pp. 627–638.
12. Tam, A. C. (1986). Applications of photoacoustic sensing techniques, *Rev. Mod. Phys.*, **582**, p. 381.
13. Yang, J. M., Maslov, K., Yang, H. C., Zhou, Q. F., Shung, K. K., and Wang, L. V. (2009). Photoacoustic endoscopy, *Opt. Lett.*, **34**, pp. 1591–1593.
14. Yang, J. M., Maslov, K., Yang, H. C., Zhou, Q. F., and Wang, L. V. (2009). Endoscopic photoacoustic microscopy, *Proceedings SPIE 7177, Photons Plus Ultrasound: Imaging and Sensing*, p. 71770N.
15. Yang, J. M., Chen, R., Favazza, C., Yao, J. J., Li, C., Hu, Z. L., Zhou, Q. F., Shung, K. K., and Wang, L. V. (2012). A 2.5-mm diameter probe for photoacoustic and ultrasonic endoscopy, *Opt. Express*, **20**, pp. 23944–23953.
16. Yang, J.-M., Li, C., Chen, R., Zhou, Q., Shung, K. K., Wang, L. V. (2014). Catheter-based photoacoustic endoscope, *J. Biomed. Opt.*, **19**, p. 066001.
17. Yang, J. M., Li, C. Y., Chen, R. M., Rao, B., Yao, J. J., Yeh, C. H., Danielli, A., Maslov, K., Zhou, Q. F., Shung, K. K., and Wang, L. V. (2015). Optical-resolution photoacoustic endomicroscopy in vivo, *Biomed. Opt. Express*, **6**, pp. 918–932.

18. Yuan, Y., Yang, S. H., and Xing, D. (2010). Preclinical photoacoustic imaging endoscope based on acousto-optic coaxial system using ring transducer array, *Opt. Lett.*, **35**, pp. 2266–2268.

19. Yang, Y., Li, X., Wang, T. H., Kumavor, P. D., Aguirre, A., Shung, K. K., Zhou, Q. F., Sanders, M., Brewer, M., and Zhu, Q. (2011). Integrated optical coherence tomography, ultrasound and photoacoustic imaging for ovarian tissue characterization, *Biomed. Opt. Express*, 2, pp. 2551–2561.

20. Li, C. Y., Yang, J. M., Chen, R., Zhang, Y., Xia, Y. N., Zhou, Q. F., Shung, K. K., and Wang, L. V. (2013). Photoacoustic endoscopic imaging study of melanoma tumor growth in a rat colorectum in vivo, *Proceedings SPIE 8581, Photons Plus Ultrasound: Imaging and Sensing*, p. 85810D.

21. Li, C. Y., Yang, J. M., Chen, R. M., Yeh, C. H., Zhu, L. R., Maslov, K., Zhou, Q. F., Shung, K. K., and Wang, L. V. (2014). Urogenital photoacoustic endoscope, *Opt. Lett.*, **39**, pp. 1473–1476.

22. Yang, J. M., Favazza, C., Chen, R. M., Yao, J. J., Cai, X., Maslov, K., Zhou, Q. F., Shung, K. K., and Wang, L. V. (2012). Simultaneous functional photoacoustic and ultrasonic endoscopy of internal organs in vivo, *Nat. Med.*, **18**, pp. 1297–1302.

23. Lloyd-Jones, D., Adams, R., Carnethon, M., De, S. G., Ferguson, T. B., Flegal, K., Ford, E., Furie, K., Go, A., Greenlund, K., Haase, N., Hailpern, S., Ho, M., Howard, V., Kissela, B., Kittner, S., Lackland, D., Lisabeth, L., Marelli, A., McDermott, M., Meigs, J., Mozaffarian, D., Nichol, G., O'Donnell, C., Roger, V., Rosamond, W., Sacco, R., Sorlie, P., Stafford, R., Steinberger, J., Thom, T., Wasserthiel-Smoller, S., Wong, N., Wylie-Rosett, J., Hong, Y., Writing, G. M., Amer Heart Assoc Stat, C., and Stroke Stat, S. (2009). Heart disease and stroke statistics-2009 update: A report from the American heart association statistics committee and stroke statistics subcommittee, *Circulation*, **119**, pp. 480–486.

24. Cheruvu, P. K., Finn, A. V., Gardner, C., Caplan, J., Goldstein, J., Stone, G. W., Virmani, R., and Muller, J. E. (2007). Frequency and distribution of thin-cap fibroatheroma and ruptured plaques in human coronary arteries: A pathologic study, *J. Am. Coll. Cardiol.*, **50**, pp. 940–949.

25. Nissen, S. E., and Yock, P. (2001). Intravascular ultrasound—novel pathophysiological insights and current clinical applications, *Circulation*, **103**, pp. 604–616.

26. Bouma, B. E., Tearney, G. J., Yabushita, H., Shishkov, M., Kauffman, C. R., Gauthier, D. D., MacNeill, B. D., Houser, S. L., Aretz, H. T., Halpern, E. F., and Jang, I. K. (2003). Evaluation of intracoronary stenting by intravascular optical coherence tomography, *Heart*, **89**, pp. 317–320.

27. Waxman, S., Dixon, S. R., L'Allier, P., Moses, J. W., Petersen, J. L., Cutlip, D., Tardif, J. C., Nesto, R. W., Muller, J. E., Hendricks, M. J., Sum, S. T., Gardner, C. M., Goldstein, J. A., Stone, G. W., and Krucoff, M. W. (2009). In vivo validation of a catheter-based near-infrared spectroscopy system for detection of lipid core coronary plaques initial results of the SPECTACLE study, *J. Am. Coll. Cardiol. Img.*, **2**, pp. 858–868.

28. Sethuraman, S., Aglyamov, S. R., Amirian, J. H., Smalling, R. W., and Emelianov, S. Y. (2007). Intravascular photoacoustic imaging using an IVUS imaging catheter, *IEEE Trans. Ultrasonics, Ferroelectrics, and Frequency Control*, **54**, pp. 978–986.

29. Sethuraman, S., Aglyamov, S. R., Amirian, J. H., Smalling, R. W., and Emelianov, S. Y. (2006). Development of a combined intravascular ultrasound and photoacoustic imaging system, *Proc. SPIE*, **6086**, p. 60860F.

30. Karpiouk, A. B., Wang, B., and Emelianov, S. Y. (2010). Development of a catheter for combined intravascular ultrasound and photoacoustic imaging, *Rev. Sci. Instrum.*, **81**, p. 014901.

31. Jansen, K., van der Steen, A. F. W., van Beusekom, H. M. M., Oosterhuis, J. W., and van Soest, J. (2011). Intravascular photoacoustic imaging of human coronary atherosclerosis, *Opt. Lett.*, **36**, pp. 597–599.

32. Wu, M., Jansen, K., Springeling, G., van der Steen, A. F. W., and van Soest, J. (2014). Impact of device geometry on the imaging characteristics of an intravascular photoacoustic catheter, *Appl. Opt.*, **53**, pp. 8131–8139.

33. Wei, W., Li, X., Zhou, Q. F., Shung, K. K., and Chen, Z. P. (2011). Integrated ultrasound and photoacoustic probe for co-registered intravascular imaging, *J. Biomed. Opt.*, **16**, p. 106001.

34. Bai, X. S., Gong, X. J., Hau, W., Lin, R. Q., Zheng, J. X., Liu, C. B., Zeng, C. Z., Zou, X., Zheng, H. R., and Song, L. (2014). Intravascular optical-resolution photoacoustic tomography with a 1.1 mm diameter catheter, *PLOS ONE*, **9**, p. e92463.

35. Hsieh, B. Y., Chen, S. L., Ling, T., Guo, L. J., and Li, P. C. (2010). Integrated intravascular ultrasound and photoacoustic imaging scan head, *Opt. Lett.*, **35**, pp. 2892–2894.

36. Bui, N. Q., Hlaing, K. K., Nguyen, V. P., Nguyen, T. H., Oh, Y. O., Fan, X. F., Lee, Y. W., Nam, S. Y., Kang, H. W., and Oh, J. (2015). Intravascular ultrasonic–photoacoustic (IVUP) endoscope with 2.2-mm diameter catheter for medical imaging, *Comput. Med. Image Grap.*, **45**, pp. 57–62.

37. Ji, X. R., Xiong, K. D., Yang, S. H., and Xing, D. (2015). Intravascular confocal photoacoustic endoscope with dual-element ultrasonic transducer, *Opt. Express*, **23**, pp. 9130–9136.

38. Li, Y., Xiaojing Gong, X. J., Liu, C. B., Lin, R. Q., Hau, W., Bai, X. S., and Song, L. (2015). High-speed intravascular spectroscopic photoacoustic imaging at 1000 A-lines per second with a 0.9-mm diameter catheter, *J. Biomed. Opt.*, **20**, p. 065006.

39. Wang, P., Ma, T., Slipchenko, M. N., Liang, S. S., Hui, J., Shung, K. K., Roy, S., Sturek, M., Zhou, Q. F., Chen, Z. P., and Cheng, J. X. (2014). High-speed intravascular photoacoustic imaging of lipid-laden atherosclerotic plaque enabled by a 2-kHz Barium Nitrite Raman laser, *Sci. Rep.*, **4**, p. 6889.

40. Hui, J., Yu, Q. H., Ma, T., Wang, P., Cao, Y. C., Bruning, R. S., Qu, Y. Q., Chen, Z. P., Zhou, Q. F., Sturek, M., Cheng, J. X., and Chen, W. B. (2015). High-speed intravascular photoacoustic imaging at 1.7 μm with a KTP-based OPO, *Biomed. Opt. Express*, **6**, pp. 4557–4566.

41. Piao, Z. L., Ma, T., Li, J. W., Wiedmann, M. T., Huang, S. H., Yu, M. Y., Shung, K. K., Zhou, Q. F., Kim, C. S., and Chen, Z. P. (2015). High speed intravascular photoacoustic imaging with fast optical parametric oscillator laser at 1.7 μm, *Appl. Phys. Lett.*, **107**, p. 083701.

42. Cao, Y. C., Hui, J., Kole, A., Wang, P., Yu, Q. H., Weibiao Chen, W. B., Sturek, M., and Cheng, J. X. (2016). High-sensitivity intravascular photoacoustic imaging of lipid-laden plaque with a collinear catheter design, *Sci. Rep.*, **6**, p. 25236.

43. Sethuraman, S., Amirian, J. H., Litovsky, S. H., Smalling, R. W., and Emelianov, S. Y. (2007). Ex vivo characterization of atherosclerosis using intravascular photoacoustic imaging, *Opt. Express*, **15**, pp. 16657–16666.

44. Sethuraman, S., Amirian, J. H., Litovsky, S. H., Smalling, R. W., and Emelianov, S. Y. (2008). Spectroscopic intravascular photoacoustic imaging to differentiate atherosclerotic plaques, *Opt. Express*, **16**, pp. 3362–3367.

45. Wang, B., Su, J. L., Amirian, J., Litovsky, S. H., Smalling, R., and Emelianov, S. (2010). Detection of lipid in atherosclerotic vessels using ultrasound-guided spectroscopic intravascular photoacoustic imaging, *Opt. Express*, **18**, pp. 4889–4897.

46. Wang, B., Karpiouk, A., Yeager, D., Amirian, J., Litovsky, S., Smalling, R., and Emelianov, S.(2012). Intravascular photoacoustic imaging of lipid in atherosclerotic plaques in the presence of luminal blood, *Opt. Lett.*, **37**, pp. 1244–1246.

47. Wang, B., Karpiouk, A., Yeager, D., Amirian, J., Litovsky, S., Smalling, R., and Emelianov, S. (2012). In vivo intravascular ultrasound-guided photoacoustic imaging of lipid in plaques using an animal model of atherosclerosis, *Ultrasound Med. Biol.*, pp. 2098–2103.

48. Jansen, K., Wu, M., van der Steen, A. F. W., and van Soest, G. (2013). Lipid detection in atherosclerotic human coronaries by spectroscopic intravascular photoacoustic imaging, *Opt. Express*, **21**, pp. 21472–21484.

49. Jansen, K., van der Steen, A. F. W., Wu, M., van Beusekom, H. M. M., Springeling, G., Li, X., Qifa Zhou, Q. F., Shung, K. K., de Kleijn, D. P. V., van Soest, G. (2014). Spectroscopic intravascular photoacoustic imaging of lipids in atherosclerosis, *J. Biomed. Opt.*, **19**, p. 026006.

50. Li, X., Wei, W., Zhou, Q. F., Shung, K. K., and Chen, Z. P. (2012). Intravascular photoacoustic imaging at 35 and 80 MHz, *J. Biomed. Opt.*, **17**, p. 106005.

51. Li, R., Slipchenko, M. N., Wang, P., and Cheng, J. X. (2013). Compact high power barium nitrite crystal-based Raman laser at 1197 nm for photoacoustic imaging of fat, *J. Biomed. Opt.*, **18**, p. 040502.

52. Allen, T. J., Hall, A., Dhillon, A. P., Owen, J. S., and Beard, P. C. (2012). Spectroscopic photoacoustic imaging of lipid rich plaques in the human aorta in the 740 to 1400 nm wavelength range, *J. Biomed. Opt.*, **17**, p. 061209.

53. Karpiouk, A. B., Wang, B., Amirian, J., Smalling, R. W., and Emelianov, S. Y. (2012). Feasibility of in vivo intravascular photoacoustic imaging using integrated ultrasound and photoacoustic imaging catheter, *J. Biomed. Opt.*, **17**, p. 096008.

54. Zhang, J., Yang, S. H., Ji, X. R., Zhou, Q., and Xing, D. (2014). Characterization of lipid-rich aortic plaques by intravascular photoacoustic tomography, *J. Am. Coll. Cardiol.*, **64**, pp. 385–390.

55. Wang, B., Yantsen, E., Larson, T., Karpiouk, A. B., Sethuraman, S., Su, J. L., Sokolov, K., and Emelianov, S. Y. (2009). Plasmonic intravascular photoacoustic imaging for detection of macrophages in atherosclerotic plaques, *Nano Lett.*, **9**, pp. 2212–2217.

56. Yeager, D., Karpiouk, A., Wang, B., Amirian, J., Sokolov, K., Smalling, R., and Emelianov, S. (2012). Intravascular photoacoustic imaging of exogenously labeled atherosclerotic plaque through luminal blood, *J. Biomed. Opt.*, **17**, p. 106016.

57. Qin, H., Zhou, T., Yang, S. H., Chen, Q., and Xing, D. (2013). GdIII-gold nanorods for MRI and PAI dual-modality detection of macrophages in atherosclerotic inflammation, *Nanomedicine*, **8**, pp. 1611–1624.

58. Qin, H., Zhao, Y., Zhang, J., Pan, X., Yang, S. H., and Xing, D. (2016). Inflammation-targeted gold nanorods for intravascular photoacoustic imaging detection of matrix metalloproteinase-2 (MMP2) in atherosclerotic plaques, *Nanomed. Nanotechnol.*, **12**, pp. 1765–1774.

Chapter 5

Photoacoustic Viscoelasticity Imaging Technique

5.1 Introduction

Elasticity and viscosity are important parameters that are closely connected to the mechanical properties of materials. Alterations of mechanical properties are often associated with pathological states in biological tissue [1–5]. Viscoelastography, an imaging technique that is typically implemented using existing medical imaging techniques, can map the biomechanical distribution in biological tissue.

5.2 Photoacoustic Elastography

Elastography, mapping Young's modulus E distribution in biological tissue, is based on the elasticity contrast. In elastography, tissue deformation is induced by a static or dynamic load and imaged. If the stress distribution σ is known, the deformation ε can be converted to an image of elasticity by

$$E = \frac{\sigma}{\varepsilon}. \tag{5.1}$$

Biomedical Photoacoustics
Sihua Yang and Da Xing

ISBN 978-981-4774-58-1 (Hardcover), 978-0-203-70365-6 (eBook)
www.jennystanford.com

Wang et al. implemented photoacoustic (PA) elastography by measuring the local stress and strain using a linear-array PA computed tomography system [6, 7] (Fig. 5.1). A 10 ns pulsed laser beam at 680 nm was used for PA excitation with a 20 Hz pulse repetition rate. Light was first coupled into a fiber bundle. The fiber bundle was then split into two rectangular light bars mounted on each side of a linear ultrasonic transducer array (21 MHz center frequency, 256 elements), which detected the generated PA waves. PA signals were sampled at 84 MHz. For each laser pulse, one quarter of the 256 ultrasonic array elements were used for detecting PA signals. Acquired with four laser pulses, the full data set was used to reconstruct a cross-sectional PA image, yielding a frame rate of 5 Hz. In the PA elastography system, an aluminum compression plate larger than the object exerted a small axial compressive force on the object (Fig. 5.1a). An imaging window slightly larger than the ultrasonic transducer probe was opened at the center of the compression plate (Fig. 5.1b). A piece of fully stretched polymethylpentene (TPX) plastic membrane was attached to the bottom of the compression plate to provide uniform and uniaxial force to the object while passing the illumination laser beam. The compression plate was adjusted by a manual translation stage to provide precise compression to the object against a rigid object holder. The total displacement of the object surface was read from the translation stage. The object and the object holder were placed on a high-precision digital weighing scale. The compression stress applied to the object was calculated from the difference in the scale readings before and after compression.

$$\sigma = \frac{g(m_{\mathrm{a}} - m_{\mathrm{b}})}{A} \tag{5.2}$$

Here, g is the acceleration of gravity, m_{a} and m_{b} are the scale readings before and after compression, and A is the area on which the compression force is applied.

A mouse leg was imaged in vivo by PA elastography using the system [6]. The mouse leg was imaged before and after applying an external compression force of 12 mN (Figs. 5.2a,b). A displacement image was obtained by cross-correlating the PA images before and after compression, using the image pixels with

PA signal amplitudes above the noise level. A raw strain image was then obtained by numerically differentiating the axial displacements, assuming that the applied stress was uniaxial (Fig. 5.2c). The raw strain image was then superimposed on the structural PA image (Fig. 5.2e). The regions of tissue with larger strains were softer than regions with smaller strains and thus were thought to have more fat. The PA elastography was validated by ultrasound elastography using the same linear-array imaging probe, which showed a similar distribution of strains (Fig. 5.2d,f). In ultrasound elastography, structural ultrasound images were acquired simultaneously with structural PA images before and after compression. The displacement and strain images in the ultrasound elastography were computed using the same data processing method as in the PA elastography. The average strains over the entire cross-sectional image were 0.84 ± 0.49% in PA elastography and 0.82 ± 0.29% in ultrasound elastography.

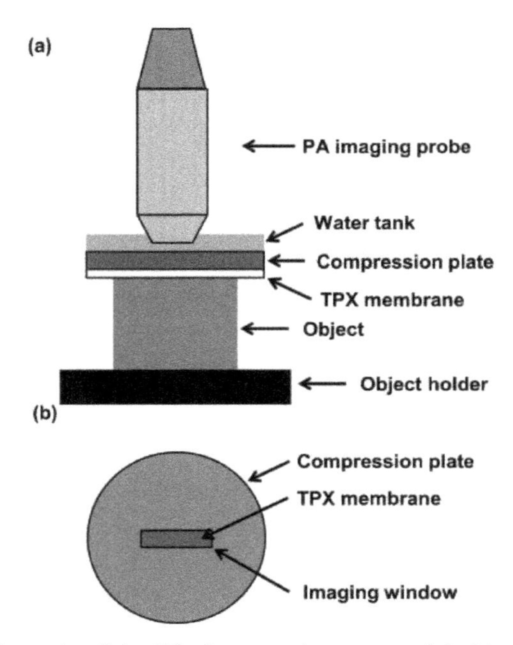

Figure 5.1 Schematic of the PA elastography system: (a) side view of the PA elastography system and (b) top view of the compression plate with the imaging window at the center. Reproduced with permission from Ref. [6].

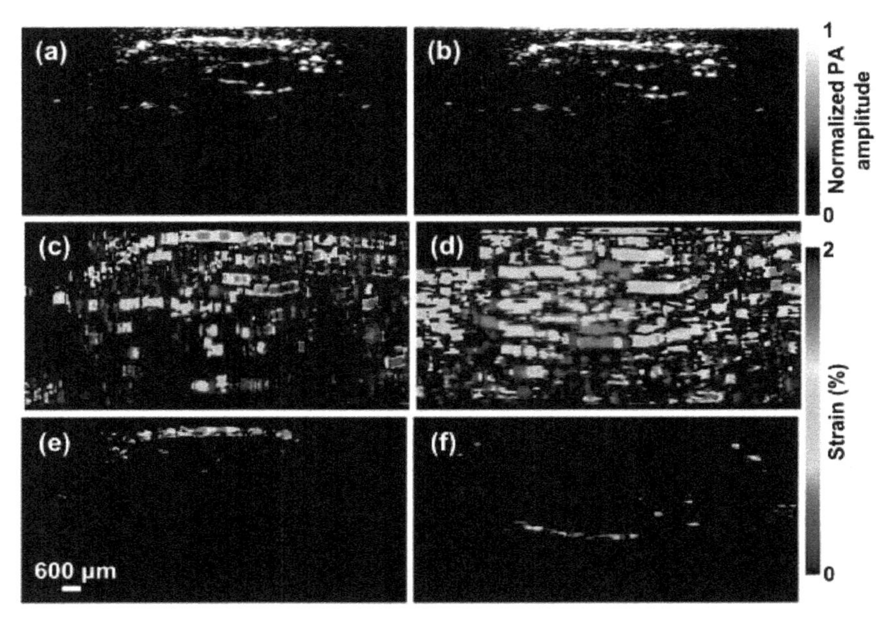

Figure 5.2 Strain measurement of a bilayer gelatin phantom by PA elastography: (a) and (b) PA images of a bilayer gelatin phantom mixed with 50 μm microspheres acquired (a) before and (b) after compression; (c) displacement image obtained from (a) and (b); (d) average displacement versus depth. The data was fitted by a linear function for each layer.

Then, the right arm of a healthy human volunteer was imaged [7]. The biceps muscle was chosen because the volunteer would have sufficient control of the arm to avoid motion artifacts and could maintain the same arm position and same elbow angle of 90° during the experiment. The compression system and the PA imaging probe were on the top of the stress sensor. Different loadings of 0.0, 2.5, 5.0, 7.5, and 10.0 kg were applied. At each loading, a B-scan PA image of a cross section of the stress sensor and arm was obtained. In the B-scan PA image, three layers of structures were resolved, including the skin layer, the blood vessels, and a muscle layer (Fig. 5.3a). Then, an axial compression force was exerted by moving the compression plate down along the z-axis. Another B-scan PA image of the same cross section of the stress sensor and arm was obtained. At each loading, a map of Young's modulus was calculated based on the method described above (Fig. 5.3a–e). With the system, we were able to obtain the Young's modulus values of the bicep up to 6 mm deep, within which the SNR was sufficiently high to calculate the displacements. We also calculated the averaged Young's modulus values for each layer (Fig. 5.3f). The skin had an average Young's modulus value of 15.9 kPa, and it stayed invariant with increasing loadings of the arm. The Young's modulus of the muscle layer increased linearly with the loading applied. The in vivo results in human arm demonstrated the capability of PA elastography.

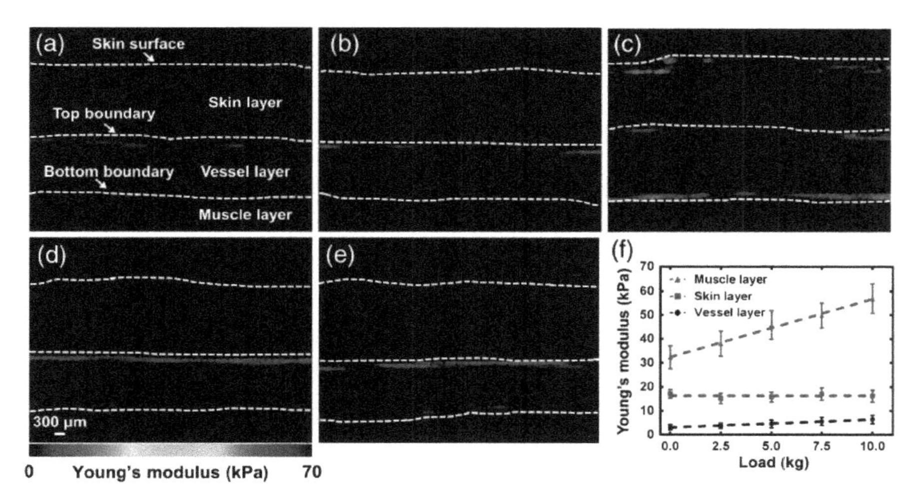

Figure 5.3 PA elastography of a human biceps muscle in vivo. PA elastography of the human biceps muscle in vivo at different loadings: (a) 0.0, (b) 2.5, (c) 5.0, (d) 7.5, and (e) 10.0 kg. The skin layer, blood vessel boundaries, and skeletal muscle can be observed. (f) Young's modulus value averaged in each layer as a function of loading. Reproduced with permission from Ref. [7].

5.3 Photoacoustic Viscoelasticity Imaging

Most biological tissues show viscoelastic properties and rheological behavior. It is not comprehensive and inaccurate to just use the elastic property to express the intrinsic characteristics of biological tissues [8, 9]. Many soft solids, such as cartilage, bone, tendon, and muscle, require both elasticity and viscosity simultaneously to describe their mechanical behaviors [10, 11]. Therefore, a method to characterize viscoelasticity of tissues will make a great progress in medical applications and clinical research. Due to the damping effect resulting from viscoelasticity of tissues, PA wave would produce a phase lag behind the excitation laser. Different tissue types or pathologies with corresponding viscoelastic properties will lead to different phase lags of the PA wave. Therefore, PA viscoelasticity imaging (PAVEI) can be obtained based on the contrast with phase delay [12–17].

5.3.1 Method

Here, we theoretically deduced the mathematical relationship between the PA phase delay and the viscosity-elasticity ratio [12, 13]. Isotropic viscoelastic structure is radiated with an intensity-modulated CW laser. The laser intensity is given by $I = 1/2I_0(1+\cos\omega t)$, where I_0 is the time-averaged laser intensity and ω is the modulation frequency. Light absorption by the absorber results in a sinusoidal temperature variation in the form of $T = T_0 e^{i\omega t}$ due to the nonradiative transition and then causes thermal expansion and shrinkage as well as PA wave generation based on the thermoelastic mechanism. Because of the sinusoidal cyclical changes of the laser intensity, the PA wave whose dominant frequency is equal to the modulated frequency is excited periodically. In the above process, the cyclical heating in the local region induces the thermal stress, and strain generates due to the stress in the form of the force-produced PA waves. Owing to the damping effect of the biological viscoelasticity, the strain also alternated periodically but would be out of phase with the stress.

The radial thermal stress caused by a periodically variational point heat source in the spherical coordinate system can be expressed as [12]

$$\sigma = -\frac{4G}{r}\frac{\partial \Phi}{\partial r} + \rho\frac{\partial^2 \Phi}{\partial t^2}, \tag{5.3}$$

where G is Green's function of thermal conduction, Φ is the thermoelastic displacement potential and is defined by $\Delta\Phi = \alpha T(1+\nu)/(1-\nu)$, α is the expansion coefficient, and ν is

Poisson's ratio. In the condition of temperature $T = T_0e^{i\omega t}$, the stress σ can be rewritten as

$$\sigma = -E\alpha T_0 e^{i\omega t} / (1 - \nu), \tag{5.4}$$

where E is Young's modulus and ω is the modulation frequency. In the rheological Kelvin-Voigt model, the constitutive equation in terms of a stress–strain relationship can be expressed as:

$$\sigma = E\varepsilon + \eta\dot{\varepsilon}. \tag{5.5}$$

Combining Eqs. (5.4) and (5.5), we have

$$\varepsilon(t) = \varepsilon_A e^{i\omega t + \delta}, \tag{5.6}$$

$$\delta = \arctan\frac{\eta\omega}{E}, \tag{5.7}$$

where $\varepsilon_A = (-E^2\alpha T)/[(E^2 + \eta^2\omega^2)\cos\delta(1 - \nu)]$ is the amplitude of the complex strain. Equation (5.7) indicates the phase delay that the strain response lags behind the stress. From Eq. (5.7), we can know the relationship between the phase delay δ and the viscosity-elasticity ratio η/E.

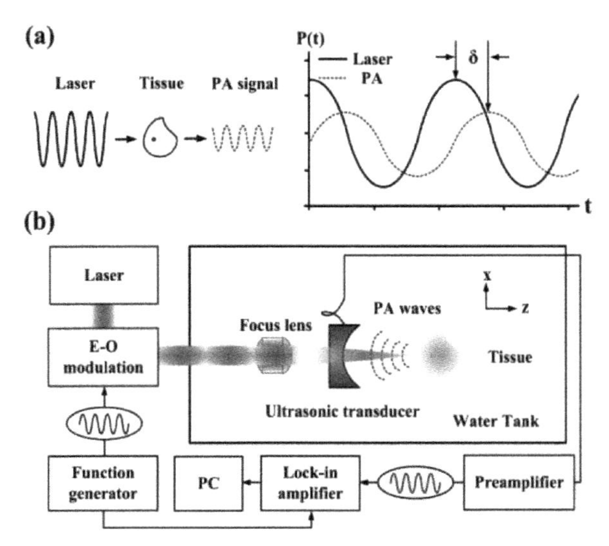

Figure 5.4 Principle (a) and the schematic setup (b) of the PA viscoelasticity imaging. Reproduced with permission from Ref. [13].

5.3.2 PAVEI System

Figure 5.4b shows the schematic setup of PAEI system [13]. A fiber-coupled CW laser was used as the excitation source operating at the wavelength of 808 nm. The intensity of the CW laser was modulated by an electro-optic modulator at 50 kHz with a 90% modulation depth by applying a 4.5 V sinusoidal signal from a function generator. This sinusoidal signal was simultaneously used as the reference signal, which was input to the lock-in detector. The transmitted laser was focused by the focus lens and then illuminated the target with a focus on about 100 μm. The measured time average laser power density on the surface was in the American National Standards Institute safety limit of 200 megawatts/cm^2. A customized hollow bowl-shaped ultrasound transducer was used to receive the PA signal. The transducer consists of a piezoelectric ceramic, and has a normal bandwidth of 20 KHz with a center frequency of 50 kHz. The diameter of the hole was 1 cm and the outer diameter of the hollow transducer was 4 cm. The pre-amplified PA signal was detected by a lock-in detector (SR830, Stanford Research Systems). During data acquisition, the sample was placed in a container and acoustically coupled with distilled water. A computer was used to analyze the amplitude and phase delay of the signal, which controlled the mechanical scanner simultaneously.

The difference between the PA signal and the reference signal can be detected by the lock-in detector. Figure 5.5 shows the basic detection process of the lock-in detector [13]. The PA signal $E_1\sin(\omega t + \alpha)$ and the reference signal $E_2\sin(\omega t + \beta)$ are fed into the signal channel input and become the two inputs to the multiplier stage. The multiplier output is

$$E(t) = E_1E_2\frac{\cos(\beta - \alpha)}{2} - E_1E_2\cos\frac{(2\omega t + \beta + \alpha)}{2}. \tag{5.8}$$

The amplitudes of the second harmonic and the dc voltage are both proportional to the amplitude of the PA signal E_1. To concentrate on the dc component, the multiplier output is fed into a low-pass filter (LPF) so that it makes the second harmonic signal strong attenuation. The lock-in detector uses phase-sensitive detection to produce a pair of quadrature-demodulated outputs "X" and "Y," where "X" and "Y" signals are the LPF product of the PA signal and the reference signal. The "X" and "Y" signals are in phase component and quadrature component of the PA signal compared to the reference signal, which are different in phase with a shift of $\pi/2$ [18, 19]. The final output R is a dc voltage directly proportional to the PA amplitude, which can be calculated by

$$R = \sqrt{X^2 + Y^2}, \tag{5.9}$$

and the phase delay can be simultaneously given as

$$\delta = \tan^{-1} Y/X. \tag{5.10}$$

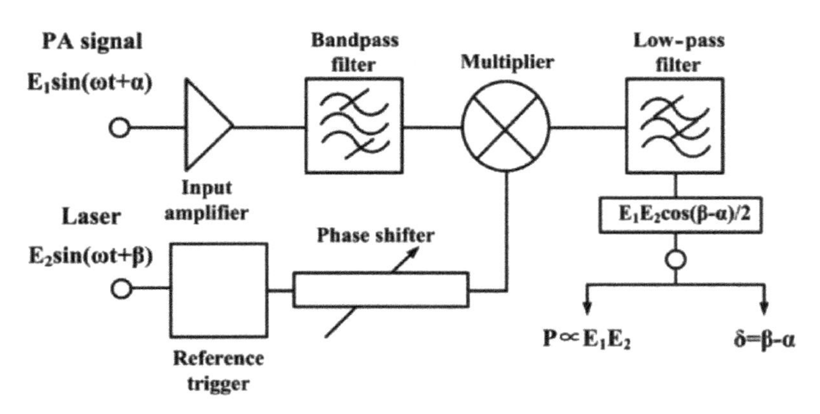

Figure 5.5 Detection process of amplitude and phase of PA signal based on the lock-in measurement. Reproduced with permission from Ref. [13].

Actually, the phase measured by the above-mentioned system is the phase difference δ_m between the PA signal and the reference signal used to modulate the laser. Except for the phase delay δ in Eq. (5.10), the phase difference δ_m also includes the relaxation time δ_t for the nonradiative transition and the delay time δ_s caused by system. Therefore, we have

$$\delta_m = \delta_t + \delta_s + \delta. \tag{5.11}$$

In practical experiments, the repetition frequency is 5×10^4 Hz and the order of relaxation time δ_t is 10^{-11} s [12], the phase delay $\delta >> \delta_t$, so δ_t can be neglected. For delay time δ_s caused by the system is very difficult to accurately measure, but in the same system delay time δ_s is stable and invariable, so it is a constant. So the magnitude difference of δ_m is only related to δ in the experiment.

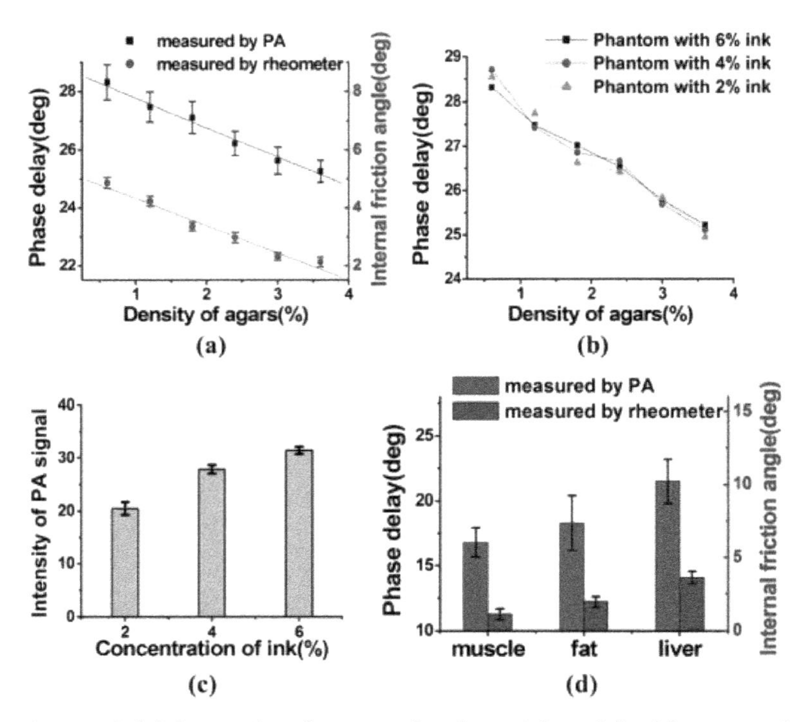

Figure 5.6 (a) Comparison between the phase delay of the PA wave and the internal friction angle measured by rheometer. The values of phase delay were averaged over 16 tests. (b) Phase delay obtained by PA measurement from agars with different absorption coefficients. (c) PA intensity of the agar phantoms with different proportions of ink. (d) Comparison between the phase delay of the PA wave and the internal friction angle measured by rheometer for biological tissues. Reproduced with permission from Ref. [12].

We used different-density agar (0.6%, 1.2%, 1.8%, 2.4%, 3.0%, and 3.6%) to simulate the biological tissues with different viscoelastic properties to verify the dependence of the PA phase delay on the viscoelastic properties. Figure 5.6a shows the phase delay measured by the PA experiment and the phase delay measured with a rheometer. From this, we can see that the variation tendency of phase delay measured by the PA experiment coincides with the measured internal friction angle using the rheometer. According to the fitting lines parallel in general of the two group data consistent with the trend, it is indicated that the phase delay depends on viscoelastic properties. The intercept between the fitting curves of the two sets of data is due to the delay time δ_s caused by the system and the relaxation time δ_t of the nonradiative transition. In addition, the influence on the PA phase delay of the absorption coefficient is also taken into account. The samples with the same density and different absorption coefficients were prepared by mixing 2% and 4% of the ink in the agar solution respectively. Figure 5.6b shows that there is almost no difference in phase delay for samples with different absorption coefficients, which indicates that the absorption coefficient does not affect the phase delay but determines the PA intensity as shown in Fig. 5.6c.

5.3.3 Medical Applications

The PAVE imaging technique has a wide range of biological application. Especially, it can be used as an alternative method for charactering diseases whose mechanical properties possess obvious changes. Moreover, in view that PAVE imaging is a derivative method for the PA technique, it has the in vivo imaging capability and can be easily integrated into the current techniques such as PA microscopy and PA endoscopy. Xing's group has developed the PAVE method for tumor detection, atherosclerosis characterization, and related vascular endoscopy [13–15].

5.3.3.1 PAVEI for tumor detection

As one of the most deadly diseases, tumors may be accompanied by the change of surface mechanical structure during their progression [20, 21]. Thus, PAVEI is suitable to detect these diseases. Our team first developed a viscoelastic imaging technique for the detection of tumors in vivo [13]. The in vivo experiment with murine EMT6 tumor was designed to confirm the feasibility of the dual-parameter imaging for biomedical applications. In the experiment, sodium pentobarbital (40 mg/kg; supplemental, 10 mg/kg/h) was administered to keep the mouse motionless. The mouse was secured with a clamp on the two-dimensional scanning stage, mouse back was smeared with ultrasound coupling fluid, and was attached to the glass slide on the bottom of the water tank. The optical absorption and viscoelasticity images of murine EMT6 tumor are shown in Fig. 5.7. The optical absorption image shown in Fig. 5.7a could visualize the tumor profile, and the tumor tissue possesses a relatively higher optical absorption than that of the surrounding tissue. The viscoelastic image shown in Fig. 5.7b suggested the location of the tumor with sufficiently high contrast to the surrounding normal tissue according to the fact that tumors are stiffer than normal tissues [22]. In fact, while the phase delay from the surrounding tissue was around 55.6°, the phase delay in the tumor region was around 41.9°. The similarity of dual-parametric images was greatly caused by the similar distribution of laser viscoelasticity and absorption in the tissue. However, these converse contrast images could provide synthetic information for pathological diagnosis. Figure 5.7c shows the photograph of the back of a BALB/c mouse, the tumor within the dashed frame was used for the in vivo study and the tumor morphology in the photo can correspond with the dual-parameter imaging results. Figure 5.7d shows the mean value and standard deviation of the PA amplitudes and phase delay within and outside of the dashed oval circle, which corresponded to the tumor area and normal tissues. This result stated that the tumor can be well distinguished from the normal tissue with the dual-parametric imaging method. Ultimately, the sections of tumor and normal tissue were sliced along the normal and tumor region. For each segment, cross sections were stained with hematoxylineosin stain and are shown in Figs. 5.7e,f. The tumor tissue showed densely staining nuclei compared to the normal tissue. The results indicated that the method has the capability to provide biomechanical and optical absorption information about biological tissues for accurate medical diagnosis.

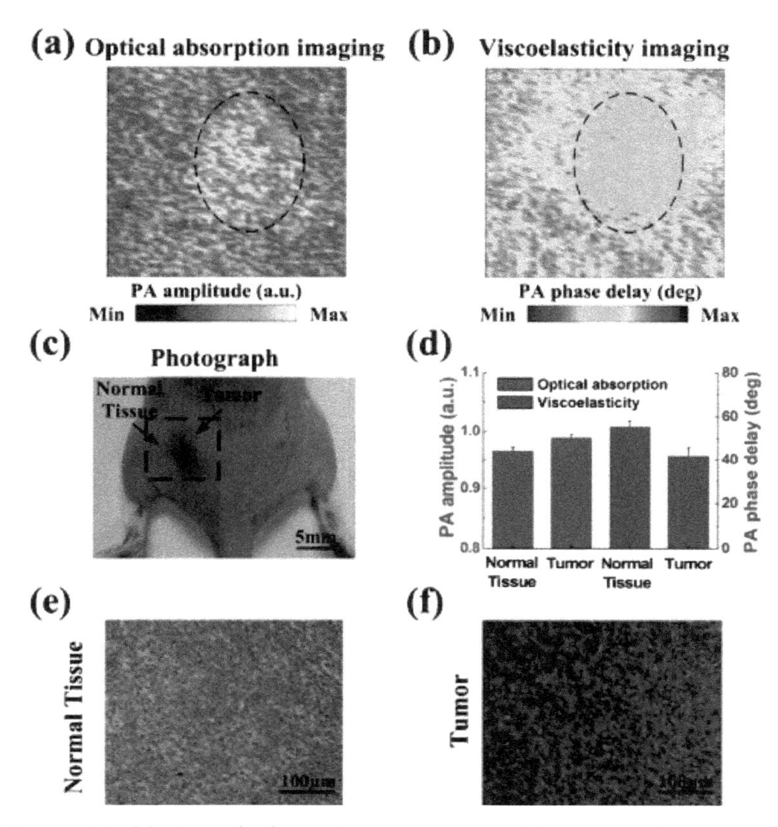

Figure 5.7 (a) Optical absorption image and (b) viscoelasticity image of the tumor. (c) Photograph of the tumor and the dashed frame is the scanning area. (d) The averaged amplitude and phase delay of the PA signal from the marked tumor region compared to the normal tissue. (e), (f) Hematoxylineosin(HE) stain of the normal tissue and tumor tissue. Reproduced with permission from Ref. [13].

5.3.3.2 PAVEI for atherosclerosis characterization

Acute myocardial infarction is the leading cause of death worldwide, which makes the intravascular atherosclerosis detection extremely important. Observations from animal and human models of atherosclerosis suggest that plaque development is initiated by lipid accumulation in the arterial wall extracellular matrix, leading to the activation of inflammation and intimal fibrosis [23, 24]. Atherosclerotic plaques with large lipid necrotic cores and thin fibrous caps are more susceptible to rupture, which is responsible for the development of the majority of acute cardiovascular events [25]. A plaque's stability is related to its morphology and histological composition. Hence, accurate identification of plaque morphology and components may allow the detection of vulnerable plaques before they rupture. The progression of atherosclerotic lesions is a complex process influenced by mechanical and biological factors. Rather than the degree of stenosis, the risk of acute coronary events is largely dependent on the mechanical properties and plaque morphology [26–30]. Therefore, noninvasive measurement of the mechanical properties of the arterial walls is useful for diagnosing atherosclerosis. A comprehensive understanding of the mechanical properties of the plaques will not only advance our understanding of the pathology of atherosclerotic disease but also provide critical information for medical or surgical treatment and for evaluating the efficacy of therapeutic interventions. Reliable techniques that are capable of characterizing the mechanical properties of plaque components may cover the clinically relevant diagnostic values. We applied the PAVEI technique for precisely evaluating the viscoelastic properties of atherosclerotic plaques, which yields new insights into biomechanical diagnosis of cardiovascular diseases [14–16].

Plaque-mimicking phantoms were studied to verify the ability of PAVEI to identify plaque lipids, as shown in Fig. 5.8 [16]. Lipid, as the main material in early plaque lesions, was mixed with the gelatin in different concentrations to simulate atherosclerotic plaques. The relation between the lipid content and phase delay is shown in Fig. 5.8a. Four circular targets are clearly shown in the PAVEI images acquired from the gelatin phantoms in Fig. 5.8b. As predicted theoretically, the observed phase delay of the PA signals increased significantly with the increasing density of the lipid.

A representative example was used to demonstrate the ability of a PAVEI system to retrieve characteristics from an atherosclerotic plaque, as shown in Fig. 5.9 [16]. Figure 5.9a was acquired from the atherosclerotic plaques found in a 4-month-old rabbit. The phase delay that was used to characterize plaque composition changed with the compositional change at the lumen surface of the plaque, as shown in Fig. 5.9b. Figure 5.9c presents the histological results, which shows the surface composition changes of the plaque; the lipid components obviously increased, and the collagen decreased significantly. We calculated the integrated optical density (IOD) per stained area (pixels) (IOD/area) in the histological results using the Image-Pro Plus program (IPP; Media Cybernetics, Inc., Bethesda, Maryland, USA) for a semi-quantitative analysis of the lipid content of the plaque [31]. We compare the average phase delay values within three regions of interest with histological results that are independent from the measurement. In histology the plaque type in region I was determined to be healthy from histology (0.024 ± 0.003 IOD/area); and an average phase delay of 31.46 ± 4.48° was observed from PAVEI. Region I exhibited a relatively shorter average phase delay compared with the area shown in region II, which had more lipid (0.102 ± 0.005 IOD/area), as indicated by histology, and region II showed an averaged phase delay of 60.40 ± 11.20°. In comparison, a different trend in the advanced fatty region (region III) was observed, where the average phase delay increased to 66.14 ± 12.65° with an increase in lipid (0.112 ± 0.006 IOD/area). In this manner, the lipid composition of the plaque surface (i.e., the vessel lumen) was characterized by PAVEI.

Figure 5.10a shows the region of the arterial lumen in a healthy rabbit the atherosclerotic lesions found in 3- and 5-month-old rabbits, respectively. In Fig. 5.10b, PAVEI images from the surface of the advanced arterial plaques show abundant lipid-rich structures differing in appearance from the luminal elastic lamina that dominate the images of healthy vessels. The average phase delay (30.52 ± 3.24°) obtained from images of the lumen surface in the healthy region is significantly lower compared with the atherosclerotic regions. Meanwhile, the average phase delay (56.40 ± 9.79°) from the plaque found in the 3-month-old rabbit is much lower than that (70.35 ± 10.27°) found in the 5-month-old rabbit. Therefore, the phase delay from regions of plaque increased in a linear fashion with the age of the rabbits, representing the overall burden of plaque in the aorta. The lipid accumulation is clearly shown in Fig. 5.10c. According to histological results, a dense knot of lipid led to an overall increase in phase delay compared to the surrounding area. The lipid content increased from a basal average IOD/area of 0.018 ± 0.002 in a healthy artery to 0.108±0.005 IOD/area in the fatty artery after 3 months of a high-fat/high-cholesterol diet. After 5 months of this diet, plaque had accumulated over large areas of the lumen surface of the aorta, and the 0.147 ±0.008 IOD/area value of the lipid plaque showed a significant difference compared with other groups. The increase in the average viscosity-elasticity ratio was predominantly due to the increase in the amount of accumulated lipid and the decrease of collagen content (Fig. 5.10d), which plays an important role in the development of vulnerable plaque.

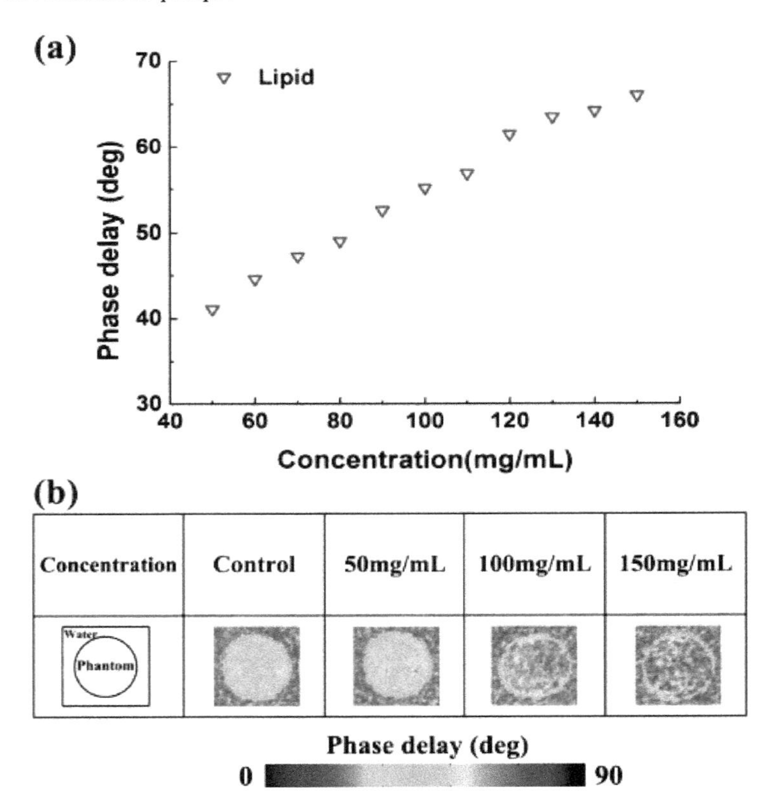

Figure 5.8 (a) Relation between the lipid concentration and phase delay; (b) PAVEI of phantoms containing various concentrations of lipid. Reproduced with permission from Ref. [16].

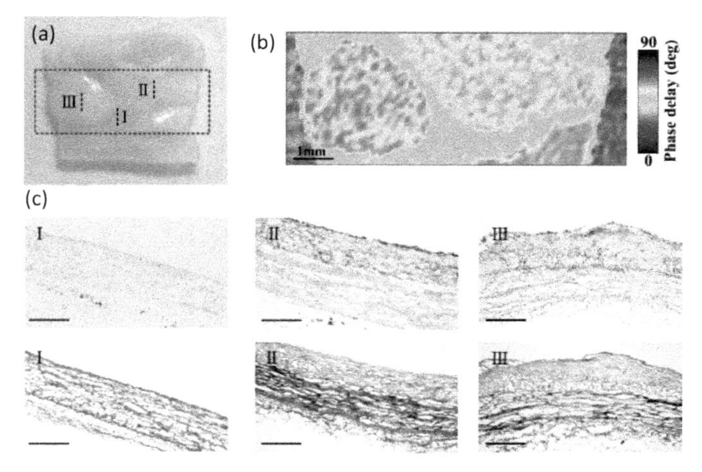

Figure 5.9 (a) Photo of the atherosclerotic arterial wall. The red dashed rectangle indicates the detection area. (b) PAVEI of the region of interest. The lipid-rich plaque with clearly demarcated borders in the color map is corroborated by the accompanying gross pathology photograph. (c) The oil red O and Masson's trichrome stain images of the area in the dashed rectangle. Scale bars = 300 μm. Reproduced with permission from Ref. [16].

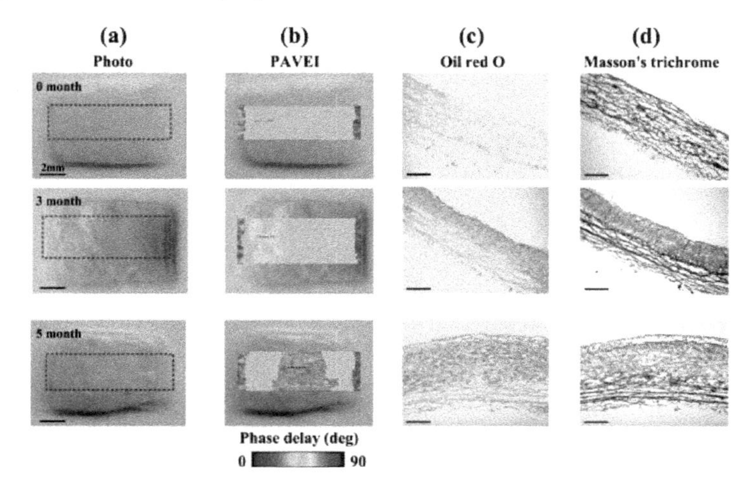

Figure 5.10 (a) Photos and (b) representative PAVEI images acquired from the luminal surface of the rabbit arteries, with healthy lumen (top row), early atherosclerotic plaque (middle row), and atherosclerotic plaque with higher lipid content (bottom row); (c) the oil red O and (d) Masson's trichrome stain images of the area in the dashed rectangle. Scale bars = 300 μm. Reproduced with permission from Ref. [16].

5.3.4 Integrated PA and PAVEI (PA-PAVEI) for Structural and Mechanical Features Characterization of Atherosclerosis

As a derivative method for PA technique, PAVEI can be easily integrated into the current PA techniques such as PA endoscopy. The combined PA-PAVEI technique could provide structural and biomechanical information about plaques and thus has received much attention in biomedical fields [14]. The schematic setup of the PA-PAVEI is shown in Fig. 5.11. A quasi-continuous laser (DS20HE-1064D/R, PHOTONICS) operating at 1064 nm with the pulse width of 22 ns and the repetition frequency of 25 kHz was used as the excitation source. A microscope objective was used to focus the collimated laser and made it illuminate on the sample surface. Between two ultrasound transducers (UT), the sample was settled on a two-dimensional motor scanning platform. UT1 has center frequency of 25 KHz and holds hollow bowl-shape style. UT2 has center frequency of 3.5 MHz (U8518056, Olympus) or 75 MHz (U8424009, Olympus). To obtain PA phase and temporal PA amplitude at the same time, the generated PA signals were acoustically coupled with distilled water and divided into two paths. On one hand, PA signals detected by UT1 were transferred to a low-pass low-noise preamplifier (SR552, Stanford Research Systems), then calculated by a lock-in detector (SR830, Stanford Research Systems) to resolve the PA phase. On the other hand, PA signals detected by UT2 were first transferred to a wide-bandwidth low-noise amplifier (Ha2, Precision Acoustics LTD) and obtained by a data acquisition system (NI PCI-5124, National Instruments). Both the PA phase and temporal PA amplitude were recorded and analyzed on a computer that was simultaneously used for controlling the motorized scanner with a custom program written by LABVIEW (National Instruments, USA) software. At each step of the scan, time-averaged laser intensity on the tissue surface was controlled rationally within the American National Standard Institute's safety limit (100 mJ/cm^2).

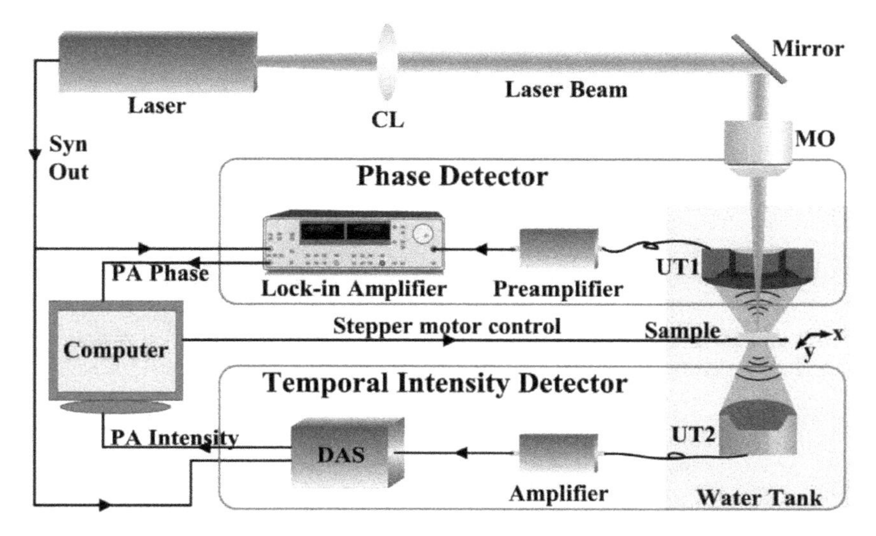

Figure 5.11 Schematic of IMS-PAI system. CL: collimating lens, MO: microscope objective, UT: ultrasonic transducer, DAS: data acquisition system. Reproduced with permission from Ref. [14].

An atherosclerotic tissue with a fatty streak obtained from a rabbit fed with a high-fat/high-cholesterol diet for 15 weeks was tested here to demonstrate the feasibility of PA-PAVEI for biomedical application [14]. Laser was illuminated on the inside surface of the sample and a high frequency UT (75 MHz) was used to image the thin-layer structure of vascular wall. Figure 5.12a shows the photograph of the atherosclerotic tissue; the tissue within the red dashed frame was used for ex vivo detection. The en-face PA viscoelasticity and absorption images shown in Fig. 5.12b distinguished the morphology of the fatty streak and corresponded well with the sample. The fatty tissue (main component is lipid) indicated a higher viscoelasticity and a slightly lower optical absorption than the surrounding normal tissue (main component is collagen). The result demonstrated that the viscoelasticity of lipid is greater than that of collagen [32, 33], and the optical absorption of lipid is less than that of collagen at 1064 nm [34]. The inhomogeneity of PA viscoelasticity and absorption distribution attributed most to

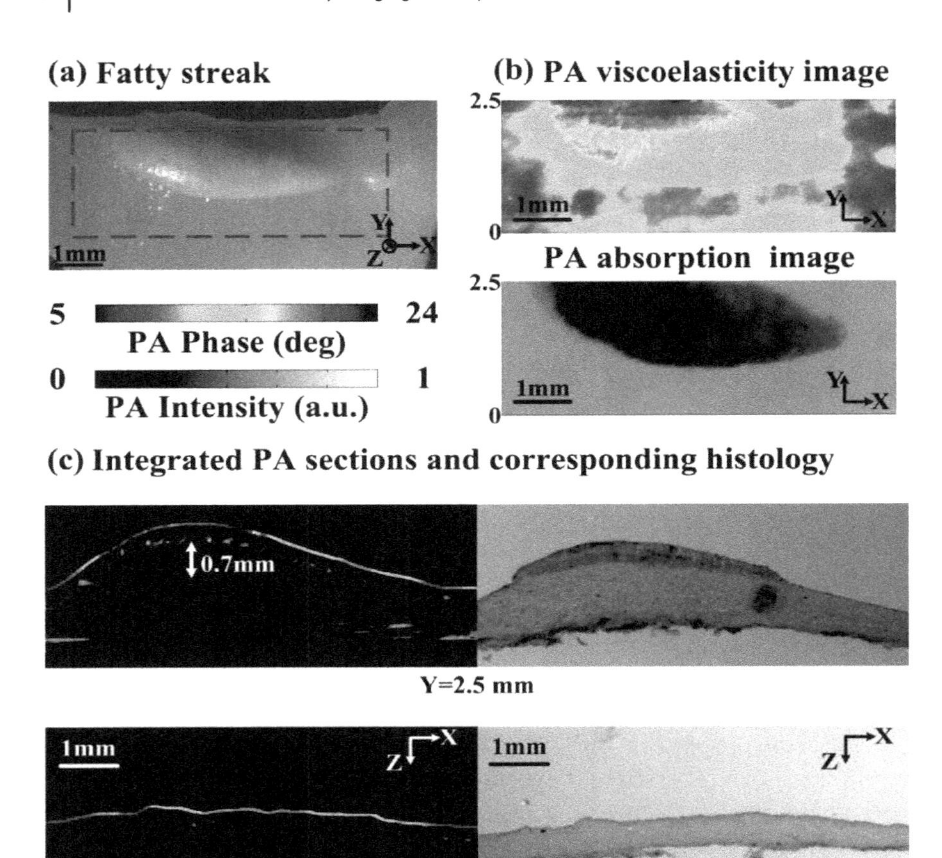

Figure 5.12 (a) Photograph of the atherosclerotic tissue with a fatty streak, region within the dashed frame is scanning area. (b) En-face PA viscoelasticity and absorption images. (c) Integrated PA sections and corresponding histology. The sections were stained with Oil red O (red) to evaluate lipid accumulation and counterstained with hematoxylin (blue) to visualize cell nuclei. Reproduced with permission from Ref. [14].

different degrees of lipid accumulation. PA absorption within the fatty streak was relatively accordant, while PA viscoelasticity exhibited a high-phase area located in the top left corner, which may be caused by the inflammation of atherosclerosis. After the PA experiment, the sample was then confirmed with cross-sectional Oil red O staining. The integrated PA sections and corresponding histology at different Y positions are shown in Fig. 5.12c. In the fatty streak, the intima thickening resulting from lipid accumulation revealed high viscoelasticity and dense Oil red O staining. The most severe intima-media thickening at Y = 2.5 mm was about 0.7 mm, which is highly consistent with the histology. Meanwhile, the imaging depth of the PA-PAVEI system for atherosclerosis characterization was about 1.5 mm. To visualize cell nuclei, hematoxylin was used for counterstaining slices. Grainy staining in clusters beneath the endothelial layer showed a sign of foam cell infiltration. These integrated PA sections allowed complementary visualization of en-face viscoelasticity distribution and in-depth structural anatomy for the fatty streak, where the distortion of viscoelasticity distribution may be an early warning of plaque rupture, and the degree of intima thickening relates to the lesion extent. The experiment result demonstrated the feasibility of accurate medical evaluation of atherosclerosis with the PA-PAVEI.

5.3.5 PAVE Endoscopy (PAVEE) for Atherosclerosis Characterization

To realize the clinical application of the PAVE technique, we developed PAVE endoscopy (PAVEE) [15] for the potential applications of intraluminal tissues such as blood vessels, esophagus, and cervical. The experimental imaging system of the PAVEE is shown in Fig. 5.13. The excitation source was a quasi-continuous laser (DS20HE-1064D/R, PHOTONICS), whose pulse width was 22 ns, the wavelength was 1064 nm and the repetition frequency was 65 kHz. The collimated laser was focused by a microscope objective and coupled into an optical fiber. The far-end of the optical fiber passed through a hollow motor and ultrasound transducer (65 kHz central frequency) to fire laser on a customized parabolic reflector, which could simultaneously reflect and focus the laser. The reflector was fixed on a rotating motor, so that a 360° cross-sectional scanning can be realized. The ultrasonic transducer and scanning mirror were held in a nickel tube (16 mm in out diameter), and a window on the side of the nickel tube was provided for transmitting laser and detecting the PA signal. All the components were mounted in a motorized call-back device for longitudinal scanning. Each angle of the scan was 0.225°, and the time-averaged laser intensity exposure to the inner surface of the vessel sample was limited well within the American National Standard Institute's safety limit (100 mJ/cm^2) [35]. In the data acquisition process, the distance between the laser focus and the transducer remains strictly consistent. The generated PA signals were acoustically coupled with distilled water and detected with an ultrasonic transducer and then sent to a low-noise low-pass preamplifier (SR522, Stanford Research System), and calculated by a lock-in detector (SR830, Stanford Research Systems) to resolve the phase lag behind the laser. The phase data acquisition and analysis were operated on a computer which controlled the motorized scanner with a Labview program at the same time.

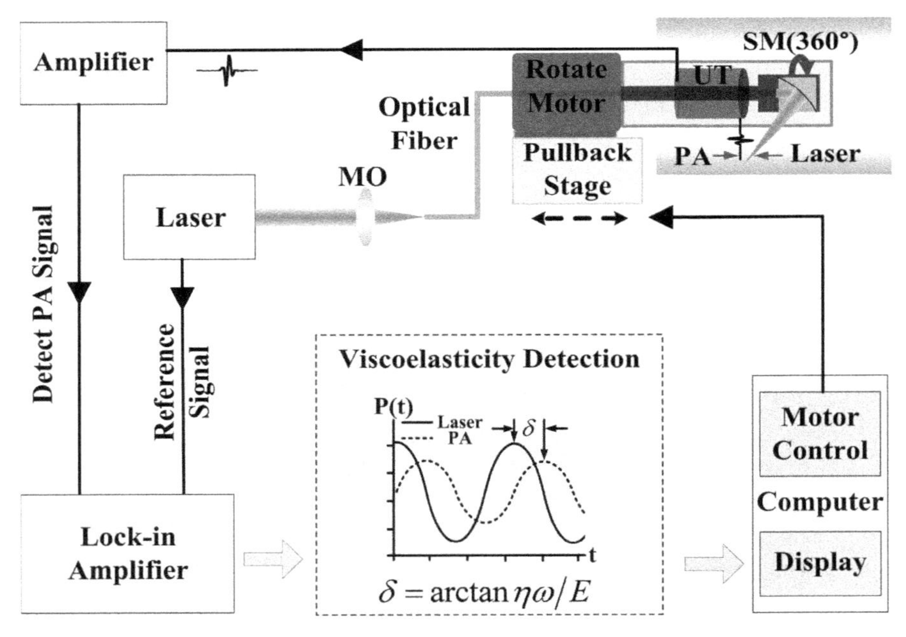

Figure 5.13 Schematic of the experimental imaging setup for PAVEE. MO: microscopic objective; UT: ultrasound transducer; SM: scanning mirror. Reproduced with permission from Ref. [15].

Ex vivo experiments were conducted on the vessel harvested from rabbits fed with high-cholesterol diet for 3 months. The vessel-shaped sample with a 19 mm luminal diameter was fixed in the agar preparation with normal and atherosclerotic segments in the large area of the atherosclerotic plaque segment, and then the PAVEE imaging was taken in the sample cavity. Three-dimensional (3-D) PAVEE image and en-face viscoelasticity distribution are shown in Fig. 5.14a. The static longitudinal PAVEE image gives the overall viscoelastic evaluation of the lumen. The atherosclerotic tissue revealed an inhomogeneity of PA viscoelasticity distribution, which was relatively homogeneous in normal tissue. This was mainly due to the different levels of atherosclerosis lipid accumulation. As expected, the lipid-rich region had a higher viscoelasticity leading to a higher PA-phase. The perfect combination of the experimental results and the pathologic principles is not only to demonstrate the accuracy of the detection of atherosclerosis but also to provide the mechanical characteristics of the medicine. The PAVEE results were validated and correlated with the sample morphology and the associated en-face PA viscoelasticity image. After the PA experiment, the specimen was cross-sectional sliced and stained with Oil red. Figure 5.14b shows the cross-sectional PAVEE section and histological staining at Z = 0.8 mm, and they exhibit good consistency. The lipid-rich plaques in atherosclerotic segment indicated high viscoelasticity and dense oil red staining. The normal-looking segment had an average phase value of 18.4° and a phase fluctuation of 2°. This phase instability was caused by the unapparent early atherosclerosis, because the normal progression and the atherosclerosis segments were harvested from the same vessel. The atherosclerotic segment had a phase ranging from 17.25° to 23.43°. The experiment shows that the PAVEE can be sensitive to distinguish between atherosclerosis, and the feasibility of accurate medical assessment. In general, different viscosities and elasticity of different materials, so for atherosclerotic plaque, normal part and lipid viscoelastic part of the larger changes will occur, resulting in different phase delay, so that can be through the PAVEE imaging system detection of plaque.

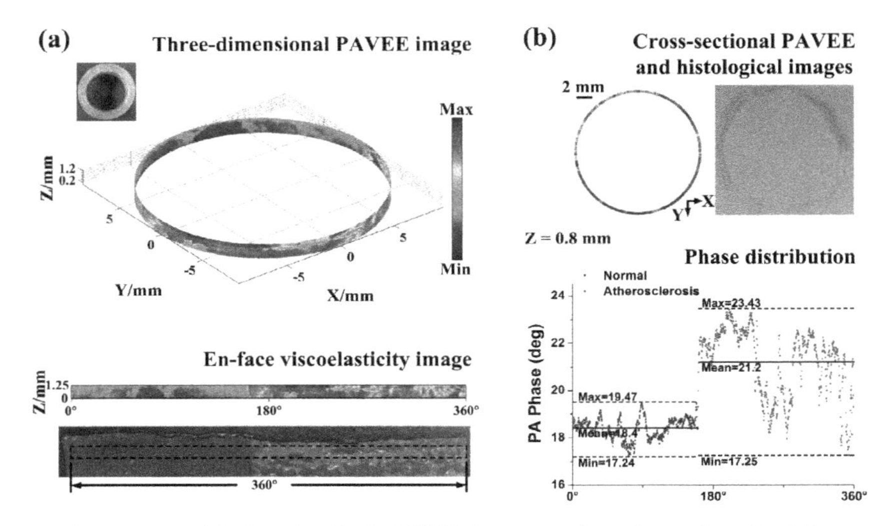

Figure 5.14 (a) Longitudinal PAVEE image and en-face viscoelasticity image viewed from inside over a 360° field of the sample. In the flattened tissue photograph, region within the dashed frame is scanning area. (b) PAVEE section and corresponding histology and phase distribution at z = 0.8 mm. The section was stained with Oil red (red) targeting lipid. Reproduced with permission from Ref. [15].

5.4 Discussion and Conclusion

The accuracy of the measurement system was limited by the time constant of the lock-in amplifier and the signal-to-noise ratio (SNR). A longer time constant will improve the performance but reduce the scanning speed and SNR. This PAVEI system was well applied to the ex vivo detection of lipid accumulation in atherosclerosis. However, because of the long acquisition time and the large diameter of the transducer, the in vivo application of intravascular viscoelasticity imaging is still limited. In the future, a small-diameter focused transducer will be employed to highly improve the imaging quality and accelerate the imaging speed of the system, which is an attractive prospect for the in vivo intravascular viscoelasticity evaluation of plaque. In addition, in PAVEI, the phase delay was related to the viscosity-elasticity ratio instead of the elasticity and viscosity alone. So a new method would be introduced to quantify elasticity and viscosity.

In conclusion, PAVEI characterizes the mechanical properties of tissue, which makes up for the deficiencies in the existing PA imaging with respect to providing structural information. PAVEI can be implemented on existing PA imaging systems, as an additional function, to provide more comprehensive information about the tissue's mechanical and functional information. Further, the PA technique can potentially measure elasticity concurrently with other functional parameters, including the oxygen saturation of hemoglobin, which may provide more comprehensive information for disease diagnosis and treatment evaluation. Noninvasive imaging of viscoelasticity distribution expands the functionality of PA technique and is expected to find potential applications in clinical practice, such as tumor detection and arterial plaque assessment.

References

1. Ophir, J., Cespedes, I., Ponnenkanti, H., Yazdi, Y., and Li, X. (1991). Elastography: A quantitative method for imaging the elasticity of biological tissues, *Ultrason. Imag.*, **13**, pp. 111–134.
2. Muthupillai, R., Lomas, D. J., Rossman, P. J., Greenleaf, J. F., and Ehman, R. L. (1995). Magnetic resonance elastography by direct visualization of propagating acoustic strain waves, *Science,* **269**, pp. 1854–1857.
3. Jamin, Y., et al. (2015). Exploring the biomechanical properties of brain malignancies and their pathologic determinants in vivo with magnetic resonance elastography. *Cancer Res.,* **75**, pp. 1216–1224.
4. Deffieux, T., et al. (2015). Investigating liver stiffness and viscosity for fibrosis, steatosis and activity staging using shear wave elastography, *J. Hepatol.,* **62**, pp. 317–324.
5. Chen, S. G., et al. (2013). Assessment of liver viscoelasticity by using shear waves induced by ultrasound radiation force, *Radiology,* **266**, pp. 964–970.
6. Hai, P. F., Yao, J. J., Li, G., Li, C. Y., and Wang, L. V. (2016). Photoacoustic elastography, *Opt. Lett.,* **41**, pp. 725–728.
7. Hai, P. F., Zhou, Y., Gong, L., and Wang, L. V. (2016). Quantitative photoacoustic elastography in humans, *J. Biomed. Opt.,* **21**, pp. 066011-1-5.

8. Kovach, I. S. (1996). A molecular theory of cartilage viscoelasticity, *Biophys. Chem.*, **59**, pp. 61–73.
9. Mow, V. C., Ratcliffe, A., Poole, A. R. (1992). Cartilage and diarthrodial joints as paradigms for hierarchical materials and structures, *Biomaterials*, **13**, pp. 67–97.
10. Lerner, R. M., Huang, S. R., and Parker, K. J. (1990). "Sonoelasticity" images derived from ultrasound signals in mechanically vibrated tissues, *Ultrasound Med. Biol.*, **16**, pp. 231–239.
11. Fatemi, M., and Greenleaf, J. F. (1998). Ultrasound-stimulated vibro-acoustic spectrography, *Science*, **280**, pp. 82–85.
12. Gao, G. D., Yang, S. H., and Xing, D. (2011). Viscoelasticity imaging of biological tissues with phase-resolved photoacoustic measurement, *Opt. Lett.*, **36**, pp. 3341–3343.
13. Zhao, Y., Yang, S. H., Chen, C. G., and Xing, D. (2014). Simultaneous optical absorption and viscoelasticity imaging based on photoacoustic lock-in measurement, *Opt. Lett.*, **39**, pp. 2565–2568.
14. Chen, C. G., Zhao, Y., Yang, S. H., and Xing, D. (2015). Integrated mechanical and structural features for photoacoustic characterization of atherosclerosis using a quasi-continuous laser, *Opt. Express*, **23**, pp. 17309–17315.
15. Chen, C. G., Zhao, Y., Yang, S. H., and Xing, D. (2015). Mechanical characterization of intraluminal tissue with phase-resolved photoacoustic viscoelasticity endoscopy, *Biomed. Opt. Express*, **6**, pp. 4975–4980.
16. Zhao, Y., Yang, S. H., Chen, C. G., and Xing, D. (2016). Mechanical evaluation of lipid accumulation in atherosclerotic tissues by photoacoustic viscoelasticity imaging, *Opt. Lett.*, **41**, pp. 4522–4525.
17. Zhao, Y., Chen, C. G., Liu, H. W., Yang, S. H., and Xing, D. (2016). Time-resolved photoacoustic measurement for evaluation of viscoelastic properties of biological tissues, *Appl. Phys. Lett.*, **109**, pp. 203702-1-5.
18. Fang, H., Maslov, K., and Wang, L. V. (2007). Photoacoustic Doppler flow measurement in optically scattering media, *Appl. Phys. Lett.*, **91**, pp. 264103-1-3.
19. Fang, H., Maslov, K., and Wang, L. V. (2007). Photoacoustic Doppler effect from flowing small light-absorbing particles, *Phys. Rev. Lett.*, **99**, pp. 184501-1-4.
20. Araujo, R. P., and McElwain, D. L. S. (2004). A linear-elastic model of anisotropic tumour growth, *Eur. J. Appl. Math.*, **15**, pp. 365–384.

21. Lyshchik, A., Higashi, T., Asato, R., Tanaka, S., Ito, J., Mai, J. J., Pellot-Barakat, C., Insana, M. F., Brill, A. B., Saga, T., Hiraoka, M., and Togashi, K. (2005). Thyroid gland tumor diagnosis at US elastography, *Radiology,* **237**, pp. 202–211.

22. Paszek, M. J., Zahir, N., Johnson, K. R., Lakins, J. N., Rozenberg, G. I., Gefen, A., Reinhart-King, C. A., Margulies, S. S., Dembo, M., Boettiger, D., Hammer, D. A., and Weaver, V. M. (2005). Tensional homeostasis and the malignant phenotype, *Cancer Cells,* **8**, pp. 241–254.

23. Lusis, A. J. (2001). Atherosclerosis, *Nature,* **407**, pp. 233–241.

24. Glass, C. K., and Witztum, J. L. (2001). Atherosclerosis: The road ahead, *Cell,* **104**, pp. 503–516.

25. Virmani, R., Burke, A. P., Farb, A., Kolodgie, F. D. (2006). Pathology of the vulnerable plaque, *J. Am. Coll. Cardiol.,* **47**, pp. C13–C18.

26. De Korte, C. L., Carlier, S. G., Mastik, Doyley, F., Van Der Steen, A. F. W., Serruys, P. W., and Bom, N. (2002). Morphological and mechanical information of coronary arteries obtained with intravascular elastography, Feasibility study in vivo, *Eur. Heart J.,* **23**, pp. 405–413.

27. Sadat, U., Teng, Z., and Gillard, J. H. (2010). Biomechanical structural stresses of atherosclerotic plaques, *Expert Rev. Cardiovasc. Ther.,* **8**, pp. 1469–1481.

28. Teng, Z., Brown, A. J., Calvert, P. A., Parker, R. A., Obaid, D. R., Huang, Y., Hoole, S. P., West, N. E., Gillard, J. H., and Bennett., M. R. (2014). Coronary plaque structural stress is associated with plaque composition and subtype and higher in acute coronary syndrome: The BEACON I (biomechanical evaluation of atheromatous coronary arteries) study, *Circ. Cardiovasc. Imaging,* **7**, pp. 461–470.

29. Finn, A. V., Nakano, M., Narula, J., Kolodgie, F. D., and Virmani, R. (2010). Concept of vulnerable/unstable plaque, *Arterioscler. Thromb. Vasc. Biol.,* **30**, pp. 1282–1292.

30. Stone, G. W., Maehara, A., Lansky, A. J., de Bruyne, B., Cristea, E., Mintz, G. S., Mehran, R., McPherson, J., Farhat, N., Marso, S. P., Parise, H., Templin, B. H., White, R., Zhang, Z., and Serruys, P. W. (2011). PROSPECT Investigators, A prospective natural-history study of coronary atherosclerosis, *N. Engl. J. Med.,* **364**, pp. 226–235.

31. Jin, W., Jia, Y. Q., Huang, L. N., Wang, T. J., Wang, H. B., Dong, Y. H., Zhang, H. Z., Fan, M. Y., and Lv, P. Y. (2014). Lipoxin A 4 methyl ester ameliorates cognitive deficits induced by chronic cerebral hypoperfusion through activating ERK/Nrf2 signaling pathway in rats, *Pharmacol. Biochem. Behav.,* **124**, pp. 145–152.

32. Lake, L. W., and Armeniades, C. D. (1972). Structure-property relations of aortic tissue, *Trans. Am. Soc. Artif. Intern. Organs,* **18**, pp. 202–208.

33. Yoo, L., Gupta, V., Lee, C., Kavehpore, P., and Demer, J. L. (2011). Viscoelastic properties of bovine orbital connective tissue and fat: Constitutive models, *Biomech. Model. Mechanobiol.,* **10**, pp. 901–914.

34. Tsai, C. L., Chen, J. C., and Wang, W. J. (2001). Near-infrared absorption property of biological soft tissue constituents, *J. Med. Biol. Eng.,* **21**, pp. 7–14.

35. Hai, P. F., Yao, J. J., Maslov, K. I., Zhou, Y., and Wang, L. V. (2014). Near-infrared optical-resolution photoacoustic microscopy, *Opt. Lett.,* **39**, pp. 5192–5195.

Chapter 6

All-Optical Photoacoustic Technology

6.1 Introduction

Photoacoustic microscopy (PAM) is a noninvasive imaging modality, which hybrids high contrast of optical imaging and deeper imaging depth of ultrasound. PAM images optical absorption contrast based on the photoacoustic effect [1]. PAM has a wide range of application in biomedicine, such as subcutaneous microvasculature imaging, brain functional imaging, or early detection of tumors [2–4]. In addition, PAM has been used to monitor the blood oxygenation of vessels with high sensitivity and high specificity [2, 5]. At present, almost all of the PAMs still utilize the ultrasound transducer to detect the PA signals in contact fashion. Nevertheless, ultrasound waves have a strong attenuation in air. So, in order to attain sensitivity signals, the coupling media, such as water or ultrasound gel, has to been added between the sample and ultrasound transducer. This requirement of the contact detection mode is often difficult for PAM imaging of biomedical samples, which limits the practical application of PAM in ophthalmology and surgical navigation.

Biomedical Photoacoustics
Sihua Yang and Da Xing

ISBN 978-981-4774-58-1 (Hardcover), 978-0-203-70365-6 (eBook)
www.jennystanford.com

Compared with the use of an ultrasound transducer, the optical detection method can detect ultrasound signal in noncontact fashion [6], thus eliminating the requirement of the coupling media, which is attractive for in vivo imaging. Besides, optical detection also has an important advantage of potentially offering a higher imaging resolution than PAM. Due to imaging of PAM, the broadband ultrasound transducer is required to detect the broadband acoustic signal that excited by pulsed laser. Furthermore, in PAM, the axial resolution is inversely proportional to the bandwidth of the ultrasonic transducer, so its bandwidth is a critical factor to the improvement of imaging quality. However, owing to the limit of piezoelectric material and preparation technology, nearly all of ultrasound transducers usually offer finite detection bandwidth. Although the maximum bandwidth of piezoelectric ultrasonic detector can reach the center frequency of 120%, it is impossible to make a broadband ultrasound transducer that can cover from low frequency to high frequency. Due to photoelectric detector with a broadband characteristic, noncontact optical detection is presented constantly and has a wide application. In general, optical detection of ultrasound is a promising ultrasound detection modality, which can provide high detection sensitivity of ultrasound [6] and broadband width [7]. Because of the noncontact character of PAM based on the optical detection method, it could be conveniently used in many special applications such as the disease diagnosis in ophthalmology and imaging of the affected area in surgical operation.

As early as in the last century in 90s, Paul C. Beard et al. have presented a method that use polymer-film-based Fabry–Perot interferometer to sense the pressure waves, whose depth and range of photoacoustic imaging is up to centimeter level, realizing the ability to monitor tumor growth. Fabry–Perot was integrated in the optical fiber by using coating method, achieving the miniaturization of optical ultrasonic detection, which can actualize the detection of intravascular photoacoustic signals [8]. Then, some noncontact imaging methods constantly were proposed by using coherence detection approach to measure the vibration velocity and displacement of object surface. The following gives some examples:

In 1996, B. F. Pounet et al. employed heterodyne interferometer for ultrasound detection on rough surface, by mixing the speckled scatter beam and a planar coherent pump beam in photorefractive cubic crystals. The main advantage in using cubic crystals was their relatively fast response compared to ferroelectric crystals, which allowed them to work in a noisy environment with reasonable laser power. Very good sensitivity was demonstrated for the detection of small-amplitude ultrasonic surface displacements [9].

In 2004, S. A. Carp et al. exploited a modified Mach–Zehnder interferometer to measure surface displacement for photoacoustic imaging, with a temporal resolution of 4 ns and a displacement sensitivity of 0.3 nm. Compared with photoacoustic approaches based on time-resolved stress detection, POIS (pulsed optoacoustic interferometric spectroscopic imaging) was capable of noncontact imaging of phantoms with realistic tissue optical properties and provided images formed from data sets acquired from several highly scattering tissue phantoms, provided better than 200 μm resolution, and showed great promise for high-resolution noninvasive imaging of heterogeneous tissues at depths approaching 1 cm [10].

In 2007, Edward Zhang et al. proposed a noncontact scheme of multi-wavelength backward-mode planar photoacoustic scanner for 3D imaging with a spatial resolution in the tens to hundreds of micrometers range. A laser of 1550 nm nominal wavelength was used to provide the detected beam, in which a remote confocal Fabry–Perot interferometer (FPI) was used to measure displacements [11].

In 2010, Berer et al. used a two-wave mixing interferometer for remote and contactless photoacoustic imaging on solid. Picosecond laser pulses were employed to excite the semitransparent sample for broadband ultrasonic waves, which were detected utilizing a two-wave mixing interferometer. Image reconstruction was accomplished with both wave types. This experiment showed the potential of PAI for the detection of absorbing inclusions in a semitransparent host material [12].

In 2012, G. Rousseau et al. raised non-contact photoacoustic tomography (PAT) and ultrasonography (US) to obtain tissue imaging. The system used suitably shaped laser pulses and a confocal Fabry–Perot interferometer, which detected ex vivo in chicken breast and calf brain specimens for endogenous and exogenous inclusions exhibiting optical and acoustic contrasts. Inclusions down to 0.5 mm in size were detected at depths well exceeding 1 cm. The method could significantly expand the range of applications of PAT and US in biomedical imaging [6].

In 2007, G. Paltauf et al. presented a method for three-dimensional photoacoustic imaging by using a Mach–Zehnder interferometer as an acoustic line detector. The signals acquired with the interferometer correspond to line integrals over the acoustic wave field; therefore, an algorithm was served to reconstruct of a three-dimensional image. The spatial resolution of the imaging system can provide a 100–300 μm range spatial resolution, which was capable of producing three-dimensional images of objects with an overall size in the range of several millimeters to centimeters. This method is suitable for the imaging of small objects such as isolated organs of small animals [13].

Lately, Yi Wang et al. utilized a low-coherence interferometer as the acoustic detector to achieve noncontact photoacoustic imaging, obtaining photoacoustic images of the blood vessels in the excised mouse ear [14]. Cedric Blatter et al. exploited intrastate phase-sensitive optical coherence tomography (OCT) [15], which realized noncontact optical photoacoustic imaging. However, these methods could only provide bandwidth-limited ultrasound detection as well as the piezoelectric ceramic transducer. Moreover, low-coherence Michelson interferometer is required that the system had to work in the homodyne mode and detect the ambient vibrations to obtain the highest sensitivity detecting point, which seriously limited the imaging speed and stability of the system.

This chapter introduces a novel noncontact broadband all-optical photoacoustic microscopy fabricated with microchip laser and all-fiber low-coherence interferometer, which can provide broad detection bandwidth and high imaging resolution [16]. As we know, the all-fiber low-coherence interferometer is the important part of the optical coherence tomography, so the broadband all-optical photoacoustic microscopy can be conveniently integrated with OCT and other optical imaging systems (such as fluorescence imaging, second harmonic imaging, and so on) to form a multimode imaging system [17, 18].

6.2 The Principle of the Noncontact All-Optical Photoacoustic Imaging

When the pulsed laser excites the biological tissue, the absorber of organization will absorb the energy of pulse light. Due to the photoacoustic effect, it will produce photoacoustic signal. The essence of photoacoustic signal is a mechanical wave that will cause vibration in the tissue during generation and transmission. We can calculate the size of the photoacoustic signal by detecting the amplitude of tissue vibration, which can reflect the amount of light absorbed by the absorber. Michelson interferometer is a commonly used tool for modern precious measurement; the measurement accuracy can reach the pm level or even higher [15]. In biomedical imaging technology, OCT is based on the Michelson interferometer, using low-coherence light source as detecting light. Because the coherence length of the low coherent light can be very short, we can achieve interference signals of incident light and reflected light to identify the micro-level structure information about the biological tissue.

The principle of noncontact detection photoacoustic signal is shown in Fig. 6.1. The detection light through a beam splitter is divided into two beams, a beam through the lens focused on the mirror, and another beam is focused onto the sample surface through the dichroic mirror and the objective lens. The reflected light from the sample arm and the reflected light from reference arm have interfered in the beam splitter. And then the coherent signal is received by the photoelectric detector. The excitation light will pass through the same objective lens together to stimulate the biological tissue. When absorbing light energy, the biological tissue will further generate photoacoustic signal. Next, the photoacoustic signal is transmitted to the surface, which will cause the vibration of the tissue surface. The amplitude of the vibration is related to the intensity of the photoacoustic signal. According to the propagation properties of the photoacoustic signal in the medium, the relationship between the vibration displacement of the tissue and the intensity of the photoacoustic signal is as follows [19]:

$$\Delta x(t) = -\frac{1}{\rho v_a} \int p(t) dt \tag{6.1}$$

Here, v_a is the medium of sound velocity, ρ is the density of the medium, t is the laser pulse excitation time. The vibration displacement of the tissue surface changes the phase of the probe light in the Michelson interferometer. Thus, the size of the vibration displacement can be measured by detecting the magnitude of the phase change. For convenience, $\Delta x(t)$ is replaced with Δx. The relationship between the displacement and the phase of the tissue caused by the photoacoustic signal is as follows:

$$\Delta\varphi = \frac{2\pi}{\lambda_0}\Delta x \tag{6.2}$$

Here, λ_0 is the center wavelength of probe light. The intensity of the interferometer output is

$$I = I_1 + I_2 + 2\sqrt{I_1 I_2}\cos\varphi. \tag{6.3}$$

Here, I_1 and I_2 are DC, when the light through the differential photodetectors, DC will disappear. Therefore,

$$I = A\sqrt{I_1 I_2}\cos\varphi \tag{6.4}$$

Here, A is the magnification of differential photodetectors; φ includes the phase induced by the ambient vibrations, φ_0, and the phase induced by the PA signals, $\Delta\varphi$. Therefore, $\varphi = \varphi_0 + \Delta\varphi$. In order to accurately extract the PA signal from the coherence signal, the system has to be working in the homodyne mode, i.e., $\varphi_0 = k\pi \pm \pi/2$, the system has the greatest detection sensitivity for the acoustic wave [14, 16].

In order to realize the high-sensitivity detection of the photoacoustic signal, only when the system judges that the output of the differential detector will be zero, the detector will trigger the pulsed laser to excite the photoacoustic signal. And, the system starts to acquire the photoacoustic signal only when the phase produced by the surrounding environment is zero. The advantage of this is the system will form a closed loop control, which will ensure that the system is, at the highest sensitivity, collecting photoacoustic signals. The temporal logic is shown in Fig. 6.2.

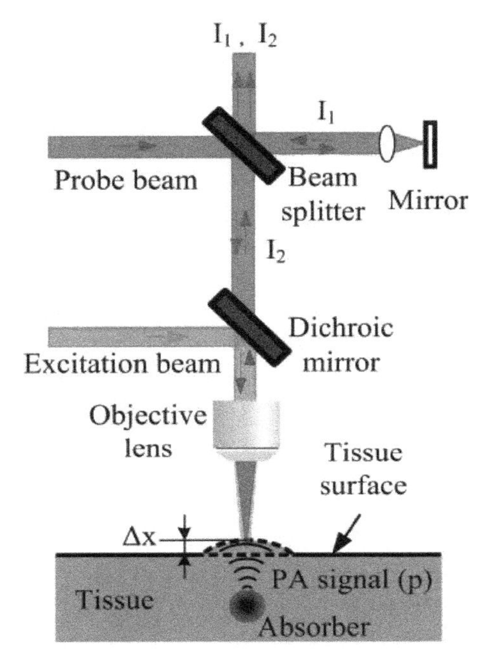

Figure 6.1 Noncontact photoacoustic signal detection schematic.

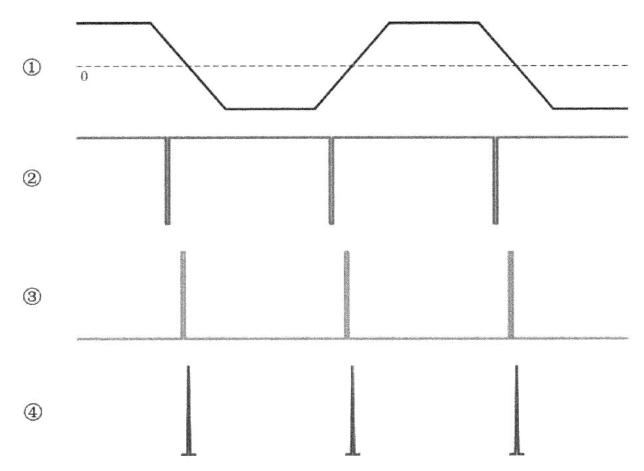

Figure 6.2 Timing diagrams for triggering the signals acquisition: ① the differential amplifier output waveform figure; ② pulse laser external trigger signal; ③ pulse laser synchronization output signal; ④ displacement of the organization surface by photoacoustic signal. Reproduced with permission from Ref. [16].

6.3 The Noncontact All-Optical Photoacoustic Microscopy

6.3.1 Noncontact All-Optical Photoacoustic Microscopy

The schematic of the noncontact broadband all-optical photoacoustic microscopy setup is shown in Fig. 6.3. The detection system of photoacoustic signal consists of the all-fiber low-coherence interferometer, which contains a super luminescent diode with a central wavelength of 1310 nm and a spectral bandwidth of 45 nm, an optical circulator and a 2 × 2 fiber coupler. The light outputted from the coupler is focused by the lens to the mirror as the reference arm of the Michelson interferometer and the other through the dichroic mirror is focused by the objective lens onto the sample surface as the sample arm. That the sample arm's optical path difference is equal to the reference arm's optical path difference made the optical signal detection system working in the zero-difference mode. A microchip laser (HLX-I-F005, Horus Laser) at 532 nm was used to excite PA signals in the sample.

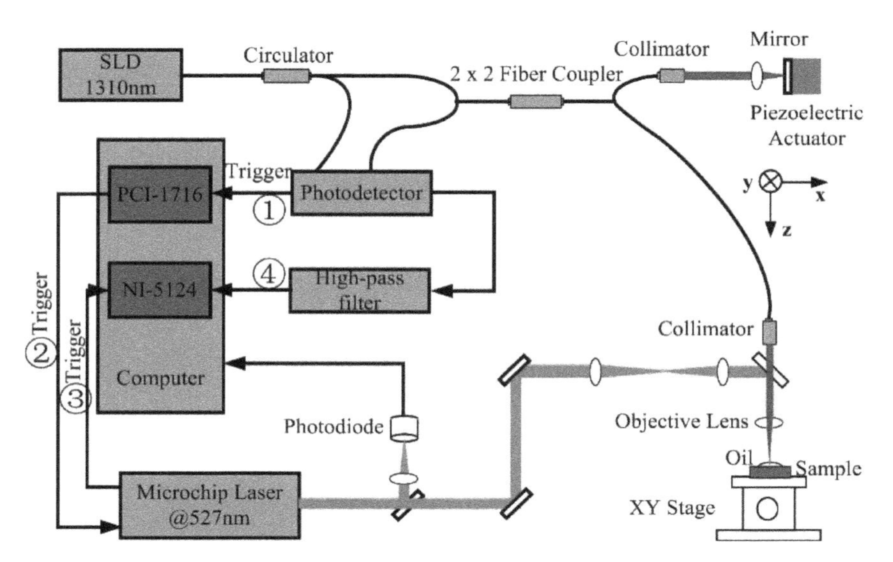

Figure 6.3 The schematic of the all-optical integrated photoacoustic microscopy setup. Reproduced with permission from Ref. [16].

The pulse width of the excitation laser was ~10 ns. The excitation laser trigger was locked at the zero point by the control card (PCI-1716, Advantech). A trigger signal from the microchip laser was sent to the data acquisition card (NI5124, National Instruments) simultaneously for sampling the PA signals. In order to reduce the influence of environmental signals, the collected signal through cut-off frequency of 30 KHz high-pass filter. And the photodiode detects the output laser pulse energy to correct the jitter caused by the laser energy jitter of the photoacoustic signal. To further improve the detection sensitivity of the system, the imaged area of the sample was covered with a thin layer of mineral oil. Figure 6.4 shows the PA signal obtained from hair: Fig. 6.4a is the displacement of photoacoustic signal, and Fig. 6.4b is the first derivative of the PA signal.

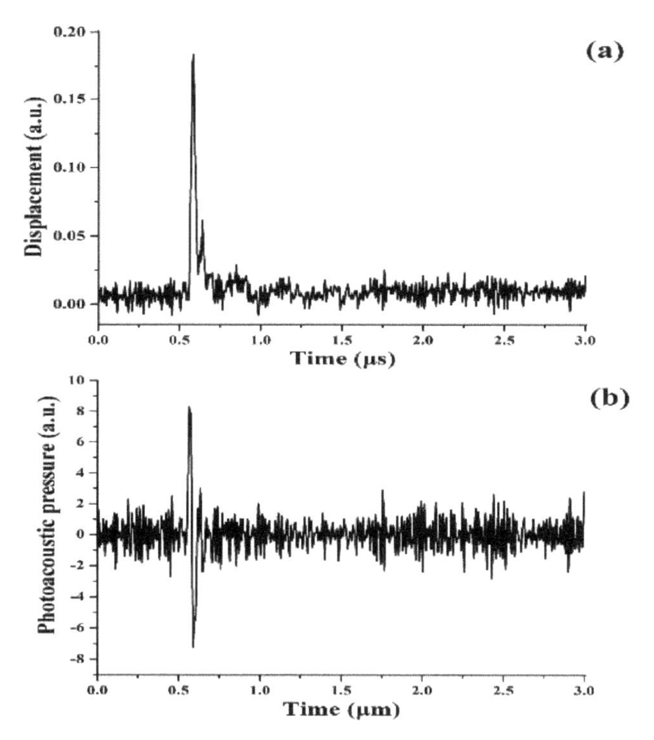

Figure 6.4 Photoacoustic signals detected from a hair sample: (a) transient displacement and (b) corresponding transient pressure. Reproduced with permission from Ref. [14].

6.3.2 The Bandwidth of the Noncontact Photoacoustic Microscopy

The detection bandwidth of the noncontact photoacoustic microscopy was measured by detecting the PA signal from the red ink film plated on a transparent thin polyethylene film. The spectrum width of the signal is obtained by Fourier transform of the detected photoacoustic signal, and the bandwidth of the standard signal is regarded as the detection bandwidth of the system. The PA signal (no averaging) of the red ink film is shown in the inset of Fig. 6.5a. The Fourier transform of the PA signal spectra is shown in Fig. 6.5a, and the –6 dB bandwidth is about 67 MHz, which is close to the bandwidth of photodetector (75 MHz). According to the formula $R_A = 0.88 \cdot V_s/\Delta f$ [20], the axial resolution of the noncontact photoacoustic microscopy system is estimated to be 19.7 μm. Figure 6.5b is the B-scan image in the *x*-*z* plane of the red ink film; the full width half magnitude (FWHM) of the profile along *z* direction is ~20.6 μm, which matches well with theoretical resolution. Generally, the detection bandwidth of the noncontact photoacoustic microscopy is limited by the photodetector. Employing wider bandwidth photodetector and shorter laser pulse could further improve the detection bandwidth of the photoacoustic microscopy.

6.3.3 The Lateral Resolution of the System

The lateral resolution is a key parameter of an imaging system. To quantify the lateral resolution of the noncontact photoacoustic microscopy, we imaged two ~6 μm diameter crossed carbon fibers using a scanning step of 3 μm at rate of 40 Hz. Figure 6.6a shows a maximum amplitude projection (MAP) image of the two carbon fibers. Figure 6.6b shows the profile of the vertical fiber along the *x* axis, demonstrating that the lateral resolution of our system is at least as fine as 11 μm. The signal-to-noise ratio (SNR), measured from Fig. 6.6a, is as high as 19 dB.

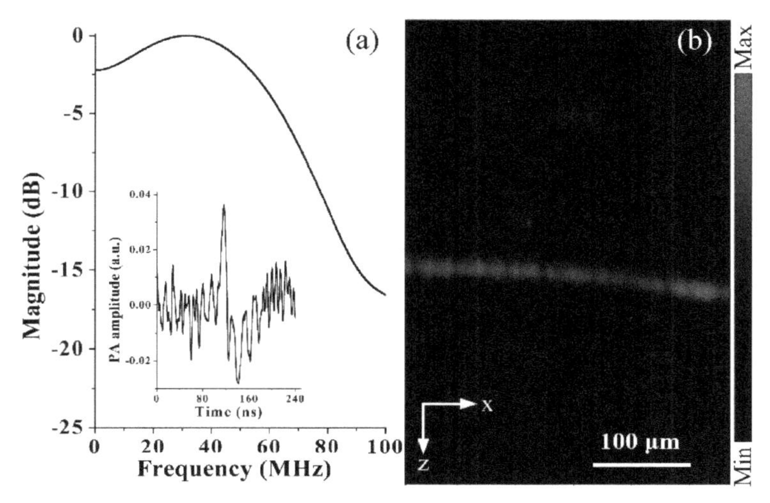

Figure 6.5 Detection bandwidth of the noncontact BD-AO-PAM. (a) Spectrum of the PA signal. (b) B-scan image in the *x*–*z* plane of the red ink film. Reproduced with permission from Ref. [16].

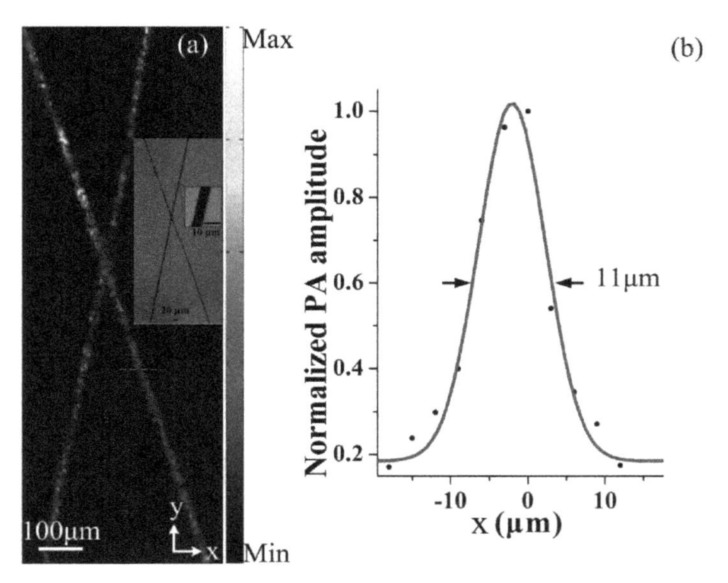

Figure 6.6 Lateral resolution of the noncontact photoacoustic microscopy. (a) PA image of two crossed 6 μm diameter carbon fibers. (b) Distribution of the PA amplitude (dots) along the dashed line in (a). Reproduced with permission from Ref. [16].

6.3.4 Mimicking and in vivo Experimental Results

6.3.4.1 Mimicking experiment

In order to verify the imaging capacity of the noncontact photoacoustic microscopy, a phantom experiment was conducted. To make the phantom, we used four strands of hair embedded within a scattering gel at depths approximately between 0 and 1 mm. The scattering gel was made of agar mixed with 1% Intra lipid, which gave a scattering background similar to typical highly scattering tissue. Figure 6.7a shows a MAP image of the phantom, and the edge of the hairs is distinct. The depth of the hair in the scattering gel, *z*, was extracted using the delay time of the PA amplitude, as shown in Fig. 6.7b. The hair strands were clearly distributed from 0 to 1 mm in the scattering gel. Meanwhile, the noncontact photoacoustic microscopy has 3D imaging capability, and the result is shown in Fig. 6.7c, in which the spatial morphology of the hair strands is clearly displayed.

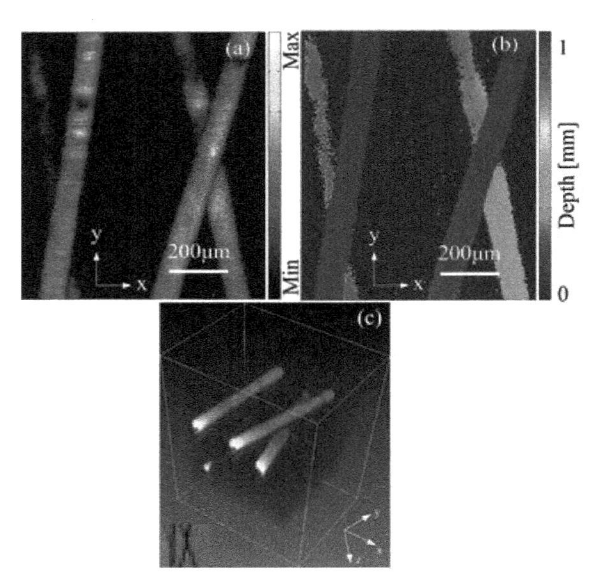

Figure 6.7 Photoacoustic image of scattering phantom; (a) Photoacoustic MAP image of the four hair strands. (b) Depth map of the scattering phantom. (c) Snapshot of a 3D animation showing the four hair strands in (a). Reproduced with permission from Ref. [16].

6.3.4.2 In vivo experiment

To further demonstrate the ability of the noncontact photoacoustic microscopy, we used it to image blood vessel of a BALB/c mouse ear in vivo. The hair was removed from the ear with commercial hair remover (Payven Depilatory China). Sodium pentobarbital (40 mg/kg; supplemental, 10 mg/kg/h) was administered to keep the mouse motionless during the experiment. The laser fluence incident on the skin surface was controlled below 12 mJ/cm^2 in the experiment [21]. Figure 6.8a shows the results of the microvasculature of mouse ear acquired from an area of 1.6 × 1.6 mm^2 (200 × 200 pixels). The diameter of the blood vessel pointed by the white arrow in the figure is about 12 μm, and the SNR is as high as 24 dB. Figures 6.8b–d show the B-scan image of blood vessel corresponding to the three white dotted line in (a), and the different size scales of blood vessel along the *z* axis can be clearly distinguished. The results suggested that the noncontact photoacoustic microscopy has the capacity to image the microvasculature with high spatial resolution in vivo.

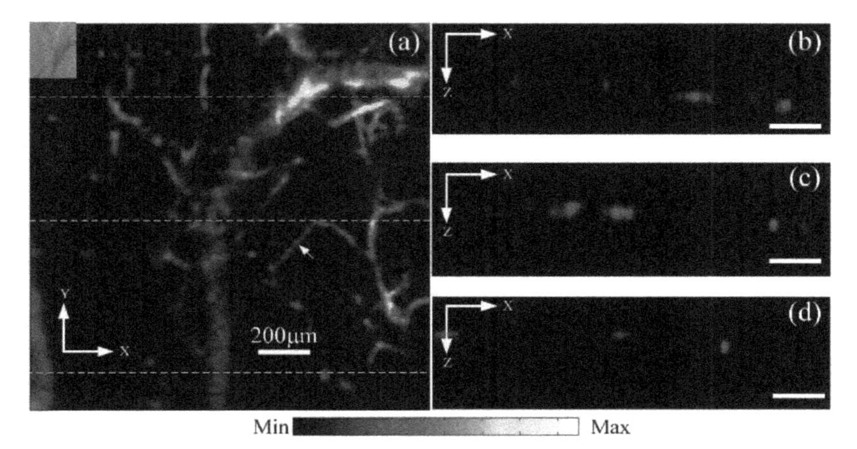

Figure 6.8 In vivo photoacoustic image of microvasculature of a mouse ear. (a) Photoacoustic MAP image of the microvasculature of a mouse ear. (b–d) B-scan image of microvasculature corresponding to the three white dotted line in (a). Reproduced with permission from Ref. [16].

In summary, a noncontact photoacoustic microscopy system, consisting of a microchip laser and an all-fiber low-coherence interferometer, was developed. The system provided broad detection bandwidth and high imaging resolution, which realized lifelike microvascular imaging on the animal model in vivo. More important, the all-optical character and noncontact fashion of the system make it easy to integrate a portable system and should be valuable for biomedical applications.

6.4 Multi-Modality Photoacoustic Imaging System

Photoacoustic microscopy is a label-fee and noninvasive biomedical imaging technology that images optical absorption contrast based on the photoacoustic effect and has shown enhanced imaging capability with more depth of field than traditional optical microscopy [22, 23]. PAM has been applied in subcutaneous functional imaging, microvasculature imaging, and early detection of tumors with high sensitivity and high specificity. However, a drawback of the photoacoustic imaging is the loss of information of the low-absorptive tissue, which is often required for proper image interpretation. Optical coherence tomography shows high sensitivity to the endogenous contrast of tissue scattering but lacks absorption information, which is a relatively mature noncontact imaging tool that detects the interference between the back-reflected light from the sample and the reference arm. What is more, optical coherence tomography has been used in clinical practice for intravascular imaging and ophthalmology, which is the branch of medicine that deals with the anatomy, physiology, and diseases of the eye, to exploit various optical scattering properties and functional information of biological tissues [24]. Because optical coherence tomography is based on the contrast of tissue optical scattering properties, optical coherence tomography technique penetration is limited to 2 mm [25]. Hence, photoacoustic imaging and optical coherence tomography can make complementarity to each other about the characterization of tissue structure and their physiological parameters; moreover, photoacoustic imaging can make up for the deficiency of optical coherence tomography in imaging depth.

On the basis of noncontact photoacoustic microscopy [16], a novel all-optical noncontact dual-mode imaging system was proposed, which can be the best co-imaging strategy of PAM and OCT in biomedicine, integrating photoacoustic microscopy and optical coherence tomography by using a single Michelson detector.

6.4.1 Optically Integrated Dual-Mode Image System of Combined Photoacoustic Microscopy and Optical Coherence Tomography

The schematic of a setup of the optically integrated dual-mode image system of combined photoacoustic microscopy and optical coherence tomography is shown in Fig. 6.9. The photoacoustic microscopy sub-system is based on the Michelson detector (the Michelson detector consist of a low-coherence light source (central wavelength of 1310 nm and spectral bandwidth of 45 nm) and an all-fiber Michelson interferometer (an optical circulator, a 2×2 fiber coupler, and a photodetector)), giving access to rapid small path length changes, such as interface displacement due to a transient PA wave, which resolves the phase difference between reference and sample arm along time. A microchip pulse laser (HLX-I-F005, Horus Laser, central wavelength of 532 nm with ~10 ns) was used to excite photoacoustic signals from the sample. The low-coherence Michelson interferometer with the highest sensitivity to the vibration, when the output of the photodetector (PDB420C, Thorlabs, USA) is crossing zero, is at the point. Only when the zero point is locked, a trigger signal from the microchip laser is sent to the data acquisition card (NI5124, National Instruments) for sampling the PA signals and the excitation laser triggered by the control card (PCI-1716, Advatech) at the same time. The imaged area of the sample was covered with a ~1 mm thin layer of mineral oil to eliminate the influence of the speckles, to further improve the PA signal detection sensitivity of the PAM sub-system, due to the rough surface of the sample and ensure the surface of the sample within the coherence length of the light source. A rapid scanning optical delay line (RSOD) realizing the depth scanning of the OCT sub-system at the reference arm, which consisted of an optical grating, a Fourier lens, a scanning galvanometer, and a gold-coated reflection mirror. Initially, when the optical path length of reference arm is equal to the optical path length from the oil surface on the sample arm, the rapid scanning optical delay line is set a proper position, the PAM sub-system starts working, and the PA signal is collected by the data acquisition card. Then the PA signal and OCT signal are alternately acquired and stored in the computer in order that the scanning galvanometer is scanning and the OCT sub-system starts working and the OCT signal is read by the data acquisition card [17].

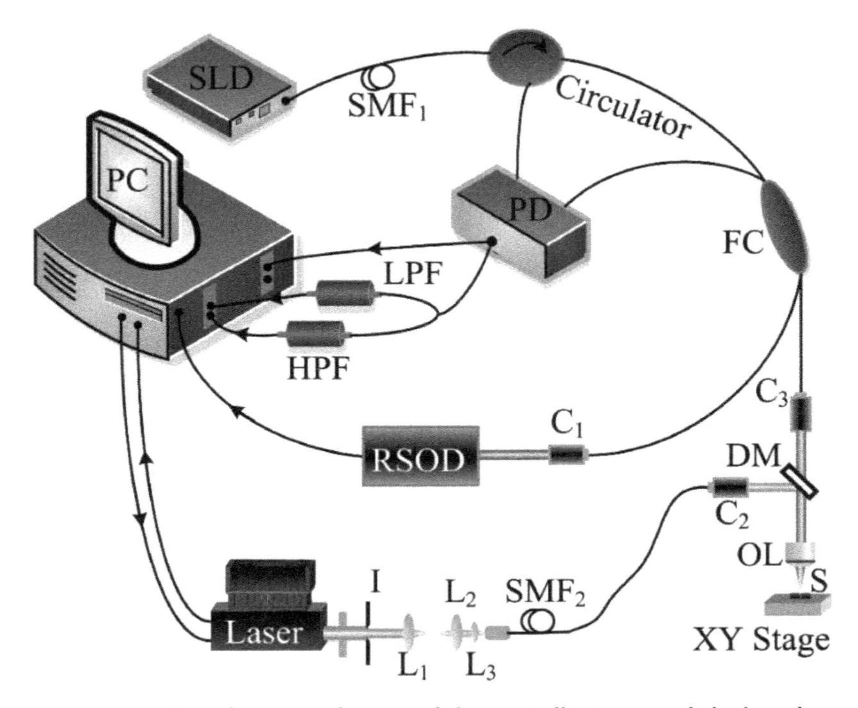

Figure 6.9 The schematic of setup of the optically integrated dual-mode PAM-OCT system combined photoacoustic microscopy and optical coherence tomography. $SMF_{1,2}$, single mode fiber; PD, photodetector; RSOD, rapid scanning optical delay line; $L_{1,2,3}$, lens; SLD, super-luminescent diode; FC, fiber coupler; LPF, low pass filter; HPF, high pass filter; $C_{1,2,3}$, collimator; SM, scanning mirror; OL, objective lens; S, sample I, iris; DM, dichroic mirror. Reproduced with permission from Ref. [17].

6.4.2 The Phantom and in vivo Experiments of the Optically Integrated Dual-Mode PAM-OCT System

The lateral resolution of the dual-mode image system combined photoacoustic microscopy, optical coherence tomography was quantified by imaging the sharp edge of a blade with a scanning step of 2 μm at rate of 40 Hz. The results are shown in Fig. 6.10. The axial resolution of PAM sub-system was estimated to be 20 μm based on the detection bandwidth of the system and the speed of sound in tissue. The lateral resolution of optical coherence tomography and photoacoustic microscopy sub-systems as defined by the full-width at half maximum (FWHM) of the line spread function was 12.5 and 13 μm, respectively. The edge spread function (ESF) was calculated along parallel to the x axis and shown by the red dotted line in Fig. 6.10a,b. Taking the derivative of the ESF yielded the line spread function (LSF), which is shown by the black line in Fig. 6.10a,b. On the surface of the sample, the small distinction can be the result of the thin layer of mineral oil. According to Raleigh criterion, the axial resolution is defined by the distance between the two peaks, so the axial resolution of optical coherence tomography sub-system was obtained by imaging the air gap between the microscope cover glass and microscope slide, the B-scan image was shown in Fig. 6.10c. The result is better than 18.4 μm in air as shown in Fig. 6.10d, which matched well with the theoretical value of 16.8 μm.

In order to verify whether the system is capable of imaging the absorption and scattering targets within the tissue-mimicking phantom and to make the phantom, six black and white human hair strands were placed; three black hair strands (yellow dots circles) are clearly imaged. Meanwhile, the three white hair strands (white dots circles) are not identified in Fig. 6.11a, which shows the photoacoustic microscopy B-scan image of the phantom. Because optical coherence tomography technique utilizes optical scattering, which can image both black and white hair regardless of the hair color, we can visualize both of them, including the surface of scattering gel in the optical coherence tomography image (Fig. 6.11b). However, the photoacoustic microscopy is based on optical absorption; the black hair induced the strong PA signals owing to its melanin concentration, but it lacked in the white hair. Figure 6.11c shows the fused photoacoustic microscopy and optical coherence tomography B-scan image: The scattering gel surface, the strong absorption black hair, and the strong scattering white hair are obviously distinguished. Figures 6.11d,e show the hair raw PA signal and optical coherence tomography signal, respectively. In Fig. 6.11e, the signals of the phantom surface and the hair boundary are clearly shown. Therefore, the dual-mode image system of combined photoacoustic microscopy and optical coherence tomography can simultaneously acquire both optical absorption and optical scattering information, indicated by the experiment results.

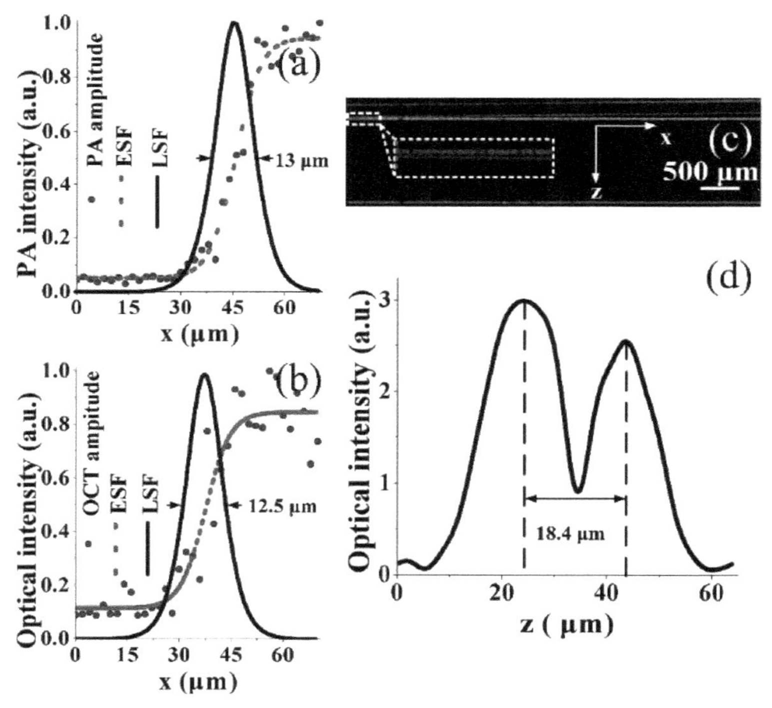

Figure 6.10 The spatial resolution including lateral resolution and axial resolution of the dual-mode PAM-OCT. LSF, line spread functions (solid black lines), extracted from the fitted ESF, edge spread functions (red dotted lines) acquired in the air. (a) PAM sub-system. (b) OCT subsystem. (c) OCT B-scan image of the air gap between the microscope cover glass and microscope slide. Inset: zoomed image of the white rectangle in (c). (d) The profile along the red dotted line in (c). Reproduced with permission from Ref. [17].

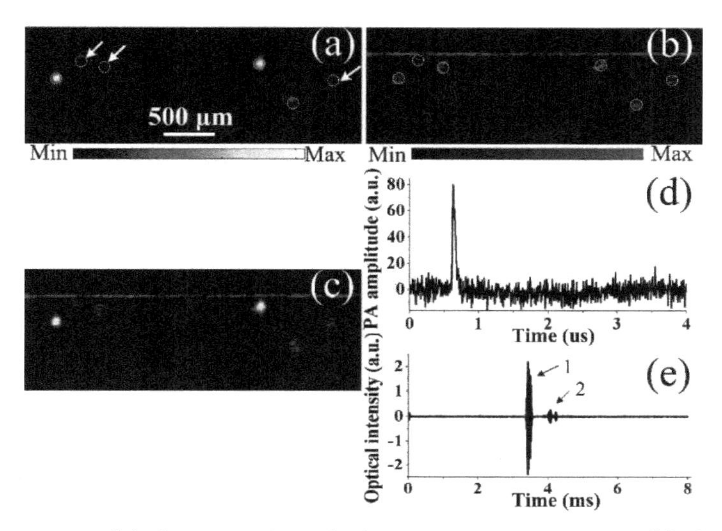

Figure 6.11 (a) Cross section of photoacoustic microscopy. (b) Cross section of optical coherence tomography. (c) Fused photoacoustic microscopy and optical coherence tomography images. The white and yellow dots represent the white and black human hair, respectively. (d) The raw photoacoustic signals. (e) Optical coherence tomography signals of hair. 1, the signal of gel surface; 2, the signals of hair. Simultaneous imaging of a scattering phantom in which six human hair strands (three white and three black) are placed. Reproduced with permission from Ref. [17].

All experimental animal procedures were carried out in accord with the guidelines of the South China Normal University. We chose to image the ear of a mouse, so that we could demonstrate the in vivo microscopic imaging ability of this dual-mode image system of combined photoacoustic microscopy and optical coherence tomography. The hair on the ear was gently removed using a human hair-removing lotion before the experiments. To keep the mouse motionless, the mouse was placed on a homemade animal holder, and sodium pentobarbital (40 mg/kg; supplemental, 10 mg/kg/h) was administered in the experiment.

Figures 6.12a,b show the MAP images of photoacoustic microscopy and optical coherence tomography, respectively. Figures 6.12c,d are B-scan images corresponding with the maximum amplitude projection of photoacoustic microscopy and optical coherence tomography in Figs. 6.12a,b, which show the benefits of combining these two modalities more clearly.

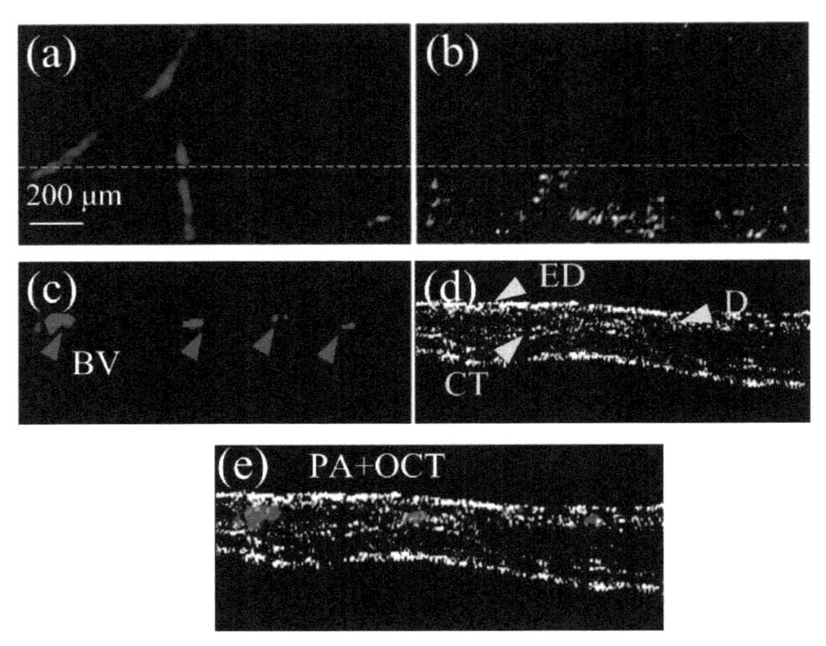

Figure 6.12 In vivo simultaneous imaging of mouse ear by using the dual-mode image system combined photoacoustic microscopy, optical coherence tomography. (a) photoacoustic microscopy maximum amplitude projection image. (b) Optical coherence tomography maximum amplitude projection image. (c) photoacoustic microscopy B-scan image. (d) Optical coherence tomography B-scan image. (e) Fused optical coherence tomography and PAM images. ED, epidermis; CT, cartilage; D, dermis; BV, blood vessel. Reproduced with permission from Ref. [17].

Figure 6.12e shows the fused cross-sectional image of photoacoustic microscopy and optical coherence tomography. The incident laser fluence on the skin surface, which is lower than the American National Standards Institute (ANSI) laser safety limit (20 mJ/cm^2), was controlled below 10 mJ/cm^2 during the in vivo experiment. The ear's thickness in the optical coherence tomography image is about 200 to 300 μm. The dermal structure and the sebaceous gland are clearly observed. In the cross-sectional optical coherence tomography, we can identify epidermis, dermis, and cartilage, indicated with yellow arrows. We have observed that photoacoustic microscopy is good at locating micro-vessels in the ear with limited surrounding tissue information. Optical coherence tomography and photoacoustic microscopy visualize different tissue structures: photoacoustic microscopy clearly maps the microvasculature, while optical coherence tomography images the tissue structure with high resolution.

These experimental results show that the PAM and OCT images of mimic phantom and biology tissues are reconstructed with high resolution through the dual-mode imaging system, which can contribute to promoting the application of PAM-OCT. Generally, the coefficients of absorption and scattering are the important parameters of the tissue, because all kinds of pathological changes would lead to the changes of the coefficient of absorption and scattering. We can abstract the information of absorption and scattering to diagnose the pathology changes of the tissue. Typically, PA and OCT can simultaneously measure the size of lipid core and the thickness of thin fibrous cap in cardiovascular disease [24, 26]. In addition, this all-optical noncontact dual-mode imaging system can be conveniently used to diagnose ophthalmology and dermatology disease due to its noncontact mode.

6.4.3 The Optically Integrated Tri-Modality Image System of Combined Photoacoustic Microscopy, Optical Coherence Tomography, and Fluorescence Imaging

Although the dual-mode image system of combined photoacoustic microscopy and optical coherence tomography is advantageous since it can simultaneously obtain two different sets of imaging contrast information. However, it still remains unable to image more information about the tissue. Consequently, some tri-modality imaging systems were proposed, such as PAM-OCT-US [27] and OCT-FL-US [28] system. These tri-modality imaging systems can simultaneously provide more different imaging contrast information than dual-modality systems and can obtain complementary tissue anatomical structure and physiological function information to precisely locate the lesion and diagnose the disease.

On the basis of the optically integrated dual-mode photoacoustic and optical coherence tomography, the tri-modality imaging system is presented that optically integrated with all-optical photoacoustic microscopy, OCT and fluorescence imaging to provide complementary information, including optical absorption, optical back-scattering, and fluorescence contrast about biological tissue. In this tri-modality imaging system, AOPAM, OCT, and FLM (AOPAM-OCT-FLM) share the same optical path and scanning system. In particular, AOPAM and OCT share the same Michelson detector to detect the photoacoustic signal and backscattering photons, moreover, AOPAM and FLM use a common laser source to induce photoacoustic signal and excite fluorescence signal [18].

The schematic of the setup of the all optically integrated tri-modality image system combined photoacoustic microscopy, optical coherence tomography and fluorescence imaging is shown in Fig. 6.13. All-optical photoacoustic microscopy (AOPAM), optical coherence tomography and fluorescence imaging share the same optical path and scanning system. In particular, all-optical photoacoustic microscopy and optical coherence tomography share the same Michelson detector to detect the PA signal and backscattering photons. Moreover, the highest sensitivity of the Michelson detector to the vibration is at the point when the output of the balanced photodetector (PDB420C, Thorlabs, USA) is crossing zero. A home-made field programmable gate array (FPGA) based analog-to-digital converter (ADC) was used to lock the zero point and externally trigger the microchip laser, and at the same time a trigger signal from the microchip laser is sent to the data acquisition card (NI5124, National Instruments) for sampling the PA signals. Finally, all-optical photoacoustic microscopy and fluorescence imaging use a common laser source to induce PA signal and excite fluorescence signal. The depth scanning of the OCT sub-system is realized by a rapid scanning optical delay line, which consists of a uniaxial scanning galvanometer (6231H, Cambridge Technology), an achromatic lens, a grating and a gold-coated mirror. A microchip pulse laser (HLX-I-F005, Horus Laser) at 532 nm with ~10 ns was focused into sample by an objective lens (10×, LSM02, Thorlabs) to excite PA signals. To further improve the PA signal detection sensitivity of the AOPAM sub-system, the imaged area of the sample was covered with a thin layer of mineral oil to eliminate the influence of the speckles due to the rough surface of the sample. The noise-equivalent pressure of the PAM sub-system is estimated to be about 16 Pa/$\sqrt{\text{Hz}}$. Owing to sharing a common exciting laser, the fluorescence signal is detected by the photomultiplier tube (PMM02, Thorlabs) and simultaneously acquired by the same data acquisition card. Two emission filters further eliminate the reflected excitation light.

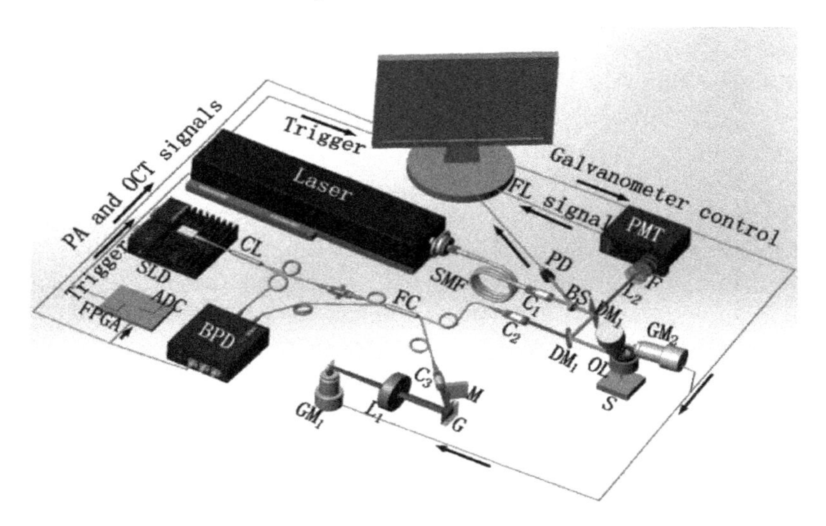

Figure 6.13 Schematic of the optically integrated tri-modality imaging system. RSOD, rapid scanning optical delay line; $L_{1,2}$, lens; $C_{1,2,3}$, collimator; $DM_{1,2}$, dichroic mirror; M, mirror; OL, objective lens; S, sample; $GM_{1,2}$, scanning galvanometer; G, grating; F, filter; SLD, super-luminescent diode; SMF, single mode fiber; PD, photodetector; BPD, balanced photodetector; PMT, photomultiplier tube; FC, fiber coupler; ADC, analog-to-digital converter. FPGA, field programmable gate array. Reproduced with permission from Ref. [18].

6.4.4 The Phantom and in vivo Experiments of Optically Integrated Tri-Modality AOPAM-OCT-FLM System

Before the experiment, the fluorescence intensity of Rhodamine B solution with 10 different concentrations was measured by the FLM sub-system and then the obtained detection limit of fluorescence was about 100 nmol/L.

Figures 6.14a,b show the MAP AOPAM and FLM images of the phantom. The veins of the leaf are visualized by both AOPAM and FLM. This is because the leaf veins have already been stained by high-concentration Rhodamine B, which absorbs the pulse energy and transfers into PA wave and emits the fluorescence. FLM only provides two-dimensional plane imaging without any cross-sectional depth information, but AOPAM and OCT can offer high-resolution depth information. Figures 6.14c,d show the B-scan AOPAM and OCT images of veins and scattering surface corresponding to the white dotted line in (a). The feasibility of the optically integrated tri-modality imaging system was validated by experiments using a turbid tissue-mimicking phantom. The scattering gel was made of agar mixed with 1% intra-lipid, which gave a scattering background similar to typical highly scattering biological tissue. To make the phantom, a piece of leaf vein was immersed in 1 mol/L Rhodamine B, which was used as optical absorption and fluorescence targets for PA and FL imaging, and then the leaf veins were air-dried and placed in a scattering gel. Obviously, the AOPAM and OCT can also image the veins clearly. Besides, the OCT can distinguish the surface of the gel due to its high scattering. Meanwhile, AOPAM and OCT have three dimensional (3-D) imaging capability; the results are shown in Figs. 6.14e,f; the spatial morphology of the veins is clearly displayed. As shown in Fig. 6.14f, OCT can simultaneously obtain the surface and veins structure (indicated by arrows). Therefore, optical absorption, scattering information, and fluorescence contrast can be acquired by the tri-modal AOPAM-OCT-FLM system, as indicated by experimental results.

In vivo imaging of the mouse ear was performed to further demonstrate the potential clinical capabilities of this optically integrated tri-modal AOPAM-OCT-FLM system. Before the experiments, by using a human hair-removing lotion, the hair on the ear was gently removed. To evaluate the capacity of the FLM sub-system in vivo, 10 nL of 10 μmol/L Rhodamine B was injected into mouse ear using a 29-gauge needle. The mouse was placed on a homemade animal holder and sodium pentobarbital (40 mg/kg; supplemental, 10 mg/kg/h) was administered to keep the mouse motionless in the experiment.

The laser fluence, which is lower than the American National Standards Institute (ANSI) laser safety limit (20 mJ/cm^2), was about 18 mJ/cm^2 at the optical focus on the ear tissue during the in vivo experiment [21].

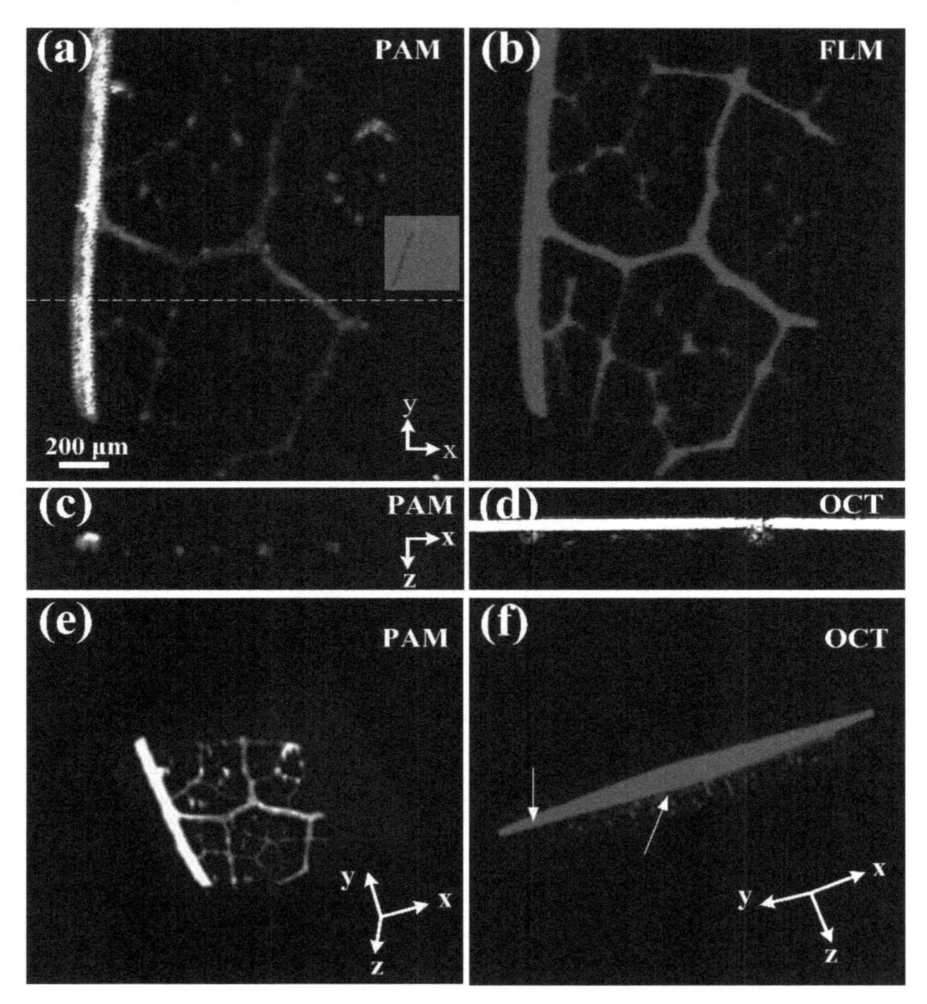

Figure 6.14 Tirmodality imaging of a scattering phantom in which Rhodamine B disseminated leaf veins: MAP image of AOPAM (a) and FLM (b). Cross section image of AOPAM (c) and OCT (d). The 3-D image of AOPAM (e) and OCT (f). Inset: Photography acquired within the imaging region. Reproduced with permission from Ref. [18].

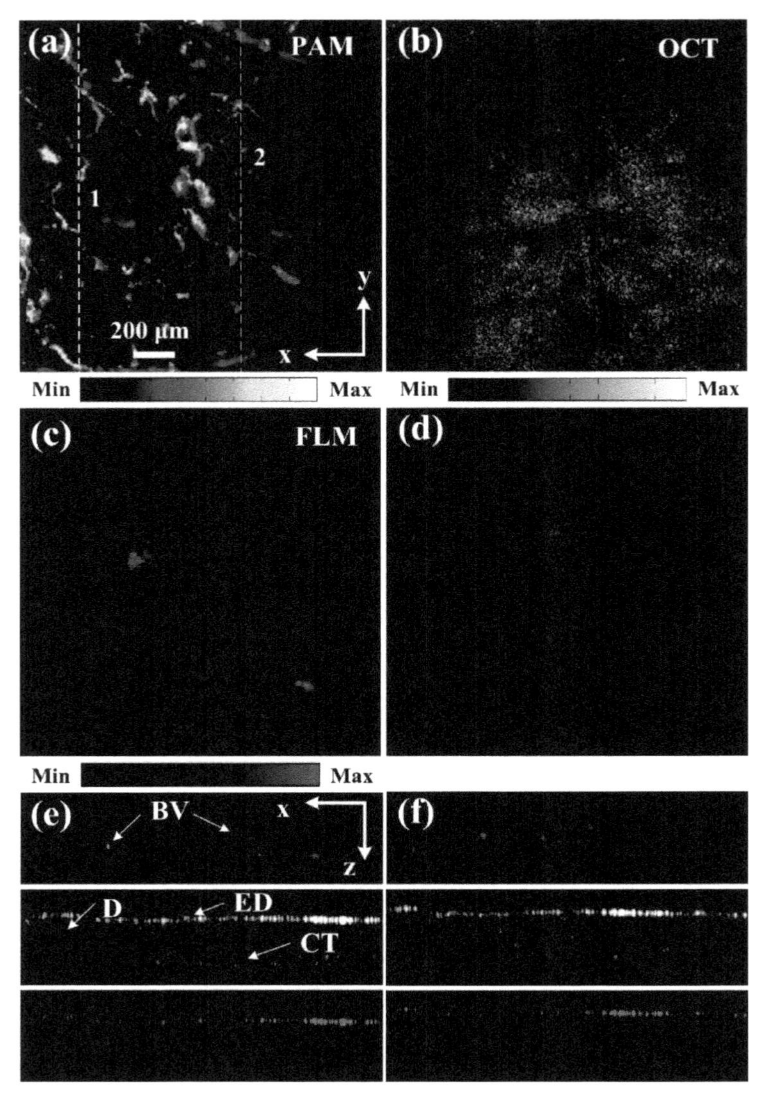

Figure 6.15 In vivo imaging of mouse ear by using the tri-modal imaging system. (a) AOPAM MAP image. (b) OCT MAP image. (c) FLM MAP image. Inset, fluorescence image acquired with fluorescence stereo microscope. (d) Fused AOPAM, OCT and FLM image. (e) AOPAM, OCT and fused AOPAM-OCT B-scan image of the white dotted line 1 in (a). (f) AOPAM, OCT and fused AOPAM-OCT B-scan image of the white dotted line 2 in (a). ED, epidermis; CT, cartilage; D, dermis; BV, blood vessel. Reproduced with permission from Ref. [18].

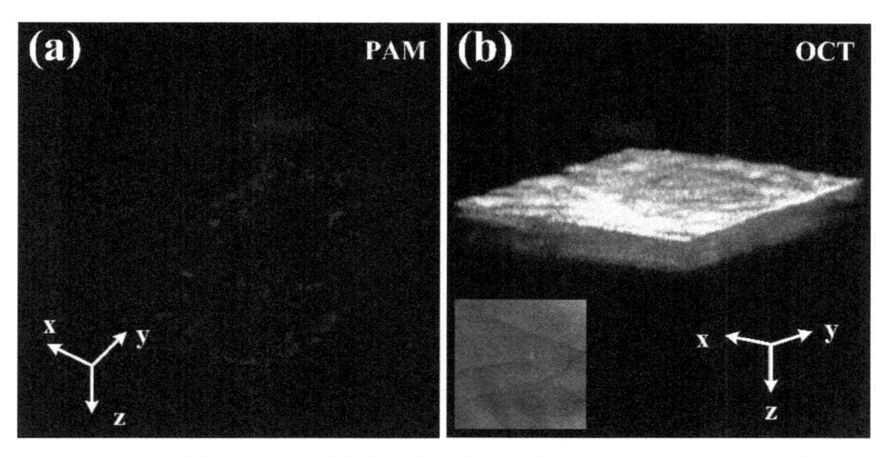

Figure 6.16 3-D image of AOPAM (a) and optical coherence tomography (b). Inset, photography acquired within the imaging region.

Figures 6.15a–c show the MAP images of AOPAM, OCT, and FLM, respectively. OCT, AOPAM, and FLM visualize different tissue components: AOPAM clearly maps the microvasculature; OCT images the tissue surface structure, while FLM indicates the orthotopical injection low-concentration Rhodamine B with high resolution. Figures 6.15e,f are B-scan AOPAM and OCT images corresponding to the white dotted line "1" and "2" in Fig. 6.15a, which shows the benefits of combining these two modalities more clearly. In the cross-sectional OCT, the epidermis, dermis, and cartilage can be easily identified, and the ear's thickness is about 200–300 μm. From the fused images, we can observe that AOPAM is good at locating micro-vessels in the ear with limited surrounding tissue information. Figure 6.15d shows the fused MAP image of AOPAM, OCT, and FLM. The microvasculature structure of the tissue and fluorescence molecular are obtained in one scan. Figures 6.16a,b show the 3D imaging of AOPAM and OCT, respectively. The network of microvasculature and anatomical structure of tissue are clearly shown.

The phantom and in vivo experiments demonstrated that the trimodality imaging system can simultaneously image the absorption, scattering, and fluorescence imaging information of the tissues, which can contribute to promoting the applications of multimodal imaging technology. The changes of the coefficient of absorption and scattering generally represent the pathological changes of tissue. This absorption and scattering information can be used to guide diagnosis and therapy. The presentation of different antibody and fluorescence dyes can also indicate the status of inflammation and cancers, or estimate the effect of drug release and therapy. In addition, this optically integrated trimodality imaging system can be conveniently used to diagnose ophthalmology disease due to its noncontact mode. Currently, because of the limit of exciting light, only a few fluorescence molecular can be used for FL imaging. A supercontinuum laser could be used to induce PA effect and exciting fluorescence. In the next, more functional imaging would be realized, such as hemoglobin oxygen saturation (SO_2), photoacoustic spectroscopy, and multiband OCT system.

The tri-modality imaging system can be easily integrated and is low-cost, which suggests that the AOPAM-OCT-FLM system has a great potential for biomedical applications. So far, the phantoms and in vivo experiments have demonstrated that the tri-modality imaging system has the capacity of imaging microvasculature and fluorescence molecular specifications with high resolution and high contrast. We have developed a novel optically integrated tri-modal AOPAM-OCT-FLM imaging system, which is based on a single Michelson detector and a single exciting light for in vivo imaging of biological tissue. Moreover, AOPAM-OCT shares the same detection system and AOPAM-FLM shares the common exciting laser. Even AOPAM-OCT-FLM shares mostly optical path.

References

1. Wang, L. V. (2009). Multiscale photoacoustic microscopy and computed tomography. *Nat. Photonics*, **9,** pp. 503–509.
2. Ermilov, S. A., Khamapirad, T., Conjusteau, A., Leonard, M. H., Lacewell, R., Mehta, K., Miller, T., and Oraevsky, A. A. (2009). Laser optoacoustic imaging system for detection of breast cancer. *J. Biomed. Opt.*, **5**, p. 024007.
3. Zhang, H. F., Maslov, K., Sivaramakrishnan, M., Stoica, G., and Wang, L. V. (2007). Imaging of hemoglobin oxygen saturation variations in single vessels in vivo using photoacoustic microscopy. *Appl. Phys. Lett.*, 90, p. 053901.

4. Yang, S. H., Xing, D., Zhou, Q., Xiang, L. Z., and Lao, Y. Q. (2007). Functional imaging of cerebrovascular activities in small animals using high-resolution photoacoustic tomography. *Med. Phys.*, **8**, pp. 3294–3301.

5. Chen, Z. J., Yang, S. H., and Xing, D. (2012). In vivo detection of hemoglobin oxygen saturation and carboxyhemoglobin saturation with multiwavelength photoacoustic microscopy. *Opt. Lett.*, **16**, pp. 3414–3416.

6. Rousseau, G., Blouin, A., and Monchalin, J. (2012). Non-contact photoacoustic tomography and ultrasonography for tissue imaging, Biomed. *Opt. Express*, **1**, pp. 16–25.

7. Hou, Y., Kim, J., Ashkenazi, S. S., Huang, S. W., Guo, L. J., and O'Donnell, M. (2007). Broadband all-optical ultrasound transducers. *Appl. Phys. Lett.*, **7**, p. 073507.

8. Jathoul, A. P., Laufer, J., Ogunlade, O., Treeby, B., Cox, B., Zhang, E., Johnson, P., Pizzey, A. R., Philip, B., Marafioti, T., Lythgoe, M. F., Pedley, R. B., Pule, M. A., and Beard, P. C. (2015). Deep in vivo photoacoustic imaging of mammalian tissues using a tyrosinase-based genetic reporter. *Nat. Photonics*, **4**, pp. 239–246.

9. Pouet, B. F., Ing, R. K., Krishnaswamy, S., and Royer, D. (1996). Heterodyne interferometer with two-wave mixing in photorefractive crystals for ultrasound detection on rough surfaces. *Appl. Phys. Lett.*, **25**, pp. 3782–3784.

10. Carp, S. A., Guerra, A., Duque, S. Q., and Venugopalan, V. (2004). Optoacoustic imaging using interferometric measurement of surface displacement. *Appl. Phys. Lett.*, **23**, pp. 5772–5774.

11. Zhang, E., Laufer, J., and Beard, P. (2008). Backward-mode multiwavelength photoacoustic scanner using a planar Fabry–Perot polymer film ultrasound sensor for high-resolution three-dimensional imaging of biological tissues. *Appl. Opt.*, **4**, pp. 561–577.

12. Berer, T., Hochreiner, A., Zamiri, S., and Burgholzer, P. (2010). Remote photoacoustic imaging on solid material using a two-wave mixing interferometer. *Opt. Lett.*, **24**, pp. 4151–4153.

13. Paltauf, G., Nuster, R., Haltmeier, M., and Burgholzer, P. (2007). Photoacoustic tomography using a Mach–Zehnder interferometer as an acoustic line detector. *Appl. Opt.*, **16**, pp. 3352–3358.

14. Wang, Yi., Li, C. H., and Wang, R. K. (2011). Noncontact photoacoustic imaging achieved by using a low-coherence interferometer as the acoustic detector. *Opt. Lett.*, **20**, pp. 3975–3977.

15. Blatter, C., Grajciar, B. Zou, P., Wieser, W., A., Verhoef, J., Huber, R., and Leitgeb, R. A. (2012). Intrasweep phase-sensitive optical coherence tomography for noncontact optical photoacoustic imaging. *Opt. Lett.*, **21**, pp. 4368–4340.
16. Chen, Z., Yang, S., Wang, Y., and Xing, D. (2015). Noncontact broadband all-optical photoacoustic microscopy based on a low-coherence interferometer. *Appl. Phys. Lett.*, **4**, p. 043701.
17. Chen, Z., Yang, S., Wang, Y., and Xing, D. (2015). All-optically integrated photo-acoustic microscopy and optical coherence tomography based on a single Michelson detector. *Opt. Lett.*, **12**, pp. 2838–2841.
18. Chen, Z., Yang, S., and Xing, D. (2016). Optically integrated trimodality imaging system: Combined all-optical photoacoustic microscopy, optical coherence tomography, and fluorescence imaging. *Opt. Lett.*, **7**, pp. 1636–1639.
19. Zhou, W., and Chen, Z. (2017). Noncontact all-optical photoacoustic microscopy based on optical coherent detection. *Acta Laser Biol. Sin.*, **25**, pp. 77–82.
20. Wang, L. V., and Hu, S. (2012). Photoacoustic tomography: In vivo imaging from organelles to organs. *Science*, **6075**, p. 1458.
21. Laser Institute of America, (2007). American National Standard for Safe Use of Lasers, ANSI Z136.1-2007 (American National Standards Institute.
22. Omar, M., Soliman, D., Gateau, J., and Ntziachristos, V. (2014). Ultrawideband reflection-mode optoacoustic mesoscopy. *Opt. Lett.*, **13**, pp. 3911–3914.
23. Hai, P., Yao, J., Maslov, K. I., Zhou, Y., and Wang, L. V. (2014). Near-infrared optical-resolution photoacoustic microscopy. *Opt. Lett.*, **17**, pp. 5192–5195.
24. Sinclair, H., Bourantas, C., Bagnall, A., Mintz, G. S., and Kunadian, V. (2015). OCT for the identification of vulnerable plaque in acute coronary syndrome. *J. Am. Coll. Cardiol. Img.*, **2**, pp. 198–209.
25. Vakoc, B. J., Fukumura, D., Jain, R. K., and Bouma, B. E. (2012). Cancer imaging by optical coherence tomography: Preclinical progress and clinical potential. *Nat. Rev. Cancer*, **5**, pp. 363–368.
26. Zhang, J., Yang, S., Ji, X., Zhou, Q., and Xing, D. (2014). Characterization of lipid-rich aortic plaques by intravascular photoacoustic tomography. *J. Am. Coll. Cardiol.*, **4**, pp. 385–390.

27. Dai, X., Xi, L., Duan, C., Yang, H., Xie, H., and Jiang, H. (2015). Miniature probe integrating optical-resolution photoacoustic microscopy, optical coherence tomography, and ultrasound imaging: Proof-of-concept. *Opt. Lett.*, **12**, pp. 2921–2924.

28. Liang, S., Ma, T., Jing, J., Li, X., Li, J., Shung, K. K., Zhou, Q., Zhang, J., and Chen, Z. (2014). Trimodality imaging system and intravascular endoscopic probe: Combined optical coherence tomography, fluorescence imaging and ultrasound imaging. *Opt. Lett.*, **23**, pp. 6652–6655.

Chapter 7

Nanoprobes as Contrast Agents for Biomedical Photoacoustic Imaging

Currently, fluorescence imaging (FLI), magnetic resonance imaging (MRI), positron emission tomography (PET), X-ray computed tomography (CT), and ultrasound imaging (USI) have achieved great progress in biomedical imaging. These modalities have shown great capabilities and have been extensively used in disease diagnosis and prognosis. However, they suffer from certain limitations such as system complexity and temporal resolution in MRI, limited imaging depth in optical imaging, or limited spatial and temporal resolution in PET and SPECT. Therefore, there is still urgent need for the further development of robust imaging modalities, capable of addressing these limitations and reducing system complexity and implementation cost.

Photoacoustic imaging (PAI) as a rapidly emerging biomedical imaging modality has many advantages, such as high spatial resolution (0.1~100 μm), ideal depth for clinical application, without ionizing radiation, in particular the capability of indicating the functional and molecular information in real-time. These advantages, which make PAI a suitable diagnostic modality for clinical applications, will have the ability to provide molecular information at clinically relevant depths with a high resolution and in real time [1–9]. Moreover, it can be easily combined with

Biomedical Photoacoustics
Sihua Yang and Da Xing

ISBN 978-981-4774-58-1 (Hardcover), 978-0-203-70365-6 (eBook)
www.jennystanford.com

ultrasound (US) imaging because both imaging modalities have a common signal detection regimen and can share some hardware components. Therefore, it is possible to simultaneously obtain more comprehensive information on anatomical, functional, and molecular content of diseased tissues through the combination of US and PAI.

7.1 Introduction

Nowadays, PAI using the endogenous contrast agents has been widely studied and shown great potential in a variety of applications. The inherent contrast of PAI is largely determined by optical absorption coefficient of various endogenous chromophores, including hemoglobin [10–12], melanin [13–15], lipid [16–19], and myoglobin [20]. The optical absorption of endogenous chromophores is presented in Fig. 7.1.

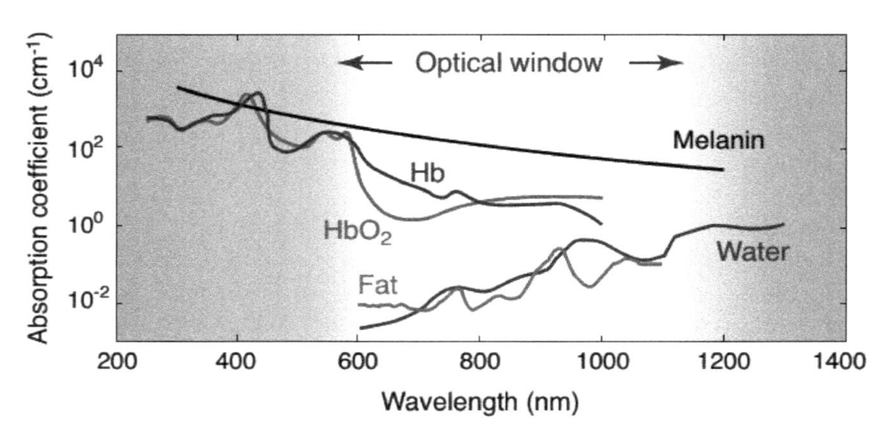

Figure 7.1 Absorption spectra of endogenous chromophores in the body [4]. The optical absorption of these endogenous chromophores is wavelength dependent; therefore, the PA signal intensity at different optical wavelengths can be used to characterize optical properties of tissue. Data for the absorption coefficient were obtained from http://omlc.ogi.edu/spectra/. The "optical window" (600–1100 nm) is the wavelength range in which tissue absorption is at a minimum. Reproduced with permission from Ref. [4].

Thus, in the previous reports, the preclinical study of PAI has shown tremendous potential in diagnosing associated diseases, such as ischemia [21], vascular disorders [22–25], cardiovascular diseases [16, 18], stroke [26], diabetes [27], epileptic seizures [28], and cancer [14, 29–32]. The endogenous chromophore-mediated PAI has its own advantages. For example, the endogenous contrast agents are inherently biologically safe. Moreover, the physiological changes of endogenous contrast agents can in real time indicate physiological and metabolic changes that differentiate pathological tissue from normal tissue, such as oxygen saturation and vascular blood volume in the body [10, 33, 34]. However, it is often difficult to precisely detect many diseases with PAI, because many diseases and physiological processes usually just provide little or insufficient endogenous contrast agents. Therefore, it is highly essential and urgently needed to develop reliable exogenous contrast agents for fully utilizing PAI potentials, as the introduction of exogenous contrast agents can greatly enhance imaging contrast and further enable cellular and molecular PAI.

7.2 Nanoprobes as Contrast Agents for PAI

Introduction of exogenous contrast agents, which can improve the imaging contrast and resolution by changing the acoustic and optical properties of the local tissue, will greatly expand PAI applications. With the development of materials science and nanotechnology, a myriad of advances have been made in terms of exogenous contrast agent development for PA molecular imaging applications, particularly nanoparticle based technologies. First, since the PA signal depends on the optical absorption of the photoabsorbers, the nanoparticles are usually designed with peak absorption in the near-infrared (NIR) region from 700 to 1100 nm, where the optical absorption of endogenous tissue is minimum. Thus, the optical attenuation of tissue is relatively low. Therefore, it is possible to increase the imaging contrast and depth by tuning the optical absorption of nanoparticles to fall in the "tissue optical window." Second, targeting molecules such as peptides, antibodies, and aptamers can be conjugated to the nanoparticle surface. Moreover, nanoparticles are small enough, which vary in diameter from 1 nm to several hundred nanometers. Owing to small structure combined with the targeting property, nanoparticles can extravasate from the vasculature, get close to and ultimately bind to specific cells or molecules, and thereby visualize events at the cellular and molecular levels through PAI [35–39]. Furthermore, nanoparticles with different composition and various shapes make it possible to adjust their absorption peak according to the required imaging depth, desired light source to be used, and delivery mechanism of nanostructures. Last, PAI can be easily combined with photoacoustic therapy as PAI-guided tumor therapeutic technique, because both techniques are based on the photoacoustic effect [40–43].

Emerging nanoprobes as contrast agents include gold nanomaterials, carbon nanomaterials, dye-related nanomaterials, and some other related nanomaterials. A great variety of PA contrast agents have been reported; however, there is no single best agent for all applications (Table 7.1).

Table 7.1 Various contrast agents explored in PAI

Materials	Contrast agents	Advantages	Disadvantages
Gold nanomaterials	Gold nanocages [44, 45]; gold nanorods [41, 42, 46–50]; gold nanostars [51, 52]; gold nanoshells [53, 54]; gold nanospheres [55]; gold nanoflowers [56, 57]; gold nanoplates [58]	Tunable physicochemical properties; theoretically safe and stable components; high photoacoustic conversion efficiency	Non-biodegradability; suboptimal photothermal stability
Carbon nanomaterials	Carbon nanotubes [59–63]; graphenes [40, 64, 65]; carbon dots [66, 67]	Large specific surface area; good photothermal stability	Non-biodegradability; heterogeneity
Dye-related nanomaterials	ICG [68, 69]; IR780 [70]; IR825 [71], etc.	Good biocompatibility and biodegradability	Easy photobleaching; short bloodstream circulation half-life

Materials	Contrast agents	Advantages	Disadvantages
Transition metal chalcogenides	CuS [72–75]; WS2 [76]; MoS2 [77]; FeS [78], etc.	High photoacoustic conversion efficiency; good photothermal stability;	Non-biodegradability; potential toxicity of heavy metals
Other related nanomaterials	Perfluorocarbon (PFC) nanodroplets [43, 79, 80]; Organic polymer-related nanoparticles [81–86]; reporter genes [87–89], etc.	Good biocompatibility; integrating multimodal imaging	Biodegradation behaviors remain unknown; potential toxicity

7.2.1 Dye-Related Nanoprobes

Dyes are the first class of PAI contrast agents. Most of them are typically small molecules at the nanometer scale and can be readily cleared by the renal system. Many of these dyes with capacity of emission fluorescence are commonly used in fluorescence imaging. Currently, it has been reported that many biocompatible dyes that absorb in the optical window, including indocyanine green (ICG) [68, 69], IRDye800CW [70], Alexa Fluor 750 [71], and methylene blue [90], have been used in PA imaging. Compared with inorganic nanomaterials, these dyes show good biocompatibility and biodegradability. Only the absorption of dyes contributes to the photoacoustic signal. Therefore, for fluorescent dyes, a lower quantum yield will result in more efficient PA signal generation, because it means that more of the absorbed energy is converted into the photoacoustic signal instead of being emitted by fluorescence.

7.2.1.1 Indocyanine green

Indocyanine green is an FDA-approved fluorescent contrast agent that can be used for both fluorescence and photoacoustic imaging. Its absorption spectrum is primary in the range of 600–900 nm and fluorescence emission spectrum is from 750–950 nm. On mass basis, ICG at 780 nm is ~7 times more absorbing than single-wall carbon nanotubes (SWNTs) and ~8500 times more absorbing than gold nanorods (GNRs) [91]. However, free ICG molecules have a half-life of only several minutes after intravenous injection into body [92, 93], which greatly limits the in vivo applications. To provide a more effective imaging system, ICG is usually conjugated with or encapsulated into nanoparticles to deliver ICG molecules to the disease site more efficiently [69, 79, 91, 94, 95]. Nanoparticles delivery ICG will significantly increase the circulation time in blood, allowing for an extended imaging period. There are several other advantages through dyes loading in nanoparticles. First, the surface of nanovehicles can be easily functionalized for specific purposes, for example, a targeting moiety can be conjugated for selective PAI. Second, a large number of ICG loaded on the nanovehicles will greatly increase the concentration of delivery dyes and enhance the photoacoustic signal intensity and thus improve imaging contrast. For example, ICG can covalently conjugate to SWNTs to deliver a higher amount of ICG molecules to the disease site [91]. As shown in Fig. 7.2a, ICG-loaded SWNTs showed 20 times higher optical absorption than plain SWNT at the peak absorption wavelength, 780 nm. After the SWNT-ICG functionalized with RGD as targeting moiety, the nanocomposites of SWNTs-ICG-RGD greatly improve the PAI contrast of targeted cells (Fig. 7.2b,c).

Finally, nanoparticles loading ICG will protect dye molecules against the effects of an aqueous medium and other destabilizing effects from the biological environment improving its stability and blood circulation time. The nanoparticles (ICG-PL-PEG) consisting of ICG and phospholipid–polyethylene glycol (PL–PEG) have been developed for enhanced PAI [69]. It inherits the strong absorption of ICG in the NIR region while show spherical and well dispersed, with a core size of approximately 18 nm (Fig. 7.3). Notably, the components of these nanoparticles, ICG and PEG, have been approved for human use. Therefore, ICG-PL-PEG nanoparticles are biocompatible and relatively nontoxic. These nanoparticles have overcome numerous limitations of ICG, such as poor aqueous stability, concentration-dependent aggregation and lack of target specifity, showing the potential for clinical application.

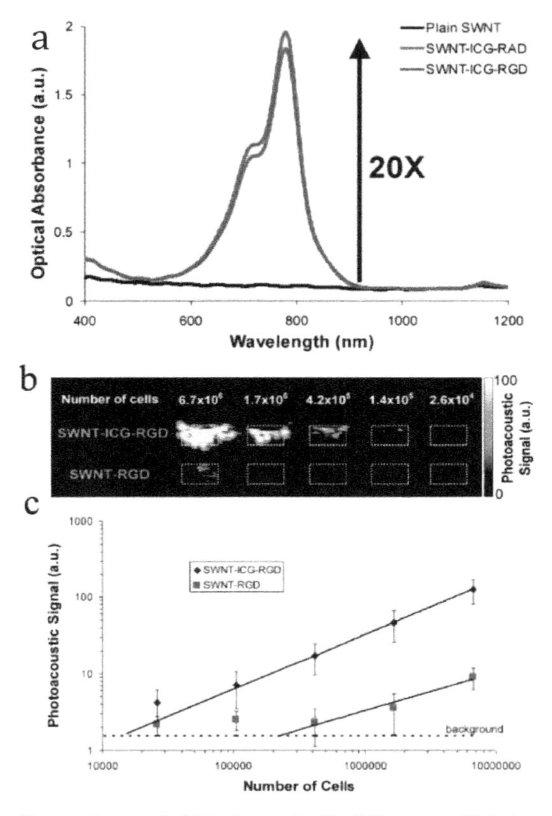

Figure 7.2 Absorption of ICG loaded SWNTs and PAI in vitro [91]. (a) Optical spectra of plain SWNT (black), SWNT-ICG-RGD (blue), and SWNT-ICG-RAD (red). ICG dye-enhanced SWNTs particles showed 20 times higher optical absorption than plain SWNT at the peak absorption wavelength, 780 nm. The similarity of SWNT-ICG-RAD and SWNT-ICG-RGD spectra suggests that the peptide conjugation does not notably perturb the photoacoustic signal. (b) Photoacoustic vertical slice image through an agarose phantom containing decreasing number of U87 cancer cells exposed to SWNT-ICG-RGD and plain SWNT-RGD particles. While 1.7×10^6 cells exposed to SWNT-RGD are barely seen on the image, a clear photoacoustic signal was observed from 1.4×10^5 cells exposed to SWNT-ICG-RGD. The signal inside the ROI (dotted white boxes) is not homogeneous due to possible aggregates of cells. (c) Quantitative analysis of the photoacoustic signals from the phantom (n = 3) showed that SWNT-ICG-RGD can visualize 20 times less cancer cells than SWNT-RGD can ($p < 0.0001$). The background line represents the average background signal in the phantom. Linear regression was calculated on the linear regime of both curves. Reproduced with permission from Ref. [91].

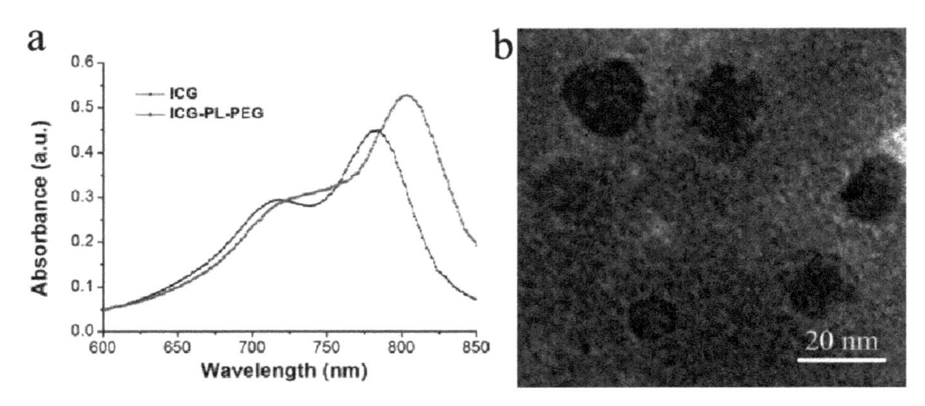

Figure 7.3 Characteristics of nanoparticles loading ICG [69]. (a) Absorption spectra of ICG and ICG-PL-PEG nanoparticles, respectively. (b) TEM imaging of ICG-PL-PEG nanoparticles. Reproduced with permission from Ref. [69].

7.2.1.2 Other dyes

Other dyes as effective contrast agents, such as IRDye800CW, an ICG derivative, can label on the protein/antibody after adding the NHS ester reactive group. In an in vivo study, IRDye800 is conjugated with cyclic peptide cyclo(Lys-Arg-Gly-Asp-Phe) (cKRGDf) that targets integrin $\alpha v\beta 3$ overexpressed by neovessels in U87MG glioblastoma tumor model, and PAI contrast of tumor region is greatly enhanced due to the accumulation of IRDye800 in tumor [34].

Another dye is Alexa Fluor 750, a common near-IR fluoescent dye. Razansky et al. showed that in small animals, Alexa Fluor 750 as contrast agent of multispectral PAI can be imaged with 25 fmol sensitivity and 150 μm spatial resolution [96]. Methylene blue (MB) is a water-soluble dye and its light absorption peak is around 670 nm. It has been used for sentinel lymph node (SLN) mapping with PAI in a rat model [90]. It should be noted that many other dyes have been used for PAI, including Evans blue (EB) [97], BHQ3, QXL680 [98], and MMPSense™ 680 [99].

Many dyes usually have excellent biosafety properties and have been approved by the FDA for human use. In addition, organic dyes have strong light absorption capacity and more efficient photoacoustic conversion efficiency. However, most organic dyes easily suffer from severe photobleaching and are susceptible to environmental factors, leading to irreproducible signals and unpredictable phenomena.

7.2.2 Gold-Based Nanoprobes

Gold-based nanoprobes, including gold nanospheres [55], gold nanorods [41, 42, 46–50], gold nanoshells [53, 54], gold nanocages [44, 45], gold nanostars [51, 52], gold nanoflowers [56, 57], and gold nanoplates [58] have attracted much attention in exogenous-mediated PAI because of their excellent physicochemical properties and photoacoustic characteristics. Their optical absorption is based on the surface plasmon resonance (SPR), in which free charges on the surface of noble metal nanoparticles oscillate in concert with the electromagnetic field, showing its optical absorption to be hundred times higher than SWNTs [42] and five orders of magnitude greater than dyes (on a per-particle basis) [100]. Moreover, the optical absorption properties can be tuned by controlling the size and morphology, to match its absorption in the NIR region. Furthermore, gold-based nanoprobes also show relative inertness, outstanding stability, and resistance to chemical/thermal denaturation, and photobleaching, which always compromise the longevity of NIR dyes.

7.2.2.1 Gold nanospheres

Gold nanospheres can be synthesized by a simple citrate synthesis method [101] to form a spherical shape. Their peek absorption is about 520 nm, which is not in the NIR region and the absorption of hemoglobin in this region is also very strong. It makes gold nanospheres difficult to apply for deep tissue imaging and distinguish nanoprobes from blood in vivo.

7.2.2.2 Gold nanorods

Among the various exogenous contrast agents for PAI, GNRs (Fig. 7.4b) have shown excellent photophysical properties and thus have been intensively explored for PAI in the past several years. First, GNRs are easy to synthesize by a seed-mediated growth and tuning of the absorption peak in NIR by manipulating the aspect ratio, which have led to multiplexing applications [102]. Second, GNRs possess strong light absorption characteristics. It has been reported that at the same molar concentration, the optical absorbance of GNRs is about 160 times and 5000 times higher than SWNTs and ICG, respectively (Fig. 7.4c) [103, 104]. Therefore, GNRs exhibit stronger photoacoustic signal than SWNTs and ICG (Fig. 7.4d). Last, the surface of GNRs can be easily modified with various functional molecules, such as antibodies, peptides, and nucleic acid for enhancing biocompatibility, targeting to the spot of interest (Fig. 7.4a) [105].

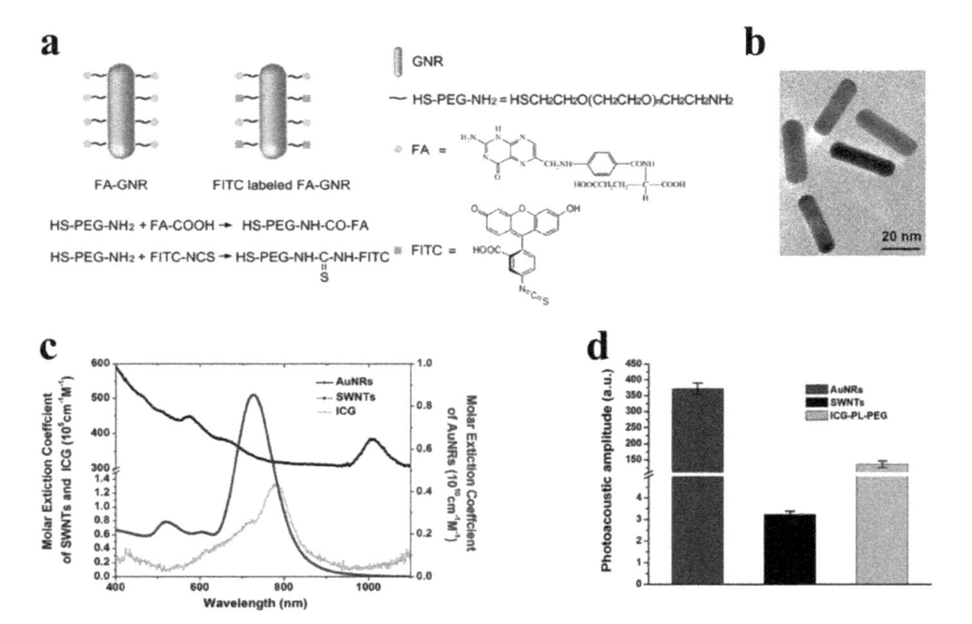

Figure 7.4 Characteristics of gold nanorods [42]. (a) Schematic drawing of functionalized AuNRs with amine-PEG-thiol (HS-PEG-NH2), folic acid (FA) and fluorescein isothiocyanate (FITC). (b) Transmission electron microscopy (TEM) imaging of AuNRs. (b) Extinction spectra of AuNRs, SWNTs and ICG. The vertical scale to the left is for SWNTs and ICG, while that to the right is for AuNRs. (c) Photoacoustic amplitudes of the nanoparticles based on the same molar concentrations (80 nM). The error bars indicate the standard deviation for each measurement (n = 5). Reproduced with permission from Ref. [42].

7.2.2.3 Gold nanoshells

Gold nanoshells consist of a spherical silica core surrounded by a thin layer of gold. Their morphology is shown in Fig. 7.5a. The absorption peak of the gold nanoshells in the NIR region can be tuned by changing the silica core size relative to the thickness of the gold shell. However, their absorption spectrum shows a broad optical spectrum and more significant scattering, making them less effective PAI contrast agents than gold nanorods (Fig. 7.5b) [54].

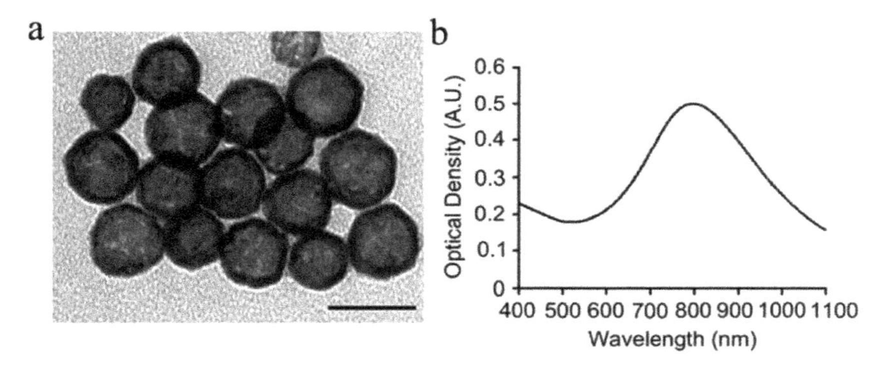

Figure 7.5 Characteristics of gold nanoshells [53]. (a) TEM image of PEG-HAuNS (Bar = 50 nm). The average outer diameter was 45 nm, and thickness of the shell was 2.5 nm. (b) Experimental absorbance spectrum of PEGHAuNS in water, which peaked at 800 nm. Reproduced with permission from Ref. [53].

7.2.2.4 Gold nanocages

Gold nanocages (GNCs) with cubical symmetry are hollow and porous. The common shape and size are shown in Figs. 7.6a–d. The surface morphology and size can be tuned by different Ag nanocubes bearing sharp corners and added amount of $HAuCl_4$ [45]. Benefiting from the strong optical absorption, hollow interior and porous structure, GNCs can also be designed as a NIR controlled-release drug delivery vehicle by coating with thermally responsive polymers on the nanocage surface [106]. In addition to their compact sizes (<50 nm), adjustable compositional and morphological properties, the absorption of GNCs based on localized SPR also can be precisely tuned throughout the visible and NIR regions (Fig. 7.6f) [107].

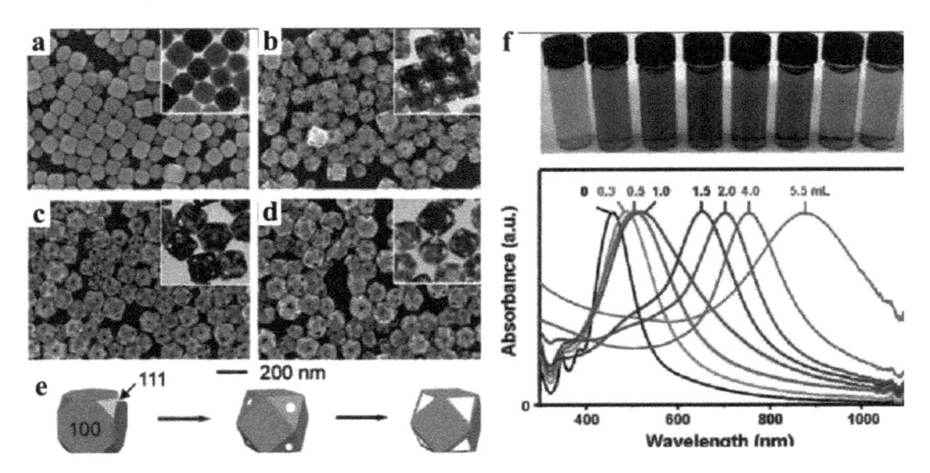

Figure 7.6 Characteristics of gold nanocages [45]. (a) Ag nanocubes with rounded corners and (b–d) product after reaction with 0.6, 1.6, and 3.0 mL of 0.1 mM $HAuCl_4$ solution, respectively. (e) Illustration summarizing morphological changes. Coloration indicates conversion of a Ag nanocube into a Au/Ag nanocage then a predominately Au nanocage. (f) Top panel, vials containing Au nanocages prepared by reacting 5 mL of a 0.2 nM Ag nanocube (edge length ≈ 40 nm) suspension with different volumes of a 0.1 mM $HAuCl_4$ solution. Lower panel, the corresponding UV-visible absorbance spectra of Ag nanocubes and Au nanocages. Reproduced with permission from Ref. [45].

7.2.2.5 Other gold-based nanoprobes

Gold nanostars (GNS) with multi-branch shape (Fig. 7.7a) have strong optical absorption in the NIR region because of their great absorption-to-scattering ratio (Fig. 7.7b) [51]. Their absorption cross section can be orders of magnitude higher than dye molecules because of a high surface-to-volume ratio. The detection limit of the photoacoustic signals of GNS in tissue phantoms was nearly at a concentration of 1 ppm Au (λ_{ex} = 767 nm; Fig. 7.7c).

Wang et al. used it for monitoring the accumulation in the sentinel lymph nodes (SLNs) of a rat with spectroscopic PAI [51]. Furthermore, the large particle surface of GNS can be modified with many functional molecules such as antibodies and ligands. The cyclic arginine-glycine-aspartic acid (RGD) peptides, which target integrin $\alpha v\beta_3$ overexpressed by neovessels in tumors, are linked on the surface of GNS, it as a novel theranostic platform on hemispherical photoacoustic imaging (HPAI) to volumetrically map the tumor angiogenesis quantitatively, suppress tumor vascularization by laser irradiation, and monitor the treatment response simultaneously [52].

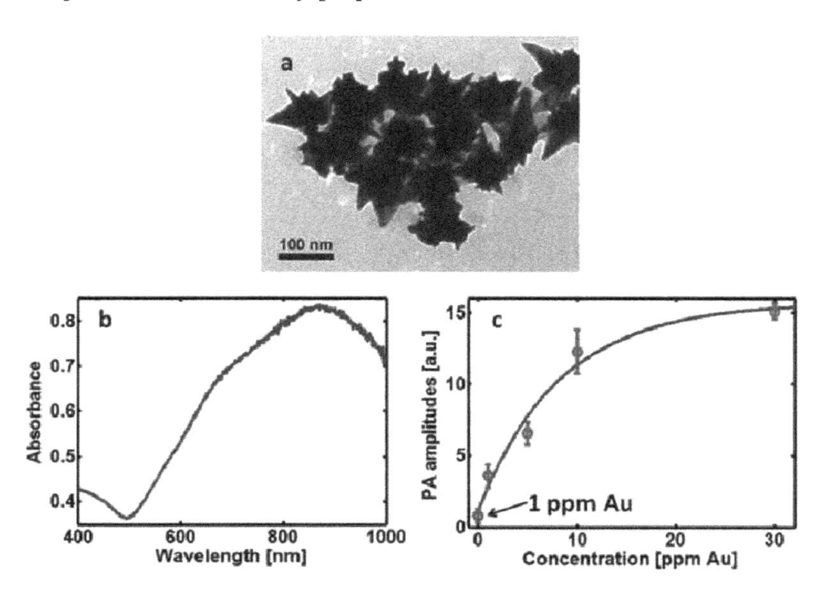

Figure 7.7 Characteristics of gold nanostars [51]. (a) TEM image of plasmon-resonant Au nanostars (NSTs, sample A). (b) Extinction spectrum of diluted sample A. The absorbance (arbitrary unit) of the Au nanostars is 0.75 at 767 nm. (c) A plot of photoacoustic (PA) amplitudes as a function of NST concentration. Reproduced with permission from Ref. [51].

Other gold-based nanoprobes such as anisotropic and branched gold nanostructures (gold nanotripods) [108] and nanoflowers [56, 57] also have been developed for PAI. However, during PAI, some gold nanomaterials are exposed to high-energy nanosecond laser pulses. They absorb the light energy and instantaneously generate substantial heat, which leads to the melting and reshaping of nanoparticles and an associated absorption reduction in NIR region [109–111]. It makes them more difficult to detect and limited the application of gold nanomaterials for persistent PAI. Gold materials' stability can be enhanced by coating them with a thin layer of silica [58]. This layer not only maintains their shape but also enhances the photoacoustic signal's amplitude [110, 111]. Moreover, silica-coated GNRs were reported to increase the uptake of gold into the cell more than fivefold [112]. A layer of poly(ethylene glycol) (PEG) is commonly modified to the surface, which can increase circulation time and decrease cytotoxicity of gold nanomaterials. Furthermore, PEG helps to stabilize nanoparticles and prevents aggregation [113]. In addition, the surface of PEG is easily linked with a variety of targeting molecular to expand the application of photoacoustic molecular imaging. As treatment for many diseases, most notably cancer, is increasingly focused on specific molecular interactions, the ability to visualize these interactions becomes a necessity [3].

7.2.3 Carbon Nanoparticles

Carbon-based nanomaterials, including carbon nanotubes [59–63], graphenes [40, 64, 65], fullerenes [114, 115], and carbon dots [66, 67], have been extensively used for pharmaceutical, biomedical, and bioimaging applications. They usually show strong absorption over a broad spectrum from UV to NIR regions. Moreover, many compounds/materials are easy to integrate with carbon-based nanomaterials for PAI and other functions due to its special surface carbon structure. In addition, carbon nanomaterials are stable against photobleaching and have low adverse effects for the whole organism [116, 117].

7.2.3.1 Carbon nanotubes

Carbon nanotubes, including SWNTs and multi-walled carbon nanotubes (MWNTs), exhibit a broad absorption spectrum. Their strong absorbance in the NIR region makes them suitable as a contrast agent for PA imaging. SWNTs can be synthesized as small as 1 nm diameter and with various aspect ratios to increase both absorption cross section and effective surface area for biomodification. PEG can be functionalized on the SWNTs surface to improve their dispersion and stability in aqueous, increase the circulation time in blood and reduce side effects. Moreover, targeting molecules such as antibody, folic acid, and RGD can be further modified on the SWNT-PEG [38, 118, 119]. RGD peptides were functionalized on the SWNTs through phospholipid-PEG (PL-PEG5000) to specific target tumor integrins, as depicted in Fig. 7.8a. SWNT–RGD as ultrasound imaging and PAI contrast agent were intravenously injected into U87MG tumor mice. 3D US and PA images of the tumor region were acquired at pre- and 4 h post-injection. As shown in Fig. 7.8b, the PA signal of tumor is greatly enhanced after injecting with SWNT–RGD, while the signal of plain SWNT injection group remains at almost plateau levels, indicating the targeting capability of SWNT–RGD and accumulation in the tumor.

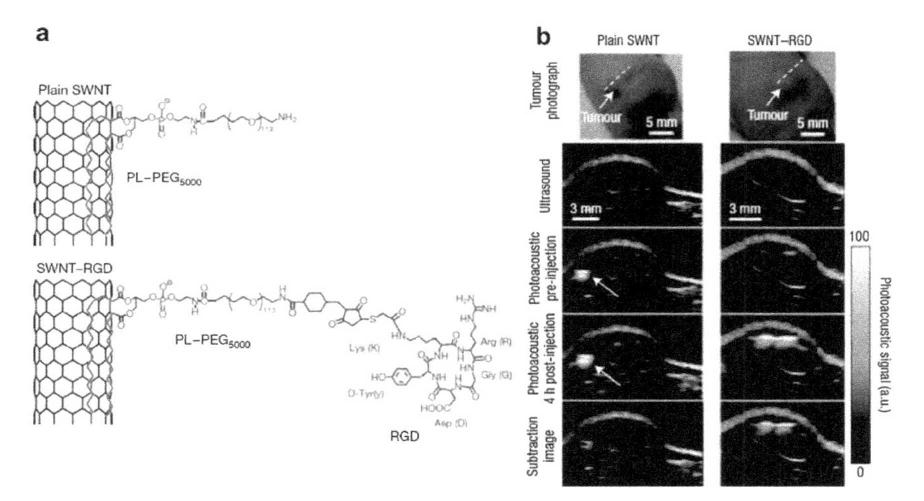

Figure 7.8 Characterization of the photoacoustic properties of single-walled carbon nanotubes [38]. (a) Illustration of plain single-walled carbon nanotubes (plain SWNT) and SWNT-RGD. The phospholipid binds to the side wall of the single-walled carbon nanotubes connecting the PEG5000 to the nanotubes. The RGD allows the single-walled carbon nanotubes to bind to tumor integrins such as $\alpha v\beta_3$. (b) B-scan US and PA images of U87MG tumor acquired along a white dotted line aided by SWNTs The US images (gray) show the skin and tumor boundaries, while PAT images (green) show optical absorption (SWNT–RGD) in the tumor. Differential images were obtained by subtraction of the pre-injection image from the 4 h post-injection image. Reproduced with permission from Ref. [38].

However, the absorption of plain SWNT at NIR wavelengths is relatively low, limiting their in vivo application. Gold-plated carbon nanotubes, termed golden carbon nanotubes (GNTs), are developed as photoacoustic and photothermal contrast agents with enhanced near-infrared contrast (~102-fold) for targeting lymphatic vessels in mice using extremely low laser fluence levels of a few mJ cm^{-2} [120]. A shortened SWNT core with a diameter of 1.5–2 nm is coated by a 4–8 nm thin gold layer. As shown in Fig. 7.9a, at the same concentration, the absorption of GNTs in the NIR region is significantly higher (85–150-fold) than the shortened SWNTs. The excellent light absorption capability makes its PA signal significantly enhanced. Figure 7.9b shows that the PA signals of GNTs are higher than the same concentration of carbon nanotubes and gold nanospheres (50–100-fold) and comparable to those of gold nanorods and nanoshells. After conjugated with antibody, the targeting GNTs can used to map the lymphatic endothelial receptor with PAI.

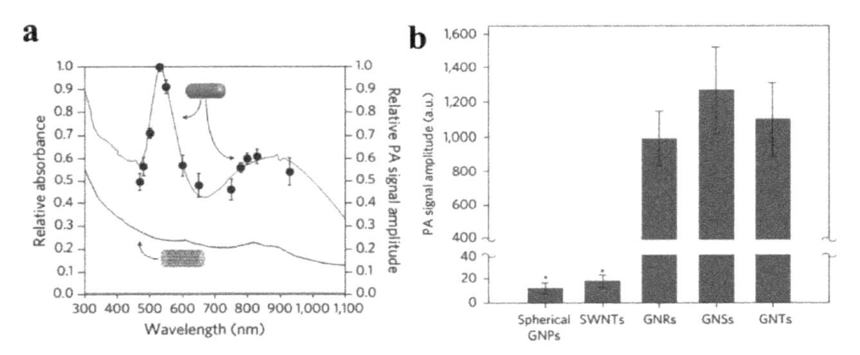

Figure 7.9 Absorption and photoacoustic properties of GNTs [120]. (a) Normalized optical spectra (left vertical axis) of GNTs in ddH_2O (red line), shortened single-walled carbon nanotubes in ddH_2O (black line) and ddH_2O only (green line) and normalized PA signal amplitudes (right vertical axis) of GNTs in ddH_2O (blue circles) at different laser wavelengths. The concentration of carbon nanotubes was 35 times higher than that of GNTs. (b) PA signal amplitudes of spherical gold nanoparticles (GNPs, 40 nm in diameter), single-walled carbon nanotubes (SWNTs, 1.7 × 186 nm), gold nanorods (GNRs, 15 × 52 nm), gold nanoshells (GNSs, 140 nm silica core with 8 nm gold shell), and GNTs (11 × 99 nm) in ddH_2O at a concentration of 1 × 10^{11} particles ml^{-1} and under a single laser pulse. Laser fluence was ~100 mJ cm^{-2}. Laser wavelengths were 532 nm for spherical GNPs and 850 nm for the remainder of the nanoparticles. The error bars represent the standard deviation in ten measurements. *$P < 0.05$, compared to GNTs. Reproduced with permission from Ref. [120].

Another strategy to improve the PAI sensitivity and contrast is the development of dye-enhanced SWNTs [121]. Many optical dyes, including ICG, QSY21, and methylene blue (MB), and cyanine dyes such as Cy5.5, and melanin have been successfully loaded on the SWNT surface (Fig. 7.10a). The optical absorption (Fig. 7.10b) and PA signal of dye-loaded SWNTs is remarkably enhanced (Fig. 7.10c,d). This strategy provides a reference for extending the application of carbon nanotubes and dye molecules in PAI.

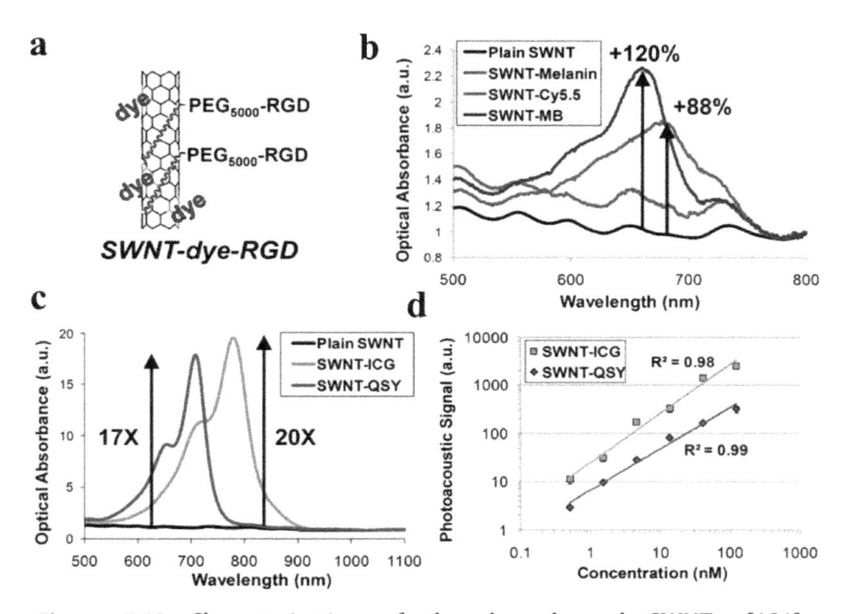

Figure 7.10 Characterization of the dye-enhanced SWNTs [121]. (a) Illustration of a SWNT-dye-RGD particle. The dye (red) is attached to the SWNT surface through noncovalent π–π stacking bonds. Phospholipid-polyethylene glycol-5000 (blue) is conjugated to the targeting peptide (RGD) on one end and to the SWNT surface on the other end. (b, c) Optical spectra of (b) plain SWNT (black), SWNT-melanin (purple), SWNT-Cy5.5 (brown), SWNT-MB (blue), (c) SWNT-QSY (red), and SWNT-ICG (green). The QSY and ICG dye-enhanced SWNT particles showed 17- and 20-times higher optical absorption than plain SWNTs at their peak absorption wavelengths, 710 and 780 nm, respectively. (d) The photoacoustic signal produced by SWNT-QSY and SWNT-ICG was observed to be linearly dependent on the particles' concentration (R^2 = 0.99 and 0.98, respectively). Reproduced with permission from Ref. [121].

7.2.3.2 Graphene

Graphene is a 2D free-standing honeycomb lattice made of a single atomic plane of graphite. Graphene, with larger surface area and lower aspect ratio compared to SWNTs, is usually oxygenated to graphene oxide (GO). The relative large GO surface can be easily conjugated with a large number of functional molecules or nanoparticles. Plain graphene and GO also only have a relative low absorption coefficient in the NIR region. Therefore, to further improve the imaging contrast, GO is usually conjugated with a strong absorbing dye such as ICG or superparamagnetic NP for enhancing absorption and achieving multi-modal imaging [40, 122]. GO can be used as a vehicle for loading dyes for quenching dye fluorescence via fluorescence resonance energy transfer (FRET), thus maximizing the PA efficiency of dyes. The absorbed light energy of fluorescent dyes mostly decays in the same radioactive way as fluorescence, and only a small part of energy is converted into thermal expansion (Fig. 7.11a).

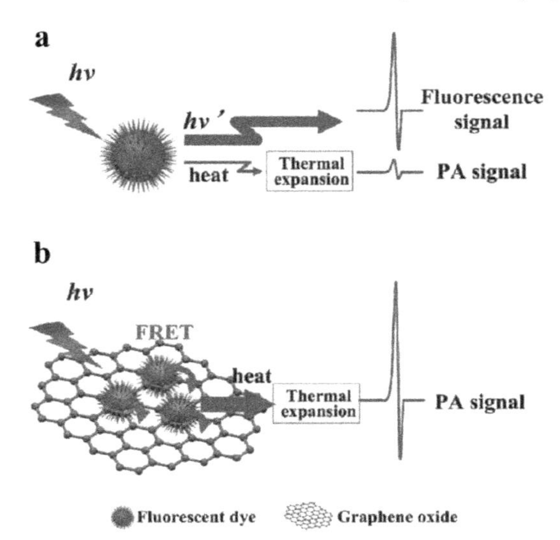

Figure 7.11 Characterization of PA enhancement process [123]. (a) Fluorescent dyes with strong absorption, strong fluorescence, and weak thermoelastic wave characteristics and (b) π-π stacking of dyes in close proximity to GO. GO quenches the dye fluorescence via FRET between dye molecules and GO. Using the pulse laser irradiation, the vast majority of absorbed light energy is converted to acoustic waves. Reproduced with permission from Ref. [123].

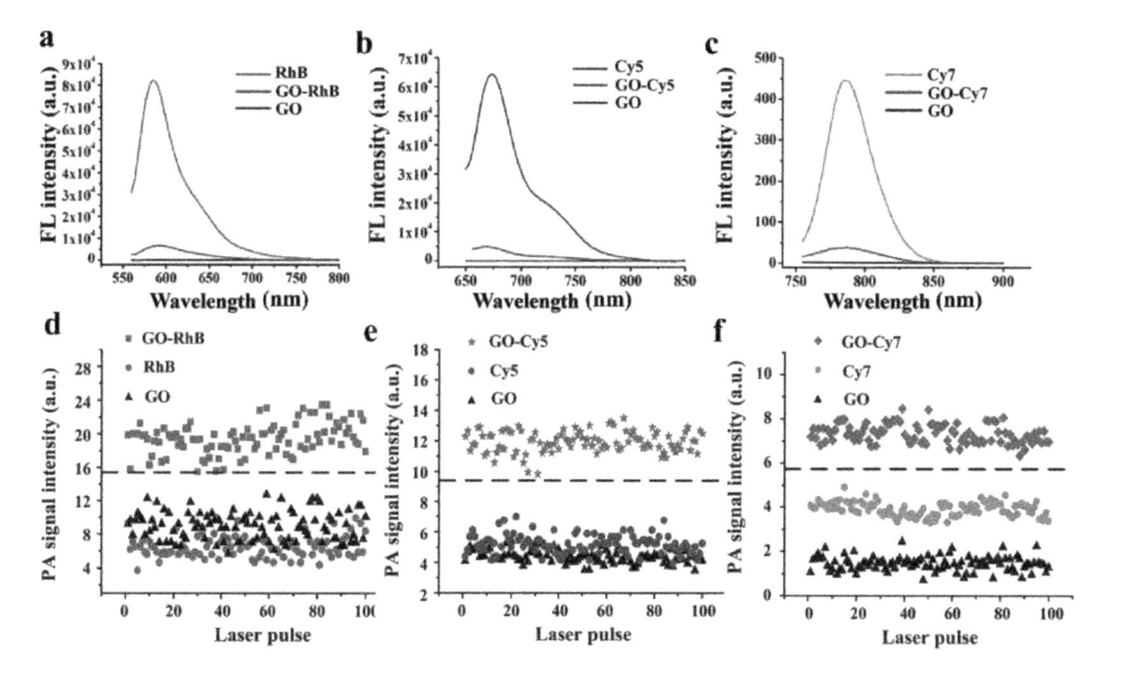

Figure 7.12 Characterization of GO-dyes [123]. (a–c) Fluorescence (FL) spectra of GO, GO–RhB, RhB, GO–Cy5, Cy5, GO–Cy7, and Cy7. (d) PA signal intensities of RhB, GO, and GO–RhB at 532 nm; (e) PA signal intensities of Cy5, GO, and GO–Cy5 at 675 nm; and (f) PA signal intensities of Cy7, GO, and GO–Cy7 at 753 nm. Dotted lines in (d), (e), and (f) indicate the sum of the separate measurements made in the dye solutions (RhB, Cy5, and Cy7) and GO particle suspension. Reproduced with permission from Ref. [123].

GO can efficiently quench the fluorescence of dyes via FRET, enabling the absorbed light energy to be mostly converted to thermal energy. Under pulse laser excitation, the absorbed light energy of GO-dyes will efficiently transform into photoacoustic shockwave (Fig. 7.11b).

Several dyes are loaded on the GO to evaluate their fluorescence quenching and PA signal enhancement. It shows that the GO efficiently quenches the fluorescence of dyes (Fig. 7.12a,b,c) and the designed fluorescence-quenching nanoprobes can produce stronger PA signals than the sum of the separate signals generated in the dye and GO (Fig. 7.12d,e,f). Therefore, this fluorescence-quenching nanoprobe based on the FRET mechanism may provide a valid strategy of designing the high-efficiency PA probes, which can expand the application of dyes and carbon materials in PAI.

7.2.4 Transition Metal Chalcogenide–Based Nanoprobes

In recent years, transition metal chalcogenides (TMC)-based nanomaterials such as copper sulfide (CuS), tungsten sulfide (WS_2), and molybdenum sulfide (MoS_2) have been newly developed for PAI, due to their low cost, low cytotoxicity, and intrinsic NIR region absorption derived from energy band transitions instead of surface plasmon resonance [124, 125]. Moreover, some semiconductor nanocrystals that show strong NIR absorption [76, 77, 126] and high photothermal transduction efficiency [124, 127, 128] have been used for photothermal therapy and shown good therapeutic effect. CuS nanoparticles with good photothermal effect and low cost have been extensively studied for PAI [72, 74, 75, 129, 130]. Common CuS nanoparticles are shown in Fig. 7.13a, and it has strong absorption in the NIR region (Fig. 7.13b). Benefiting from the strong absorption of CuS nanoparticles in 990 nm, in vivo deep PA images show that a nodule in the mouse brain is clearly imaged after intracranial injection with CuS (Fig. 7.13c–f).

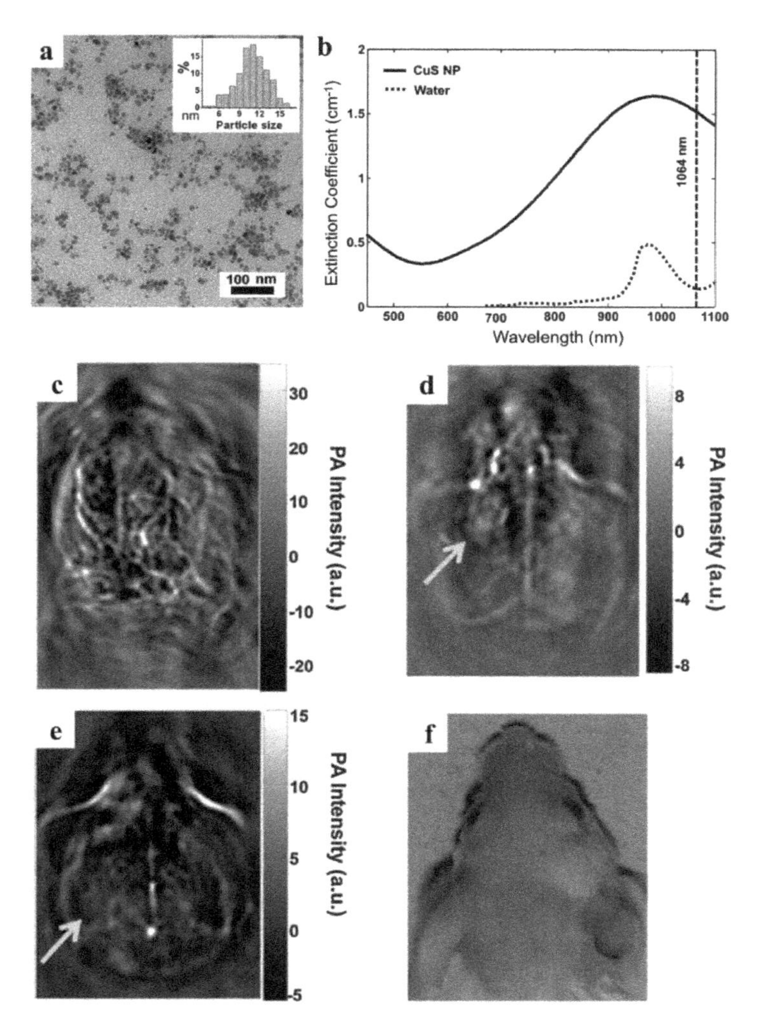

Figure 7.13 Characterization of CuS NPs [75]. (a) Transmission electron microscopy photograph of CuS NPs. Inset: Size distribution of CuS NPs. (b) Extinction coefficient spectra of 0.5 mM CuS NP aqueous solution (solid line) and pure water (dotted line). The vertical line is positioned at 1064 nm. (c) PAT of a mouse brain at a wavelength of 532 nm without CuS injection. PAT of the mouse brain at 1064 nm (d) 24 h and (e) 7 days after intracranial injection of CuS NPs solution. (f) Photograph of the imaged area of mouse head. Laser light irradiated from the top. Reproduced with permission from Ref. [75].

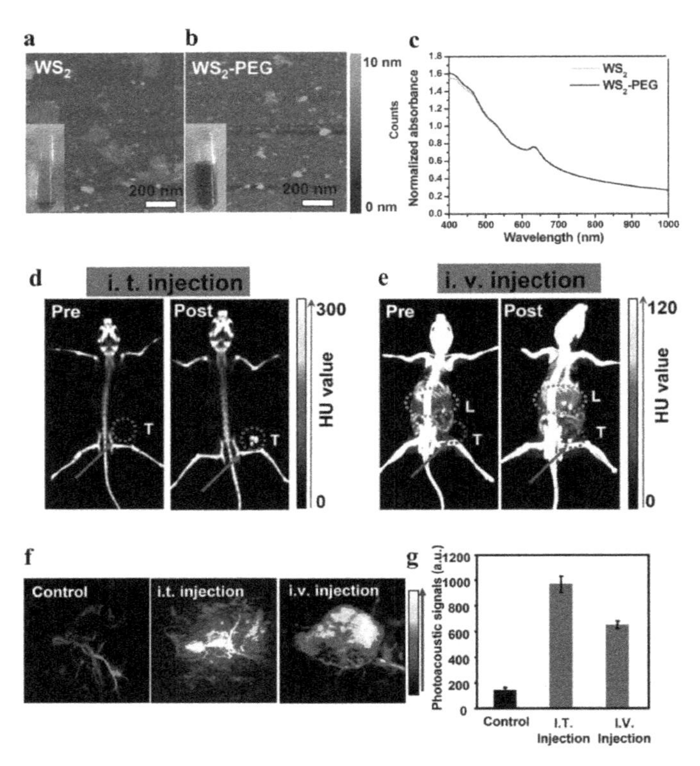

Figure 7.14 Characterization of WS_2 NPs and dual-modal imaging in vivo [76]. (a) AFM images of as-made WS_2 nanosheets and PEGylated WS_2 nanosheets. Inset: photos of as-made WS_2 and WS_2-PEG solutions in saline (0.05 mg/mL). PEGylation offered WS_2 nanosheets enhanced stability against salts. (b) Statistics of sheet thickness of WS_2 nanosheets before and after PEGylation based on AFM images. (c) UV-vis-NIR absorbance spectra of WS_2 nanosheets before and after PEGylation in water. (d) CT images of mice before and after i.t. injection with WS_2-PEG (5 mg/mL, 20 μL). (e) CT images of mice before and after i.v. injection with WS_2-PEG (5 mg/mL, 200 μL). The CT contrast was obviously enhanced in the mouse liver (green dashed circle) and tumor (red dashed circle). (f) PAT images of tumors on mice before and after i.t. or i.v. injection with WS_2-PEG. (g) Photoacoustic signals in the tumors from mice before and after i.t. or i.v. injections of WS_2-PEG solution. For PAT imaging, 20 or 200 μL of WS_2-PEG at the concentration of 2 mg/mL was i.t. or i.v. administrated, respectively. Reproduced with permission from Ref. [76].

Transition-metal dichalcogenides (TMDCs) such as MoS_2, $MoSe_2$, WS_2, and WSe_2, as a new class of two-dimensional nanomaterials with various unique electronic, optical, mechanical, and chemical properties, have attracted tremendous attention. Given the similarities in the morphology and properties between graphene and TMDCs, in particular their strong optical absorption in NIR region [131], they are very promising for biomedical applications. For example, WS_2 nanosheets are peeled off from the bulk WS_2 through Li ion insertion and ultrasonication to form water-soluble nanosheets. Then the surface of WS_2 nanosheets is modified with PEG, which greatly improves the physiological stability and biocompatibility of those nanosheets. They show a sheet shape similar to that of graphene (Fig. 7.14a,b), and their UV-vis-NIR spectrum reveals a broad NIR absorption band from 700 to 1000 nm (Fig. 7.14c). It has been demonstrated that WS_2-PEG can be used as a contrast agent in CT imaging because of the high atomic numbers of tungsten (W) (Fig. 7.14d,e) and also as PAI contrast agents for obviously enhancing the contrast of the tumor region (Fig. 7.14f). Moreover, it can serve as a powerful photothermal agent to effectively ablate cancer cells in vitro under NIR laser irradiation and finally achieve effectively high in vivo photothermal ablation of tumors.

Despite the TMC-based nanomaterials with high photothermal conversion efficiency, good photothermal stability, and strong absorbance in the NIR region, there is a lot of work to investigate their long-term toxicity to organisms and further explore their potential biomedical applications. Moreover, TMC-based nanomaterials usually contain heavy metal elements, which limits their potential for clinical translation and needs to be urgently solved.

7.2.5 Other Related Nanomaterials

7.2.5.1 Perfluorocarbon (PFC) nanodroplets

The PA signal of most of the above-mentioned PAI contrast nanoprobes is based on the thermoelastic effect, in which the nanoprobes absorb the light energy and convert it to instantaneous heat, thus resulting in rapid thermoelastic expansion. However, their PA conversion efficiency is not very high. Wilson et al. developed a new PAI contrast nanoprobe, whose photoacoustic signal generation is mainly based on vaporization, providing significantly higher signal amplitude than that from the traditionally used mechanism, thermoelastic expansion [80]. As shown in Fig. 7.15a, the photoacoustic nanoprobes consist of a droplet of liquid perfluorocarbon (PFC) and GNRs with bovine serum albumin (BSA) shell. PFC with boiling point below the body temperature (37°C) makes it easily vaporize in the presence of pulsed ultrasound with frequencies and pressures in the sub-therapeutic range [132]. Therefore, PFC has been used as a commercially available ultrasound (US) contrast agent in several formulations for diagnostic imaging [133]. The GNRs encapsulated in this nanoprobe absorb laser energy and thus generate instant heat and shock wave. It transforms the liquid PFC into gaseous phase, generating a strong PA signal through vaporization and prolonged thermoelastic expansion signal from GNRs. In addition, the generated gas will increase acoustic impedance mismatch between gas bubbles and the surrounding medium, enhancing the ultrasonic contrast (Fig. 7.15b).

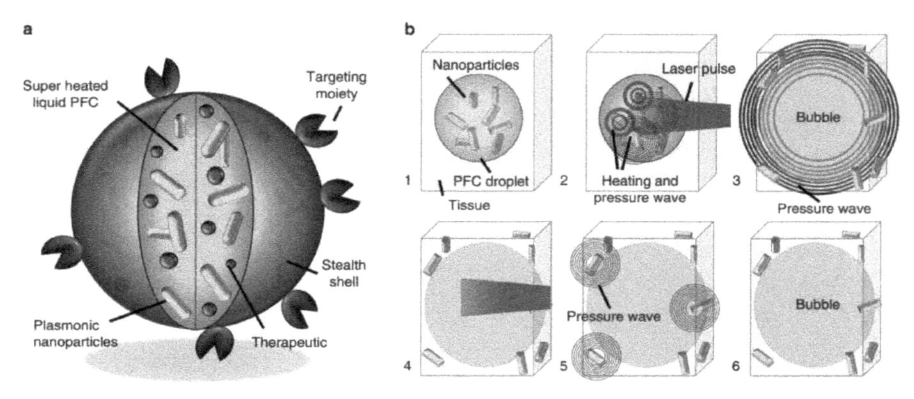

Figure 7.15 PAnDs concept and mechanisms [80]. (a) Diagram depicting the dual-contrast agent concept—photoacoustic droplet consisting of plasmonic nanoparticles suspended in encapsulated PFC (a superheated liquid at body temperature) and capped with a BsA shell. PAnDs may further contain therapeutic cargo and be surface functionalized for molecular targeting and cell–particle interactions. (b) step-by-step diagram of remote activation of PAnDs, providing photoacoustic signal via two mechanisms: vaporization of PAnDs (steps 2–3) and thermal expansion caused by plasmonic nanoparticles (steps 4–5). The resulting gas microbubble of PFC (step 6) provides us contrast due to acoustic impedance mismatch between gas and the surround environment. Reproduced with permission from Ref. [80].

On the other hand, Zhong et al. continue to develop the nanoprobe of this form by functionalization with targeting moiety on the surface of nanocomposites and loading drug molecules in it [43]. The nanocomposites show spherical morphology and the GNRs are encapsulated in them. The absorption spectrum indicates that surface modification process did not alter the specific optical properties of the GNRs, as shown in Fig. 7.16a. When the nanocomposites are irradiated with high intensity pulse laser, there is a dramatic increase in the PA signal magnitude due to the rapid, laser-triggered vaporization of nanocomposites (Fig. 7.16b). Nanocomposites are too small to be observed in the microscopic image before the nanocomposite solution is irradiated with pulse laser (Fig. 7.16c). During irradiation with laser, a large number of gas bubbles appear in the solution because of the PFC vaporization (Fig. 7.16d).

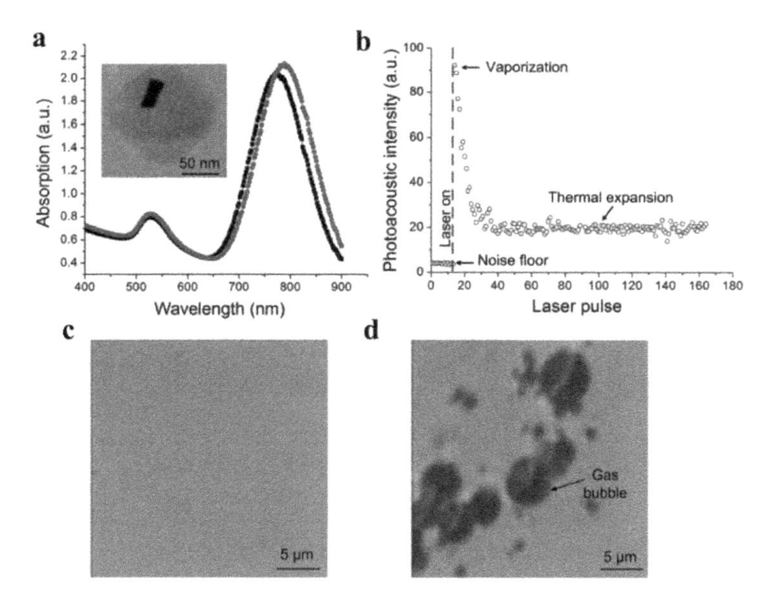

Figure 7.16 Characterization of PAnP [43]. (A) Absorption spectra of as-prepared AuNRs and AuNRs with a modified surface required for incorporation into PAnP. The red line represents the optical density spectrum of the AuNRs as synthesized, while the black line represents the spectrum of the AuNRs after organic surface modification. A typical transmission electron microscopy image of the PTX-PAnP is shown in the inset. (B) Photoacoustic signals of PAnP. (E), and gas bubbles after (F) pulsed laser irradiation. Reproduced with permission from Ref. [43].

Nanocomposites are intravenously injected into tumor-bearing mice. After pulsed laser irradiation on the tumor region, the average PA intensity begins with a very strong signal, which benefits from PFC vaporization (Fig. 7.17a,b). With the laser irradiation continued, the PA intensity decayed to a steady-state level due to the thermoelastic expansion effect of expelled GNRs and endogenous chromophores (Fig. 7.17c). In contrast, if GNRs were injected into tumor-bearing mice, the average PA intensity showed no significant change with the laser irradiation continued (Fig. 7.17d,e,f). In addition, the nanocomposites also could be US contrast agents and drug delivery vehicles for multimodal imaging-guided tumor therapy.

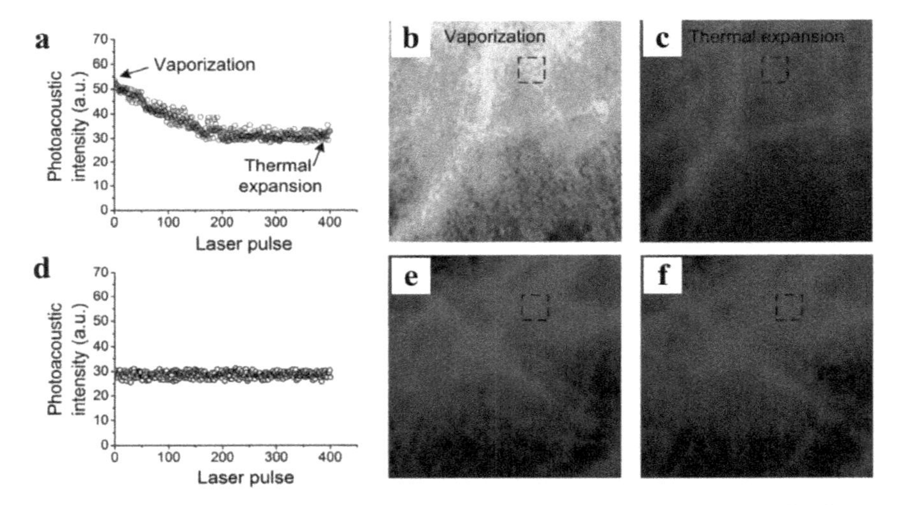

Figure 7.17 Contrast-enhanced photoacoustic imaging in vivo [43]. (a) Graph depicting the average photoacoustic intensity within the region of interest, indicated by boxes in panel B and panel C. (b) Photoacoustic image of the peak photoacoustic signal generated by PAnP vaporization. (c) Photoacoustic image representing photoacoustic signal generated from expelled gold nanorods and endogenous chromophores. (d) Graph displaying changes of photoacoustic signals after an injection of AuNRs-FA. Photoacoustic images of mouse tumor immediately after the laser was turned on (first laser pulse), (e) and at the end of the laser pulsing (last laser pulse), (f). Reproduced with permission from Ref. [43].

In the same strategy, dye molecules [79, 134, 135] or other nanoparticles [136] loaded in PFC droplets for enhanced PAI, US imaging and fluorescent imaging have been reported. However, this form of nanocomposite is not suitable for long-term continuous monitoring because the liquid-to-gas transition is too quick and irreversible. Moreover, the nanocomposites encapsulate PFC and other nanoparticles result in their relatively large size, which still poses challenges on mobility and bloodstream circulation.

7.2.5.2 Organic polymer-related nanoparticles

Organic polymers such as polypyrrole (PPy), polydopamine (PDA), and polyaniline (PANI) have also been developed as PAI contrast agents [81–86]. Compared with small-molecule organic dyes, polymer-based PAI contrast agents have much better photothermal stability, so that they can be used as long-term stable PAI contrast agents for continuous imaging. Polypyrrole (PPy) nanomaterials can be synthesized by chemical oxidation polymerization of water-soluble polymers in the metal cation aqueous solution [137]. They have received much attention in bioelectronics and biomedical applications due to their high conductivity, excellent light absorption performance, and outstanding stability [81–84]. In view of their strong NIR absorption spectrum and good photostability [138], they have the potential for being a new organic PAI contrast agent. Dai et al. used PPy nanoparticles as a biocompatible PAI nanoprobe for deep tissue imaging. The PPy nanoparticle synthesis method is shown in Fig. 7.18a. The PPy-synthesized nanoparticles showed uniform-diameter spheres with average size less than 46 nm (Fig. 7.18b) and strong absorption in the NIR region (Fig. 7.18c). The high absorption coefficient in the NIR region suggests that PPy nanoparticles were suitable for the PAI of deep tissues.

A mouse cerebral cortex was imaged by PAI before and after injection of PPy NPs. Compared to the brain PAI image with endogenous contrast agent (Fig. 7.19a), the image acquired 5 min of post-injection with PPy NPs shows the clearer brain vasculature (Fig. 7.19b). The contrast improvement is more obvious in the differential image (Fig. 7.19c) obtained by subtracting the pre-injection image (Fig. 7.19d) from the post-injection image (Fig. 7.19e). At 1 h post-injection, the vasculature PAI image still remains bright, indicating the large amount of PPy NPs circulating in the blood (Fig. 7.19c,e). This long-circulating property revealed that PPy NPs could be as a stable PAI contrast agent for long-term imaging, while it is difficult for dye-related nanoprobes.

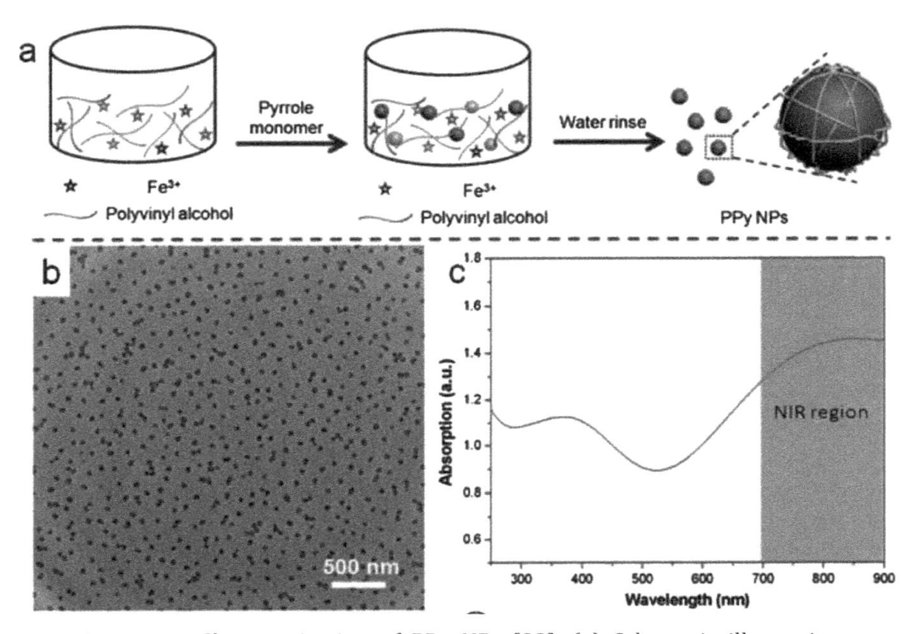

Figure 7.18 Characterization of PPy NPs [83]. (a) Schematic illustration of the formation of PPy NPs in an aqueous dispersion of water-soluble polymer–metal cation complexes; (b) the PPy NPs had a uniform diameter of average size of 46 nm based on the TEM image; (c) UV-vis-NIR extinction spectrum recorded for 50 mg mL^{-1} dispersion of PPy NPs in water. Reproduced with permission from Ref. [83].

As is well known, melanin is a natural existing pigment that is present in many organisms. There have been extensively studies in PAI in which melanin was used as an endogenous contrast agent [13–15]. Cheng's group reported that melanin was used for the successful preparation of the water-soluble melanin nanoparticles (MNP) for multimodal imaging of tumors; MNP not only provided their native optical absorption for PAI, but also actively chelated to metal ions ($64Cu^{2+}$, Fe^{3+}) for PET and MRI with a high loading capacity and stability utilizing their intrinsic chelating function [139]. These nanoparticles could be further functionalized for the drug-loading investigation [140]. Furthermore, melanin can directly co-precipitate with Fe^{3+} and Fe^{2+} ions (molar ratio at 2:1) under alkaline conditions to prepare melanin-coated magnetic nanoparticles for multimodal-imaging [141]. On the other hand, dopamine (DA) as a melanin-like mimic of mussel adhesive proteins could self-polymerize at alkaline pH values to adhere on the surface of many nanomaterials [86, 142–144]. Repenko et al. developed water-soluble dopamine-based polymers via nickel-catalyzed Kumada cross-coupling for photoacoustic imaging [145]. Dopamine was polymerized to PDA and then converted into melanin upon further oxidation. This nanopolymer as a PAI contrast agent showed good contrast performance and detection depths of more than 10 mm [85]. In these nanocomposites, PDA provides a number of advantages. First, PDA has shown its stability and biocompatiblility in vivo [144], which makes it is suitable for biomedical applications. Moreover, the existence of functional groups (i.e., catechol and amine) on the surface of PDA provided available binding sites [146–149]. Furthermore, PDA showed strong NIR absorption and high photothermal conversion efficiency (40%) [150], which suggested that it has the potential for theranostic applications.

As a biocompatible material, PANI has been used for studying cellular proliferation [151, 152]. Its strong absorption of NIR light makes it capable of generating a substantial amount of heat energy that can be used for cancer-cell ablation [153–155]. PANI nanoparticles as PAI contrast agents to guide photothermal therapy have also been reported [156].

Semiconducting polymer nanoparticles (SPNs) consisting of semiconducting π-conjugated polymer (SPs) as a new photostable fluorescent nanoprobe have been used for biomedical applications [157]. SPs are completely organic materials and have no heavy metal ion–induced toxicity to living organisms [158]. In particular, poly(cyclopentadithiophene-alt-benzothiadiazole) (SP1) and poly(acenaphthothienopyrazine-alt-benzodithiophene) (SP2) with molecular structure shown in Fig. 7.20a have been used for preparing multifunctional SPNs (Fig. 7.20b) [159]. The two formulations of SPN1 and SPN2 showed NIR absorption peak at 660 and 700 nm, respectively. Both SPNs could efficiently generate photoacoustic signals under pulsed laser irradiation (Fig. 7.20c).

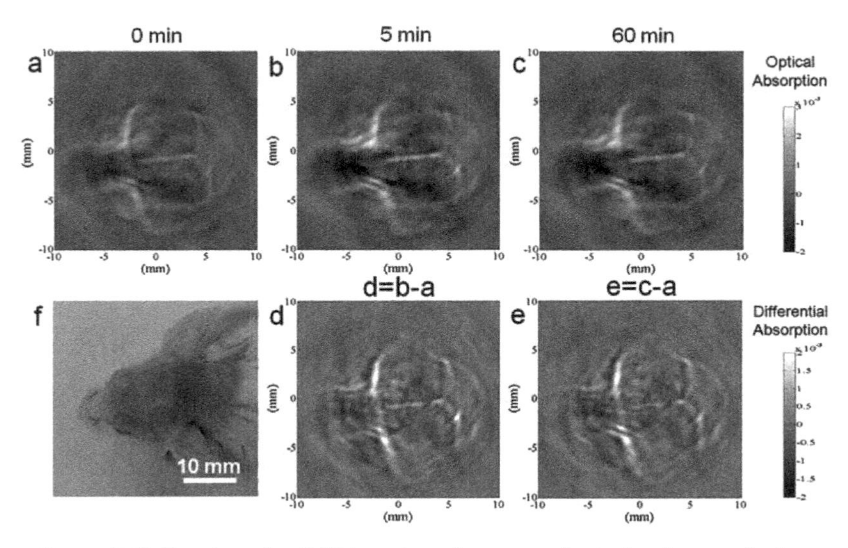

Figure 7.19 Non-invasive PAT imaging of a mouse brain in vivo employing PPy NPs and NIR light at a wavelength of 808 nm [83]. Photoacoustic image acquired (a) before, (b) 5 min after, and (c) 60 min after the intravenous injection of PPy NPs. (d and e) Differential images that were obtained by subtracting the pre-injection image from the post-injection images (image d = image b - image a; image e = image c - image a). (f) Photograph of the mouse brain obtained before the data acquisition for PAT. Reproduced with permission from Ref. [83].

Notably, compared to the extensively studied photoacoustic contrast nanoagents, SWNTs and GNRs, the photoacoustic amplitude of SPN1 was 5.2- and 7.1-times higher than SWNTs and GNRs, respectively, at the same mass concentrations. However, at equal molar concentrations, GNRs showed the highest photoacoustic amplitude because of their larger molar mass and molar extinction coefficient (5.0 × 107, 6.0 × 106, and 3.0 × 109 M^{-1} cm^{-1} for SPN1, SWNTs, and GNRs, respectively). Moreover, these nanoprobes can be further functionalized for molecular PAI [159, 160].

Other organic nanoparticles such as porphysome organic NPs, self-assembled from phospholipid–porphyrin, also have been developed for photothermal therapy and PAI because of their structure-dependent fluorescence self-quenching and strong light absorption [161]. However, the large size of the porphyrin-composed NPs limited them for in vivo application.

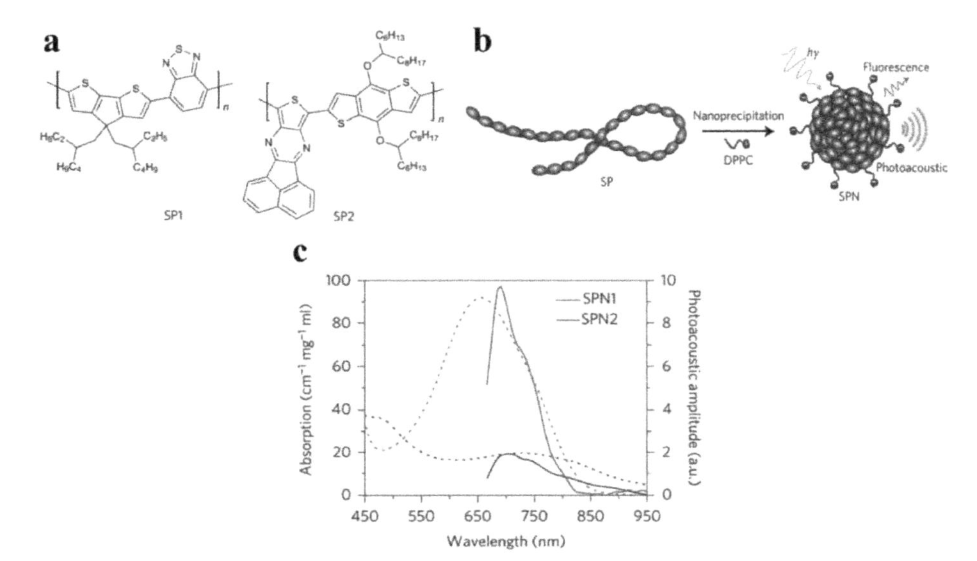

Figure 7.20 Characterization of SPNs [159]. (a) Molecular structures of SP1 and SP2 used for the preparation of SPN1 and SPN2, respectively. (b) Schematic of the preparation of SPNs through nanoprecipitation. SP is represented as a long chain of chromophore units (red oval beads). DPPC contains a short hydrophobic tail and a charged head and is illustrated as a string with a dark green ball at its end. (c) Ultraviolet–visible absorption (dashed lines) and photoacoustic spectra (solid lines) of SPNs. Reproduced with permission from Ref. [159].

7.2.6 Reporter Genes

Reporter genes as a powerful tool are very useful for molecular imaging, allowing the identification of specific cell population, analysis of transfection efficiency, and monitoring of drug treatment efficacy [87–89]. As indicators of whether a certain gene has been expressed in the cell or organism, they are typically incorporated in a regulatory sequence of another gene. Widely used reporter genes usually express visually identifiable fluorescent and luminescent proteins. For example, green fluorescent protein (GFP) gene is a commonly used reporter gene that emits green light under blue light excitation and thus provides fluorescence to identify the transfected cells as well as subcellular localization of proteins [162]. In addition, luciferase gene is another commonly used reporter gene that can be used in vivo because of its relatively long wavelength of fluorescence (<600 nm) [163]. However, its in vivo applications are restricted by the strong light-scattering and poor penetration in tissue. Although Razansky et al. has successfully applied enhanced GFP (eGFP) and mCherry fluorescent protein reporter genes as contrast agents for PAI in *Drosophila melanogaster* pupa and zebra fish [164], there are some disadvantages of using fluorescent protein reporter genes for PAI. First, fluorescent proteins are typically designed for high fluorescent yield, implying that the fraction of energy converted to heat for photoacoustic signal generation is not as high as might be possible with a low-quantum-yield species. Second, most fluorescent proteins do not have the amplification potential of enzymatic reporter genes. For an enzymatic reporter gene, one molecule of the enzyme can interact with many substrate molecules, thus providing signal amplification. Finally, fluorescent proteins are prone to laser-induced absorption bleaching, which makes long-term PAI difficult [165].

The lacZ gene, which encodes the lactose metabolizing enzyme β-galactosidase, has been used as reporter gene for PAI [166, 167]. A colorless analogue of lactose, X-gal (5-bromo-4-chloro-indolyl-β-D-galactoside) was used for β-galactosidase staining to yield a stable dark blue product (5,5′-dibromo-4,4′-dichloroindigo) which has strong light absorption between 605 and 665 nm. The gene expression can be identified by 400 μm spatial resolution subdermal PAI [166]. However, LacZ expression identified by PAI requires the injection of X-gal, while X-gal can cause skin irritation and is not efficiently metabolized in tumors, especially after intravenous injection with X-gal.

Tyrosinase (TYR) gene as a PA reporter gene does not require introduction of an enzymatic substrate and is promising for producing a very strong PA signal. Because TYR is the key enzyme to regulate melanin production in hair and skin of humans [168]. Although melanin displays maximal optical absorption at ultraviolet wavelengths, melanin has a broad spectrum of optical absorption with significant absorption at NIR wavelengths, which have strong tissue-penetrating properties [169]. Moreover, each expressed TYR will regulate the production of a number of melanin, resulting in a significant increase in PA signals. The TYR is sufficient to promote the melanin production in non-melanogenic cells [170, 171]. The PA signal of transfected cells was 10 times higher than that of wild-type cells [172]. Qin et al. reported that tyrosinase could be used as a triple-modality reporter for PET, MRI, and PAI in vivo [173]. First, melanin has a broad optical absorption spectrum and thus can be a PAI contrast agent. Second, melanin can chelate metal ions, which provides contrast for MRI [171]. Last, PET can be realized through introducing the melanin-avid probes such as N-(2-(diethylamino)ethyl)-18F-5-fluoropicolinamide (18F-P3BZA) as a PET reporter probe [170, 174]. Stritzker et al. introduced vaccinia virus-mediated tyrosinase expression system as a theranostic agent for MRI, PAI, and photothermal therapy [175]. In addition, multi-wavelength PAI [165] and deep in vivo PAI [176] with TYR gene as reporter gene have also been developed. However, the disadvantage of the reporter gene is that the melanin production process in transfected cells can potentially be toxic [177]. It is also needed to further study the toxicity of melanin production in transfected cells.

Other nanoparticles such as silver nanoparticles [178], iron oxide nanoparticles [179], and quantum dots [180] as PAI contrast agents have also been reported. However, they have not been widely studied in PAI, because of cytotoxicity [181, 182] or low PA conversion efficiency.

7.3 Biomedical Application of Nanoprobe-Mediated PAI

Nanoparticle-mediated PAI with high spatial resolution and obvious contrast can provide anatomical, functional, and molecular information of diseased loci at deep biological tissues. There have been extensive studies on tumor detection [46, 49, 119, 183], SNL mapping, tumor microenvironment monitoring [184], and tumor therapy [40–43, 62, 68, 185]. With the development of multifunctional nanoprobes, it is possible to detect disease using a single nanoplatform for multimodal imaging.

7.3.1 PAI for Diagnosis and Monitoring

Several research groups have reported various nanoparticle-mediated PAI for tumor imaging. Nanoparticles, with their small size and availability for surface modification, are very well suited for tumor imaging. First, the more permeable vasculature combined with poor lymphatic drainage of tumor tissues may allow nanoparticles (< 100 nm) to more effectively accumulate in the tumor region through a passive targeting manner, which is the so-called enhanced permeability and retention (EPR) effect [186]. In addition, the surface of nanoparticles provides a variety of functional groups for further modification. Conjugation of targeting ligands such as antibodies, peptides, and aptamers makes nanoparticles actively and specific bind to overexpressed receptors on the tumor cell membranes, thus more accumulation at the tumor site through receptor mediation [187]. Therefore, PAI can be utilized to visualize the presence of nanoparticles in the tumor, thus indicating the tumor region where the nanoparticles passively and actively accumulate. Wang's group reported gold nanocage (GNC) conjugation with [Nle4, D-Phe7]-α-melanocyte-stimulating hormone ([Nle4, D-Phe7]-MSH) as a PAI contrast agent for melanomas imaging [183]. Before injection with nanoparticles, blood vessel and melanoma images were acquired at 570 and 778 nm, respectively. Then [Nle4, D-Phe7]-MSH-conjugated GNC and PEG-GNC were intravenously injected into mouse. A series of PA images were obtained up to 6 h post injection. Melanomas before and after injection (i.e., images after injection minus images before injection) were overlaid on the images of blood vessels (Fig. 7.21). [Nle4, D-Phe7]-MSH conjugated GNC as active targeting contrast agents to melanomas (Fig. 7.21a) provided threefold higher PA signal enhancement than that of PEG–GNC (Fig. 7.21b). Nanoprobe-mediated high-resolution 3D morphological and functional PAI makes possible the detection of tumors such as early melanoma.

In addition, PAI also has been developed to map SLN, which is a very important path for cancer metastasis [188–192]. It was very promising in detecting metastatic cancer cells in the SLN for diagnosis, staging, and treatment of cancer. As shown in

Figs. 7.22a,b, an axillary region of rat which distributes SLN was monitored by PAI. With gold nanorods as SLN tracers, PA images were acquired pre- and post-injection. The PA images of vasculature were clearly observed (Fig. 7.22d). After the injection of gold nanorods on the left forepaw pad of the rat, PAI contrast of SLN gradually appeared clearly and increased (Figs. 7.22e–f). At 46 h after the injection, the SLN was clearly imaged with good contrast (Fig. 7.22g). The B-scans (Fig. 7.22c) acquired at the dotted line in Fig. 7.22g provide depth information of the SLN and the blood vessels due to the deep imaging capability of PAI. The PAI as a noninvasive approach is worthy of further development of functional nanoparticles and study on tumor metastasis through the lymph node.

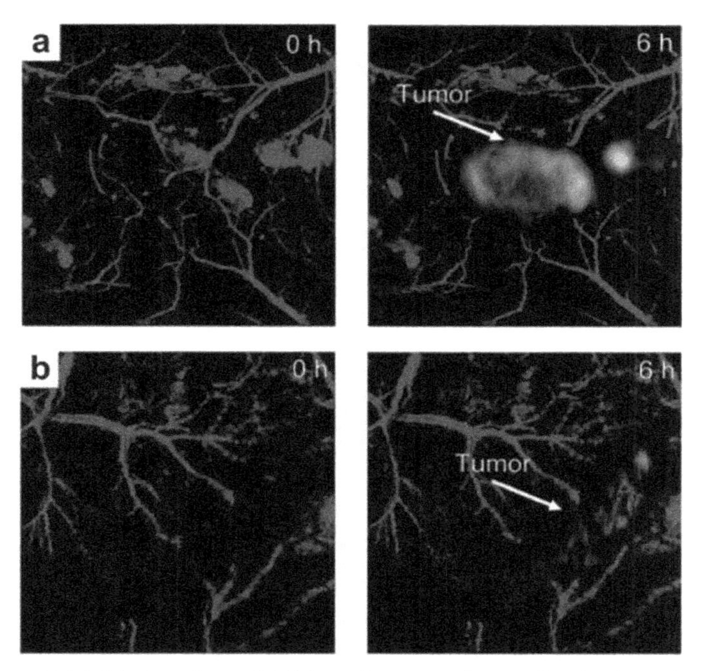

Figure 7.21 PA imaging of B16 melanoma after tail-vein injection of 100 mL GNC solution with a concentration of 10 nM [36]. Pseudo-red denotes blood vessels while pseudo-yellow denotes the melanoma in PA amplitude. (a) MIP PA images of B16 melanomas using [Nle4, D-Phe7]-α-melanocyte-stimulating hormone conjugated GNCs. (b) MIP PA images of B16 melanomas using only PEG–GNC. Reproduced with permission from Ref. [36].

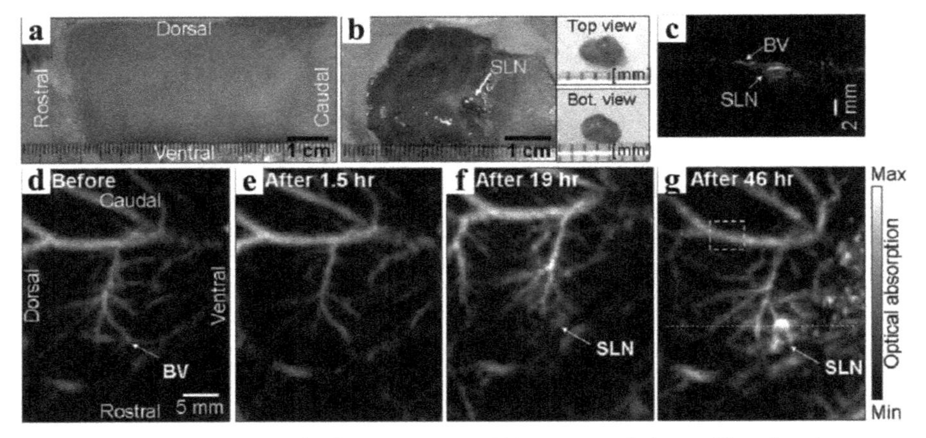

Figure 7.22 A sagittal photoacoustic MAP image of the axilla, showing the vasculature and sentinel lymph node [191]. (a) Photograph of the region of interest with hair removed before imaging. (b) Anatomical photograph of the axilla with skin and fatty tissue removed. Sentinel lymph node is indicated by an arrow. Insets are a top view and a bottom view of the dissected sentinel lymph node, respectively. (c) B-scan acquired at the dotted line in (g). (d) Photoacoustic MAP image acquired before the injection, showing the axillary vasculature. (e) Photoacoustic MAP image taken 1.5 h after the injection. (f) Photoacoustic MAP image acquired at 19 h after the injection. (g). Photoacoustic MAP image acquired at 46 h after the injection. The sentinel lymph node appears slightly in (f) and strongly in (g). Reproduced with permission from Ref. [191].

In another study, Qin et al. explored the potential application of gold nanorod–mediated PAI in atherosclerotic plaques detection [48, 193]. MMP2 (matrix metalloproteinase-2) antibody–conjugated gold nanorods (AuNRs-Abs) could specifically target MMP2, which is one of the clinical features of atherosclerotic plaques [194–196]. High-resolution photoacoustic images of the arterial structure were obtained by intravascular PAI (Fig. 7.23a), which corresponded well with the histological images (Fig. 7.23b). After labeling with AuNRs-Abs, the PA image contrast of atherosclerotic plaques containing MMP2 was significantly enhanced (Figs. 7.23c–f). Therefore, AuNRs-Abs mediated intravascular PAI can provide high-resolution morphological structure information to detect the plaque inflammation.

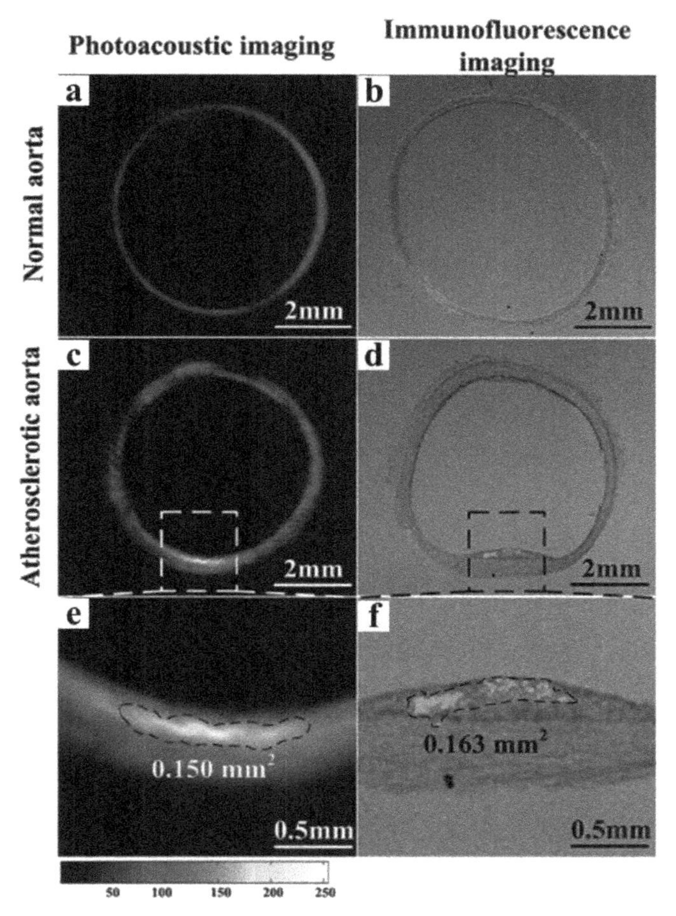

Figure 7.23 After the injection of AuNRs-Abs through the rabbit ear vein, the intravascular PAI of (a) the normal aorta and (c) the aorta containing atherosclerotic plaque was taken in situ, in which the yellow area indicates the area of expression of MMP2. The contrast/background ratio in the photoacoustic image is about 1.47. The overlay results of microscopic image and immunofluorescence result of MMP2 (b) in normal aorta and (d) in the atherosclerotic plaque were obtained soon after the PAI experiment. Images (e) and (f) are the enlarged images of the dotted rectangular areas in (c) and (d). In PAI, the area of distribution of MMP2 is 0.150 mm^2. Correspondingly, the immunofluorescence area (indicating the area of MMP2) is 0.163 mm^2. In addition, it is non-ionizing and capable of real-time imaging, making it a promising tool for use in cancer screening, diagnosis, and therapy monitoring. Reproduced with permission from Ref. [48].

Beyond the previously described studies, other researchers have investigated the utility of PAI for detecting nanoparticle-labeled micrometastasis of cancer cells within the lymphatic system [197–202], and circulating tumor cells [203, 204]. It makes possible the detection of prostate or breast cancer cells, which do not contain melanin pigments.

7.3.2 Tumor Microenvironment Monitoring

Enzyme activity, PH, reactive oxygen species (ROS) level, and metal ion concentration are very important physiological indicators in life science. For example, aberrant enzyme activity, acidosis, and hypoxia are closely related to the solid tumors growth [205–208]. As far as is known, PH values in a malignant tumor are generally lower than that in normal tissues (~7.4) [209]. Compared with the limited concentration of tumor-associated receptors, tumor acid condition as a microenvironment indicator are pervasive in almost all solid tumors and can be easily detected [210]. In particular, tumor acid condition would not only promote tumor migration and invasion [211] but also obstruct the activity of harbor p53 and p-glycoprotein, leading to the multidrug resistance (MDR) in chemotherapy [212, 213]. Therefore, in vivo real-time pH detection is promising for tumor recognition, diagnosis, monitoring, and prognosis. Some organic dyes with pH-sensitive NIR absorption shift have been exploited as PA contrast agents for tumor pH detection [184, 214, 215]. Liu et al. developed a pH-responsive albumin-based nanoprobe for in vivo ratiometric photoacoustic pH imaging [214]. Two types of NIR dyes, benzo[a]phenoxazine (BPOx) and IR825, could induce self-assembly of human serum albumin (HSA), to form albumin–dye nanocomplexes. IR825 absorbance peak at 825 nm did not change along with the decrease of pH, while the absorbance at 670–680 nm significantly increased through the protonation and intramolecular charge-transfer of BPOx (Fig. 7.24a). Therefore, the PA signal intensity at 680 significantly increased along with the decrease of PH, while PA signal intensity at 825 nm showed negligible change (Fig. 7.24b). In the pH range of 5.0–7.0, the ratiometric PA signal (PA680 nm/PA825 nm) showed excellent pH dependence, which is suitable for the PAI of tumor microenvironment (Fig. 7.24c). Then, the nanocomposites were injected into muscle and subcutaneous 4T1 tumor in situ. In the tumor region, the PA signal at 680 nm was much stronger than that at 825 nm (Fig. 7.24d), while the 680/825 PA signal ratio was obviously lower in the muscle (Fig. 7.24e). The PH values of tumor and muscle were determined to be ≈6.7 and >7.0 based on the standard calibration curve (Fig. 7.24f), which indicates that this nanoparticle could be used as a PA contrast agent for pH detection in vivo.

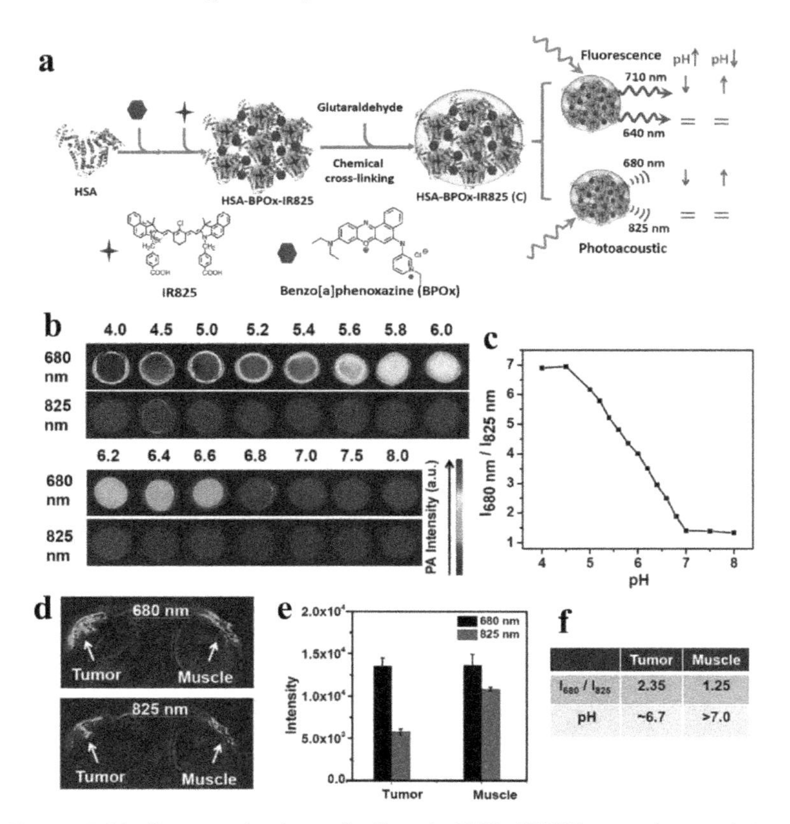

	Tumor	Muscle
I_{680}/I_{825}	2.35	1.25
pH	~6.7	>7.0

Figure 7.24 Characterization of albumin-BPOx-IR825 complex and its PAI performance [214]. (a) Illustration showing the mechanism of albumin-BPOx-IR825 complex for pH detection. PA images (b) and signal intensity ratios (c) in response to various pH values. PA images (d) and PA signals (e) of a mouse with local administration of HSA-BPOx-IR825 in its tumor (left side) or muscle (right side). (f) The quantitative analysis of pH value in the tumor and muscle. Reproduced with permission from Ref. [214].

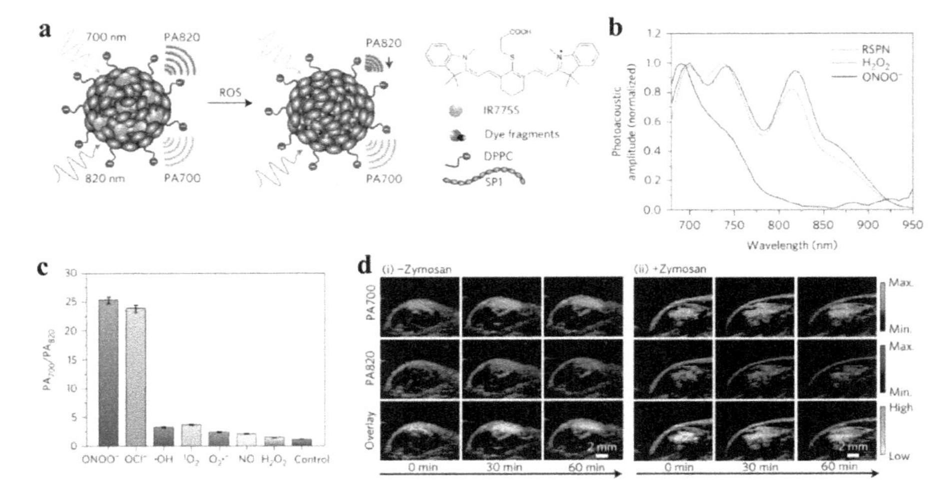

Figure 7.25 Characterization of RSPN (a) Proposed ROS sensing mechanism [159]. (b) Representative photoacoustic spectra of RSPN in the absence and presence of ROS. [RSPN] = 5 mg ml21, [ROS] = 5 mM. (c) Ratio of photoacoustic amplitude at 700 nm to that at 820 nm (PA700/PA820) after treatment with indicated ROS (5 mM). Reproduced with permission from Ref. [159].

In addition to the PH change, ROS as a fundamental component of tissue injury in most human diseases is also necessary to detect to find the etiology of disease and optimize the treatment. Rao et al. reported a PA contrast agent for real-time detection of ROS [159]. In this PAI nanoprobe, a cyanine dye (IR775S) that can sense specific ROS-mediated oxidation [216] was loaded into a semiconducting polymer nanoparticle (SPN) composed of poly(cyclopentadithiophene-altbenzothiadiazole) (SP1) (Fig. 7.25a). The PA spectrum of the prepared nanoprobe (RSPN) showed three PA peaks at 700, 735, and 820 nm. The photoacoustic peak at 735 nm decreased significantly in the presence of ONOO- and ClO-, and the peak of 820 nm almost disappeared, while the peak at 700 nm was almost unchanged (Fig. 7.25b). However, in the presence of other ROS such as •NO, •OH, 0_2•-, 10_2, and H_20_2, the photoacoustic spectrum of RSPN remained essentially unchanged (Fig. 7.25b,c). Zymosan, a structural polysaccharide of the cell wall of *Saccharomyces cerevisiae* was injected intramuscularly into the thigh of mice to generate ROS such as ONOO- and ClO-, and then the RSPN was injected into the same location. For the mice without zymosan treatment, the PA amplitude at 700 nm and 820 nm remained nearly the same over time. In contrast, the PA amplitude at 820 nm for the mice with zymosan treatment significantly decreased over time, while PA signal at 700 nm also remained nearly unchanged resulting in an increased PA700 nm/PA820 nm value. Therefore, this nanoprobe showed the potential of detecting ROS production using a dual-peak ratiometric PAI.

In other studies, the nanoparticles were developed for detecting enzyme and metal ion of physiological environment by PAI, such as matrix metalloproteinases (MMPs) [73] Cu_2+ [217], and Li^+ [218]. These emerging PAI nanoprobes provide an alternative new approach to detect disease.

7.3.3 Imaging-Guided Tumor Therapy

Among various tumor therapeutic techniques, nanoparticle-mediated PA therapy is a newly emerging tumor treatment method [40–42, 62, 68, 185]. It is based on the mechanical destruction of photoacoustic effect, in which photoabsorbers convert the light energy into acoustic waves to selectively destruct cancer cells. It is considered as one of the promising cancer treatment methods because of its cancer-specific treatment, minimal normal cell injury, and no anticancer drug resistance. Xing' group reported a variety of nanoparticles for tumor PA therapy. Among them, gold nanorods with high NIR absorption showed high-efficiency tumor killing effect, meanwhile, it also could be used as a contrast agent for PAI of tumor. Gold nanorods were functionalized with folic acid (FA) to target folate receptor-expressing cancer cells (e.g., Hela cells). HeLa cells (without FA-AuNR incubation) morphology remained intact before and after laser irradiation (Fig. 7.26a,b). In contrast, the HeLa cells incubated with FA-AuNR did not change the morphology, while after laser irradiation (Fig. 7.26c), a rough surface, blebbing, and deformation cells were observed, indicating cell damage after irradiation by the laser pulses (Fig. 7.26d).

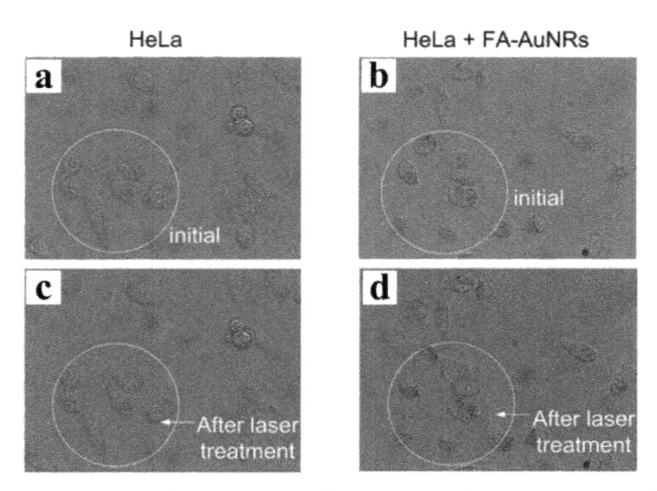

Figure 7.26 Cell viability of HeLa cells under different treatments [219]. Microscopic DIC images of the control HeLa cells (without FA-AuNR incubation) before (a) and after (b) photoacoustic treatment, showing no sign of cell damage. Image of the AuNRs-loaded HeLa cells before photoacoustic treatment (c) showing a normal morphology with an extensively attached growth pattern and smooth surface. Image of the AuNR-loaded HeLa cells after photoacoustic treatment (d) showing the same AuNR-loaded HeLa cells with a rough surface, a typical sign of cell damage. Reproduced with permission from Ref. [219].

On this basis, the PA effect as a new triggered drug release modality was further studied in tumor therapy [43]. Paclitaxel (PTX)-, perfluorohexane (PFH)-, and AuNRs-loaded nanoparticles (PTX-PAnP) were synthesized and modified with FA (PTX-PAnP-FA). Upon pulsed laser irradiation, the vaporization of PFH resulted in PTX-PAnP-FA destruction and then induced the rapid PTX release. As shown in Fig. 7.27a, the PTX released from PTX-PAnP was less than 4% of the total encapsulated drug even after 72 h, indicating sufficient stability in the physiological environment. In contrast, nanodroplets were irradiated by pulsed laser (14 mJ/cm^2), almost all drugs were released from PTX-PAnP, indicating the pulsed laser could trigger the rapid drug release (Fig. 7.27b).

In addition to the promising tumor therapeutic method, the multifunctional nanoparticles could also be a PA contrast agent for guidance therapy in a single nanoplatform. Meanwhile, with the development of new nanoparticles, PAI has been combined with many imaging modalities for multi-modal imaging, which could provide additional functional, molecular, and anatomical information.

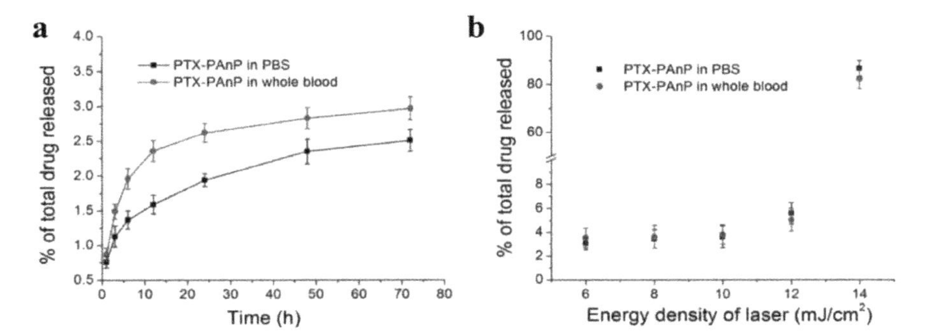

Figure 7.27 A release profile of PTX from PTX-PAnP [43]. (a) In vitro stability tests of PTX-PAnP. The fractions of PTX being detached from PTX-PAnP were determined at 37°C under gentle agitation in the presence of whole blood (red line) and 10mM PBS (black line). (b) The release profile of PTX from PTX-PAnP in whole blood (red line) and 10 mM PBS (black line) under pulsed laser irradiation (with different energy density). Reproduced with permission from Ref. [43].

Although the nanoprobes showed many advantages and functions, they have not yet been approved for clinical use, because of their potential toxicity and unclear long-term retention effect in human body [220]. Their toxicity and retention effects vary based on the difference in chemical composition, size, geometry, and surface properties of nanoparticles. It is necessary to evaluate the short-term and long-term toxicity, biodistribution, nonspecific uptake, and routes of elimination, in order to apply nanoparticles in clinical practice. Furthermore, the various nanomaterials also have their respective limitations. For example, organic dyes are extensively used in clinics, but they are rather small (<2 nm), and thus can quickly be cleared from blood and excreted by the renal system. In addition, easy photobleaching makes long-term imaging challenging. Gold nanoparticles with different sizes and morphologies have strong optical absorption and high photoacoustic conversion efficiency. However, their optical properties are highly dependent on expensive and complicated surface chemistries. Although copper-based nanoparticles synthesized in a highly stable manner and preliminary studies have shown they are safe, inexpensive, and efficient PA contrast agents, copper is known to pose neurotoxicity [221]. Therefore, there is no ideal nanoprobe to be a contrast agent for all functional PAI. With the constant advancement of the PA technology and the development of nanotechnology, developing new nanoprobes to be more suitable for photoacoustic imaging, as well as devoting to clinical transformation will be the focus of future research.

7.4 Conclusions and Future Directions

Effective photoacoustic contrast agents should significantly increase contrasts, effectively improve imaging depths, and accurately provide molecular specific information. Therefore, it is very necessary to develop most appropriate nanoprobes for specific needs, further optimization and testing in clinical trials. Contrast agents need to be prepared and used discreetly in accordance with the provisions of the Food and Drug Administration. (1) Photoacoustic contrast agents should be easily prepared and produced for a low cost due to future clinical applications. (2) The size of contrast agents has a significant impact on their distribution and targeted delivery in vivo. Large contrast agents remain in vivo longer than small ones; however, small ones have better permeability than large ones. Greater permeability makes them readily enter the bloodstream and penetrate into the desired location after injection. Dimensional control of contrast agents exerts an important control on distribution and overall curative effect of drug. (3) Agents should have excellent water solubility; otherwise they will be treated as foreign objects and rejected by the surrounding tissue. (4) They should possess proper biological compatibility and low toxicity in order to not cause adverse reactions to the tissues—the most major being damage to the endothelium, vascular system, blood–brain barrier (BBB), kidneys, and so on; (5) Contrast agents should be further optimized to actively target the specific receptors in target cells or tissues, improve the accumulation rate, and reduce the reticuloendothelial system (RES) retention by using specific biomarkers (such as peptides, antibodies, and so on), drugs, or another contrast agent. (6) They should have high stability, long half-life, and full dissolution in vivo in order to make them meet clinical requirements. (7) They should possess very good dispersibility in order to distribute evenly in the imaging region. (8) Photoacoustic contrast agents should have a high light-to-ultrasound conversion efficiency. Despite many years of research, photoacoustic molecular imaging still has a lot of issues that need to be addressed promptly. The following future directions should be discussed: (1) There are limited categories of contrast agents. Special superparamagnetism and bulk effect, the compounding of superparamagnetic nanoparticles, and their application in biomedicine will hopefully draw researchers' attention. Such methods show potential for the advance in multimodality technology of PAI, MRI, and other imaging techniques. (2) The quantitative capability of photoacoustic molecular imaging is significant. It may help us differentiate benign from malignant lesions as well as obtain absolute tissue functional information. They are perhaps determinants of molecular typing. Besides, PAI can measure the amount of a drug at its site, which will provide the theoretic foundation for ensuring the accuracy of injection doses for animals and patients. (3) The future research of contrast agents as a drug delivery system should emphasize their effectiveness and safety. The technical specifications of the system's effectiveness and safety are not established yet. How to reduce toxicity and side effects as well as improve the drug targeting, solubility, stability, slow-releasing action and how to change tissue-targeted distribution and metabolism are key issues for this research. (4) Mono-modality imaging cannot obtain all the information required. Compared to low-resolution ultrasound imaging, PAI is also able to capture information about optical absorption with high resolution. Compared to PET and SPECT, these two imaging technologies have no risk of causing injury by radiation. Compared to MRI, PAI has a quicker imaging speed. In comparison to OCT, they have longer imaging depths. Overall, PAI only demands simple equipment, is low cost, and is easy to operate. Thus, it has broad and applicable prospects. Development of PA-based multi-modal contrast agents can provide complementary information by combining information about structural and functional physiology and pathology from various imaging modalities and overcome the weaknesses of each imaging modality. (5) Unfortunately, the application of photoacoustic contrast agents in real-time imaging and their ability to confirm lesions before treatment cannot currently meet the needs of modern medicine at present. It is noteworthy that targeted drug delivery systems, advanced physical therapy, image guided surgery, heat therapy (e.g., high-intensity focused ultrasound [HIFU] or radiofrequency hyperthermia), photothermal therapy, and magnet fluid hyperthermia provide very promising treatment techniques for the future. Photoacoustic molecular imaging will aim to integrate detection, monitoring, and reasonable treatment into one platform. (6) Systematic evaluation of the stability, toxicity, biocompatibility, biodegradability, immunogenicity, and pharmacokinetics will help PA contrast agents accelerate their clinical translation for PAI. However, currently contrast agent testing are too complicated, due to their specific animal testing experiments, longer experiment durations followed by clinical trials and high research and development costs. More unfortunately, there are no unified standard, and it is rarely used in the clinic due to various differences in species, individuals, and drug delivery systems.

Therefore, the effectiveness and safety of photoacoustic molecular imaging play a key role in establishing better and more direct relations between basic research and clinical applications. Photoacoustic molecular imaging, working in tandem with powerful contrast agents, could provide groundbreaking opportunities for early diagnoses, precise lesion localization, molecular typing, drug delivery monitoring, image-guided surgery, drug-targeting therapy, and photothermic therapy of various diseases (such as cancer and cardio-cerebrovascular disorders). Photoacoustic imaging, complemented by the development of nanoprobes, will potentially bridge the gap between bench and bedside and further develop translational medicine as we know it.

References

1. Wang, L. H. V., and Hu, S. (2012). Photoacoustic tomography: In vivo imaging from organelles to organs, *Science,* **335**, pp. 1458–1462.

2. Emelianov, S. Y. Li, P. C., and O'donnell, M. (2009). Photoacoustics for molecular imaging and therapy, *Phys. Today,* **62**, pp. 34–39.

3. Luke, G. P. Yeager, D., and Emelianov, S. Y. (2012). Biomedical applications of photoacoustic imaging with exogenous contrast agents, *Ann. Biomed. Eng.,* **40**, pp. 422–437.
4. Mallidi, S., Luke, G. P., and Emelianov, S. (2011). Photoacoustic imaging in cancer detection, diagnosis, and treatment guidance, *Trends. Biotechnol.,* **29**, pp. 213–221.
5. Wang, L. V. (2008). Prospects of photoacoustic tomography, *Med. Phys.,* **35**, pp. 5758–5767.
6. Beard, P. (2011). Biomedical photoacoustic imaging, *Interface Focus,* **1**, pp. 602–631.
7. de la Zerda, A., Kim, J. W., Galanzha, E. I., Gambhir, S. S., and Zharov, V. P. (2011). Advanced contrast nanoagents for photoacoustic molecular imaging, cytometry, blood test and photothermal theranostics, *Contrast Media Mol. I,* **6**, pp. 346–369.
8. Galanzha, E. I., and Zharov, V. P. (2012). Photoacoustic flow cytometry, *Methods,* **57**, pp. 280–296.
9. Tuchin, V. V., Tarnok, A., and Zharov, V. P. (2011). In vivo flow cytometry: A horizon of opportunities, *Cytom. Part A,* **79A**, pp. 737–745.
10. Yin, G. Z., Xing, D., and Yang, S. H. (2009). Dynamic monitoring of blood oxygen saturation in vivo using double-ring photoacoustic sensor, *J. Appl. Phys.,* **106**, p. 013109.
11. Li, B. B., Qin, H., Yang, S. H., and Xing, D. (2014). In vivo fast variable focus photoacoustic microscopy using an electrically tunable lens, *Opt. Express,* **22**, pp. 20130–20137.
12. Wang, Y., Hu, S., Maslov, K., Zhang, Y., Xia, Y. N., and Wang, L. V. (2011). In vivo integrated photoacoustic and confocal microscopy of hemoglobin oxygen saturation and oxygen partial pressure, *Opt. Lett.,* **36**, pp. 1029–1031.
13. Viator, J. A., Komadina, J., Svaasand, L. O., Aguilar, G., Choi, B., and Nelson, J. S. (2004). A comparative study of photoacoustic and reflectance methods for determination of epidermal melanin content, *J. Invest. Dermatol.,* **122**, pp. 1432–1439.
14. Wang, Y. T., Xu, D., Yang, S. H., and Xing, D. (2016). Toward in vivo biopsy of melanoma based on photoacoustic and ultrasound dual imaging with an integrated detector, *Biomed. Opt. Express,* **7**, pp. 279–286.
15. He, G., Xu, D., Qin, H., Yang, S. H., and Xing, D. (2015). In vivo cell characteristic extraction and identification by photoacoustic flow cytography, *Biomed. Opt. Express,* **6**, pp. 3748–3756.

16. Zhang, J., Yang, S. H., Ji, X. R., Zhou, Q., and Xing, D. (2014). Characterization of lipid-rich aortic plaques by intravascular photoacoustic tomography ex vivo and in vivo validation in a rabbit atherosclerosis model with histologic correlation, *J. Am. Coll. Cardiol.*, **64**, pp. 385–390.

17. Chen, C. G., Zhao, Y., Yang, S. H., and Xing, D. (2015). Mechanical characterization of intraluminal tissue with phase-resolved photoacoustic viscoelasticity endoscopy, *Biomed. Opt. Express,* **6**, pp. 4975–4980.

18. Chen, C. G., Zhao, Y., Yang, S. H., and Xing, D. (2015). Integrated mechanical and structural features for photoacoustic characterization of atherosclerosis using a quasi-continuous laser, *Opt. Express,* **23**, pp. 17309–17315.

19. Wang, B., Su, J. L., Amirian, J., Litovsky, S. H., Smalling, R., and Emelianov, S. (2010). Detection of lipid in atherosclerotic vessels using ultrasound-guided spectroscopic intravascular photoacoustic imaging, *Opt. Express,* **18**, pp. 4889–4897.

20. Holotta, M., Grossauer, H., Kremser, C., Torbica, P., Volkl, J., Degenhart, G., Esterhammer, R., Nuster, R., Paltauf, G., and Jaschke, W. (2011). Photoacoustic tomography of ex vivo mouse hearts with myocardial infarction, *J. Biomed. Opt.*, **16**, p. 036007.

21. Li, Z., F., Li, H., Chen, H. Y., and Xie, W. M. (2011). In vivo determination of acute myocardial ischemia based on photoacoustic imaging with a focused transducer, *J. Biomed. Opt.*, **16, p.** 076011.

22. Sen Gupta, A. (2011). Nanomedicine approaches in vascular disease: A review, *Nanomed. Nanotechnol.*, **7**, pp. 763–779.

23. Yang, S. H., Xing, D., Lao, Y. Q., Yang, D. W., Zeng, L. M., Xiang, L. Z., and Chen, W. R. (2007). Noninvasive monitoring of traumatic brain injury and post-traumatic rehabilitation with laser-induced photoacoustic imaging, *Appl. Phys. Lett.*, **90**, p. 243902.

24. Yuan, Y., Yang, S. H., and Xing, D. (2012). Optical-resolution photoacoustic microscopy based on two-dimensional scanning galvanometer, *Appl. Phys. Lett.*, **100**, p. 023702.

25. Xu, D., Yang, S. H., Wang, Y., Gu, Y., and Xing, D. (2016). Noninvasive and high-resolving photoacoustic dermoscopy of human skin, *Biomed. Opt. Express,* **7**, pp. 2095–2102.

26. Berlis, A., Lutsep, H., Barnwell, S., Norbash, A., Wechsler, L., Jungreis, C. A., Woolfenden, A., Redekop, G., Hartmann, M., and Schumacher, M. (2004). Mechanical thrombolysis in acute ischemic stroke with

endovascular photoacoustic recanalization, *Stroke,* **35**, pp. 1112–1116.

27. Krumholz, A., Wang, L. D., Yao, J. J., and Wang, L. H. V. (2012). Functional photoacoustic microscopy of diabetic vasculature, *J. Biomed. Opt.,* **17**, p. 060502.

28. Zhang, Q. Z., Liu, Z., Carney, P. R., Yuan, Z., Chen, H. X., Roper, S. N., and Jiang, H. B. (2008). Non-invasive imaging of epileptic seizures in vivo using photoacoustic tomography, *Phys. Med. Biol.,* **53**, pp. 1921–1931.

29. Lao, Y. Q., Xing, D., Yang, S. H., and Xiang, L. Z. (2008). Noninvasive photoacoustic imaging of the developing vasculature during early tumor growth, *Phys. Med. Biol.,* **53**, pp. 4203–4212.

30. Ye, F., Yang, S. H., and Xing, D. (2010). Three-dimensional photoacoustic imaging system in line confocal mode for breast cancer detection, *Appl. Phys. Lett.,* **97**.

31. Hu, J., Yu, M. L., Ye, F., and Xing, D. (2011). In vivo photoacoustic imaging of osteosarcoma in a rat model, *J. Biomed. Optics,* **16**.

32. Xiang, L. Z., Xing, D., Guo, H., and Yang, S. H. (2009). High resolution fast digital photoacoustic CT for breast cancer diagnosis, *Acta Phys. Sin Ch. Ed.,* **58**, pp. 4610–4617.

33. Wang, X. D., Xie, X. Y., Ku, G. N., and Wang, L. H. V. (2006). Noninvasive imaging of hemoglobin concentration and oxygenation in the rat brain using high-resolution photoacoustic tomography, *J. Biomed. Opt.,* **11**.

34. Li, M. L., Oh, J. T., Xie, X. Y., Ku, G., Wang, W., Li, C., Lungu, G., Stoica, G., and Wang, L. V. (2008). Simultaneous molecular and hypoxia imaging of brain tumors in vivo using spectroscopic photoacoustic tomography, *P IEEE.,* **96**, pp. 481–489.

35. Mallidi, S., Larson, T., Aaron, J., Sokolov, K., and Emelianov, S. (2007). Molecular specific optoacoustic imaging with plasmonic nanoparticles, *Opt. Express,* **15**, pp. 6583–6588.

36. Kim, C., Cho, E. C., Chen, J. Y., Song, K. H., Au, L., Favazza, C., Zhang, Q. A., Cobley, C. M., Gao, F., Xia, Y. N., et al. (2010). In vivo molecular photoacoustic tomography of melanomas targeted by bioconjugated gold nanocages, *ACS Nano,* **4**, pp. 4559–4564.

37. Pan, D. P., Pramanik, M., Senpan, A., Yang, X. M., Song, K. H., Scott, M. J., Zhang, H. Y., Gaffney, P. J., Wickline, S. A., Wang, L. V., et al. (2009). Molecular photoacoustic tomography with colloidal nanobeacons, *Angew. Chem. Int. Edit.,* **48**, pp. 4170–4173.

38. De La Zerda, A., Zavaleta, C., Keren, S., Vaithilingam, S., Bodapati, S., Liu, Z., Levi, J., Smith, B. R., Ma, T. J., Oralkan, O., et al. (2008). Carbon nanotubes as photoacoustic molecular imaging agents in living mice, *Nat. Nanotechnol.,* **3**, pp. 557–562.

39. Galanzha, E. I., Kim, J. W., and Zharov, V. P. (2009). Nanotechnology-based molecular photoacoustic and photothermal flow cytometry platform for in-vivo detection and killing of circulating cancer stem cells, *J. Biophotonics,* **2**, pp. 725–735.

40. Qin, H., Zhou, T., Yang, S. H., and Xing, D. (2015). Fluorescence quenching nanoprobes dedicated to in vivo photoacoustic imaging and high-efficient tumor therapy in deep-seated tissue, *Small,* **11**, pp. 2675–2686.

41. Zang, Y. D., Wei, Y. C., Shi, Y. J., Chen, Q., and Xing, D. (2016). Chemo/photoacoustic dual therapy with mRNA-triggered DOX release and photoinduced shockwave based on a DNA-gold nanoplatform, *Small,* **12**, pp. 756–769.

42. Zhong, J. P., Wen, L. W., Yang, S. H., Xiang, L. Z., Chen, Q., and Xing, D. (2015). Imaging-guided high-efficient photoacoustic tumor therapy with targeting gold nanorods, *Nanomed. Nanotechnol.,* **11**, pp. 1499–1509.

43. Zhong, J. P., Yang, S. H., Wen, L. W., and Xing, D. (2016). Imaging-guided photoacoustic drug release and synergistic chemo-photoacoustic therapy with paclitaxel-containing nanoparticles, *J. Control. Release,* **226**, pp. 77–87.

44. Yang, X. M., Skrabalak, S. E., Li, Z. Y., Xia, Y. N., and Wang, L. H. V. (2007). Photoacoustic tomography of a rat cerebral cortex in vivo with au nanocages as an optical contrast agent, *Nano Lett.,* **7**, pp. 3798–3802.

45. Chen, J. Y., McLellan, J. M., Siekkinen, A., Xiong, Y. J., Li, Z. Y., and Xia, Y. N. (2006). Facile synthesis of gold-silver nanocages with controllable pores on the surface, *J. Am. Chem. Soc.,* **128**, pp. 14776–14777.

46. Huang, G. J., Yang, S. H., Yuan, Y., and Xing, D. (2011). Combining x-ray and photoacoustics for in vivo tumor imaging with gold nanorods, *Appl. Phys. Lett.,* **99**.

47. Qin, H., Zhou, T., Yang, S. H., Chen, Q., and Xing, D. (2013). Gadolinium(III)-gold nanorods for MRI and photoacoustic imaging dual-modality detection of macrophages in atherosclerotic inflammation, *Nanomedicine (UK),* **8**, pp. 1611–1624.

48. Qin, H., Zhao, Y., Zhang, J., Pan, X., Yang, S., and Xing, D. (2016). Inflammation-targeted gold nanorods for intravascular photoacoustic

imaging detection of matrix metalloproteinase-2 (MMP2) in atherosclerotic plaques, *Nanomedicine (UK)*, **12**, pp. 1765–1774.

49. Ma, Z., Qin, H., Chen, H., Yang, H., Xu, J., Yang, S., Hu, J., and Xing, D. (2017). Phage display-derived oligopeptide-functionalized probes for in vivo specific photoacoustic imaging of osteosarcoma, *Nanomedicine (UK)*, **13**, pp. 111–121.

50. Yang, S., Ye, F., and Xing, D. (2012). Intracellular label-free gold nanorods imaging with photoacoustic microscopy, *Opt. Express*, **20**, pp. 10370–10375.

51. Kim, C., Song, H. M., Cai, X., Yao, J. J., Wei, A., and Wang, L. H. V. (2011). In vivo photoacoustic mapping of lymphatic systems with plasmon-resonant nanostars, *J. Mater. Chem.*, **21**, pp. 2841–2844.

52. Nie, L. M., Wang, S. J., Wang, X. Y., Rong, P. F., Bhirde, A., Ma, Y., Liu, G., Huang, P., Lu, G. M., and Chen, X. Y. (2014). In vivo volumetric photoacoustic molecular angiography and therapeutic monitoring with targeted plasmonic nanostars, *Small*, **10**, pp. 1585–1593.

53. Lu, W., Huang, Q., Geng, K. B., Wen, X. X., Zhou, M., Guzatov, D., Brecht, P., Su, R., Oraevsky, A., Wang, L. V., et al. (2010). Photoacoustic imaging of living mouse brain vasculature using hollow gold nanospheres, *Biomaterials*, **31**, pp. 2617–2626.

54. Cole, J. R., Mirin, N. A., Knight, M. W., Goodrich, G. P., and Halas, N. J. (2009). Photothermal efficiencies of nanoshells and nanorods for clinical therapeutic applications, *J. Phys. Chem. C*, **113**, pp. 12090–12094.

55. Mallidi, S., Larson, T., Tam, J., Joshi, P. P., Karpiouk, A., Sokolov, K., and Emelianov, S. (2009). Multiwavelength photoacoustic imaging and plasmon resonance coupling of gold nanoparticles for selective detection of cancer, *Nano Lett.*, **9**, pp. 2825–2831.

56. Huang, P., Rong, P. F., Lin, J., Li, W. W., Yan, X. F., Zhang, M. G., Nie, L. M., Niu, G., Lu, J., Wang, W., et al. (2014). Triphase interface synthesis of plasmonic gold bellflowers as near-infrared light mediated acoustic and thermal theranostics, *J. Am. Chem. Soc.*, **136**, pp. 8307–8313.

57. Huang, P., Pandoli, O., Wang, X. S., Wang, Z., Li, Z. M., Zhang, C. L., Chen, F., Lin, J., Cui, D. X., and Chen, X. Y. (2012). Chiral guanosine 5′-monophosphate-capped gold nanoflowers: Controllable synthesis, characterization, surface-enhanced Raman scattering activity, cellular imaging and photothermal therapy, *Nano Res.*, **5**, pp. 630–639.

58. Luke, G. P., Bashyam, A., Homan, K. A., Makhija, S., Chen, Y. S., and Emelianov, S. Y. (2013). Silica-coated gold nanoplates as stable

photoacoustic contrast agents for sentinel lymph node imaging, *Nanotechnology,* **24**.

59. Ding, W. Z., Lou, C. G., Qiu, J. S., Zhao, Z. B., Zhou, Q., Liang, M. J., Ji, Z., Yang, S. H., and Xing, D. (2016). Targeted Fe-filled carbon nanotube as a multifunctional contrast agent for thermoacoustic and magnetic resonance imaging of tumor in living mice, *Nanomed. Nanotechnol.,* **12**, pp. 235–244.
60. Zhou, F. F., Wu, S. N., Wu, B. Y., Chen, W. R., and Xing, D. (2011). Mitochondria-targeting single-walled carbon nanotubes for cancer photothermal therapy, *Small,* **7**, pp. 2727–2735.
61. Zhou, F. F., Wu, S., Song, S., Chen, W. R., Resasco, D. E., and Xing, D. (2012). Antitumor immunologically modified carbon nanotubes for photothermal therapy, *Biomaterials,* **33**, pp. 3235–3242.
62. Zhou, F. F., Wu, S. N., Yuan, Y., Chen, W. R., and Xing, D. (2012). Mitochondria-targeting photoacoustic therapy using single-walled carbon nanotubes, *Small,* **8**, pp. 1543–1550.
63. Wen, L. W., Ding, W. Z., Yang, S. H., and Xing, D. (2016). Microwave pumped high-efficient thermoacoustic tumor therapy with single wall carbon nanotubes, *Biomaterials,* **75**, pp. 163–173.
64. Zhou, T., Zhou, X. M., and Xing, D. (2014). Controlled release of doxorubicin from graphene oxide based charge-reversal nanocarrier, *Biomaterials,* **35**, pp. 4185–4194.
65. Liu, L. J., Wei, Y. C., Zhai, S. D., Chen, Q., and Xing, D. (2015). Dihydroartemisinin and transferrin dual-dressed nano-graphene oxide for a pH-triggered chemotherapy, *Biomaterials,* **62**, pp. 35–46.
66. Huang, P., Lin, J., Wang, X. S., Wang, Z., Zhang, C. L., He, M., Wang, K., Chen, F., Li, Z. M., Shen, G. X., et al. (2012). Light-triggered theranostics based on photosensitizer-conjugated carbon dots for simultaneous enhanced-fluorescence imaging and photodynamic therapy, *Adv. Mater.,* **24**, pp. 5104–5110.
67. Ge, J. C., Jia, Q. Y., Liu, W. M., Guo, L., Liu, Q. Y., Lan, M. H., Zhang, H. Y., Meng, X. M., and Wang, P. F. (2015). Red-emissive carbon dots for fluorescent, photoacoustic, and thermal theranostics in living mice, *Adv. Mater.,* **27**, pp. 4169–4177.
68. Zhong, J. P., Yang, S. H., Zheng, X. H., Zhou, T., and Xing, D. (2013). In vivo photoacoustic therapy with cancer-targeted indocyanine green-containing nanoparticles, *Nanomedicine (UK),* **8**, pp. 903–919.
69. Zhong, J. P., and Yang, S. H. (2014). Contrast-enhanced photoacoustic imaging using indocyanine green-containing nanoparticles, *J. Innov. Opt. Heal. Sci.,* **7**.

70. Stantz, K. M., Cao, M. S., Liu, B., Miller, K. D., and Guo, L. L. (2010). Molecular imaging of neutropilin-1 receptor using photoacoustic spectroscopy in breast tumors, *Photons Plus Ultrasound Imaging Sens.*, 2010, **7564**.

71. Bhattacharyya, S., Wang, S., Reinecke, D., Kiser, W., Kruger, R. A., and DeGrado, T. R. (2008). Synthesis and evaluation of near-infrared (NIR) dye-herceptin conjugates as photoacoustic computed tomography (PCT) probes for HER2 expression in breast cancer, *Bioconjug. Chem.*, **19**, pp. 1186–1193.

72. Zhang, L. W., Gao, S., Zhang, F., Yang, K., Ma, Q. J., and Zhu, L. (2014). Activatable hyaluronic acid nanoparticle as a theranostic agent for optical/photoacoustic image-guided photothermal therapy, *ACS Nano*, **8**, pp. 12250–12258.

73. Yang, K., Zhu, L., Nie, L. M., Sun, X. L., Cheng, L., Wu, C. X., Niu, G., Chen, X. Y., and Liu, Z. (2014). Visualization of protease activity in vivo using an activatable photo-acoustic imaging probe based on CuS nanoparticles, *Theranostics*, **4**, pp. 134–141.

74. Zha, Z. B., Zhang, S. H., Deng, Z. J., Li, Y. Y., Li, C. H., and Dai, Z. F. (2013). Enzyme-responsive copper sulphide nanoparticles for combined photoacoustic imaging, tumor-selective chemotherapy and photothermal therapy, *Chem. Commun.*, **49**, pp. 3455–3457.

75. Ku, G., Zhou, M., Song, S. L., Huang, Q., Hazle, J., and Li, C. (2012). Copper sulfide nanoparticles as a new class of photoacoustic contrast agent for deep tissue imaging at 1064 nm, *ACS Nano*, **6**, pp. 7489–7496.

76. Cheng, L., Liu, J. J., Gu, X., Gong, H., Shi, X. Z., Liu, T., Wang, C., Wang, X. Y., Liu, G., Xing, H. Y., et al. (2014). PEGylated WS2 nanosheets as a multifunctional theranostic agent for in vivo dual-modal ct/photoacoustic imaging guided photothermal therapy, *Adv. Mater.*, **26**, pp. 1886–1893.

77. Yin, W. Y., Yan, L., Yu, J., Tian, G., Zhou, L. J., Zheng, X. P., Zhang, X., Yong, Y., Li, J., Gu, Z. J., et al. (2014). High-throughput synthesis of single-layer MoS2 nanosheets as a near-infrared photothermal-triggered drug delivery for effective cancer therapy, *ACS Nano*, **8**, pp. 6922–6933.

78. Yang, K., Yang, G. B., Chen, L., Cheng, L., Wang, L., Ge, C. C., and Liu, Z. (2015). FeS nanoplates as a multifunctional nano-theranostic for magnetic resonance imaging guided photothermal therapy, *Biomaterials*, **38**, pp. 1–9.

79. Hannah, A., Luke, G., Wilson, K., Homan, K., and Emelianov, S. (2014). Indocyanine green-loaded photoacoustic nanodroplets: Dual contrast nanoconstructs for enhanced photoacoustic and ultrasound imaging, *ACS Nano*, **8**, pp. 250–259.

80. Wilson, K., Homan, K., and Emelianov, S. (2012). Biomedical photoacoustics beyond thermal expansion using triggered nanodroplet vaporization for contrast-enhanced imaging, *Nat. Commun.,* **3**.
81. Song, X. J., Gong, H., Yin, S. N., Cheng, L., Wang, C., Li, Z. W., Li, Y. G., Wang, X. Y., Liu, G., and Liu, Z. (2014). Ultra- small iron oxide doped polypyrrole nanoparticles for in vivo multimodal imaging guided photothermal therapy, *Adv. Funct. Mater.,* **24**, pp. 1194–1201.
82. Liang, X. L., Li, Y. Y., Li, X. D., Jing, L. J., Deng, Z. J., Yue, X. L., Li, C. H., and Dai, Z. F. (2015). PEGylated polypyrrole nanoparticles conjugating gadolinium chelates for dual-modal MRI/photoacoustic imaging guided photothermal therapy of cancer, *Adv. Funct. Mater.,* **25**, pp. 1451–1462.
83. Zha, Z. B., Deng, Z. J., Li, Y. Y., Li, C. H., Wang, J. R., Wang, S. M., Qu, E. Z., and Dai, Z. F. (2013). Biocompatible polypyrrole nanoparticles as a novel organic photoacoustic contrast agent for deep tissue imaging, *Nanoscale,* **5**, pp. 4462–4467.
84. Pu, K., Mei, J. G., Jokerst, J. V., Hong, G. S., Antaris, A. L., Chattopadhyay, N., Shuhendler, A. J., Kurosawa, T., Zhou, Y., Gambhir, S. S., et al. (2015). Diketopyrrolopyrrole-based semiconducting polymer nanoparticles for in vivo photoacoustic imaging, *Adv. Mater.,* **27**, pp. 5184–5190.
85. Repenko, T., Fokong, S., De Laporte, L., Go, D., Kiessling, F., Lammers, T., and Kuehne, A. J. C. (2015). Water-soluble dopamine-based polymers for photoacoustic imaging, *Chem. Commun.,* **51**, pp. 6084–6087.
86. Lin, L. S., Cong, Z. X., Cao, J. B., Ke, K. M., Peng, Q. L., Gao, J. H., Yang, H. H., Liu, G., and Chen, X. Y. (2014). Multifunctional Fe_3O_4@polydopamine core-shell nanocomposites for intracellular mRNA detection and imaging-guided photothermal therapy, *ACS Nano,* **8**, pp. 3876–3883.
87. Kircher, M. F., Gambhir, S. S., and Grimm, J. (2011). Noninvasive cell-tracking methods, *Nat. Rev. Clin. Oncol.,* **8**, pp. 677–688.
88. Massoud, T. F., Singh, A., and Gambhir, S. S. (2008). Noninvasive molecular neuroimaging using reporter genes: Part II, experimental, current, and future applications, *Am. J. Neuroradiol.,* **29**, pp. 409–418.
89. Blasberg, R. G., and Tjuvajev, A. G. (2003). Molecular-genetic imaging: Current and future perspectives, *J. Clin. Invest.,* **111**, pp. 1620–1629.
90. Song, K. H., Stein, E. W., Margenthaler, J. A., and Wang, L. V. (2008). Noninvasive photoacoustic identification of sentinel lymph nodes containing methylene blue in vivo in a rat model, *J. Biomed. Opt.,* **13**.
91. de la Zerda, A., Liu, Z. A., Bodapati, S., Teed, R., Vaithilingam, S., Khuri-Yakub, B. T., Chen, X. Y., Dai, H. J., and Gambhir, S. S. (2010). Ultrahigh

sensitivity carbon nanotube agents for photoacoustic molecular imaging in living mice, *Nano Lett.,* **10**, pp. 2168–2172.

92. Desmettre, T., Devoisselle, J. M., and Mordon, S. (2000). Fluorescence properties and metabolic features of indocyanine green (ICG) as related to angiography, *Surv. Ophthalmol.,* **45**, pp. 15–27.
93. Mordon, S., Devoisselle, J. M., Soulie-Begu, S., and Desmettre, T. (1998). Indocyanine green: Physicochemical factors affecting its fluorescence in vivo, *Microvasc. Res.,* **55**, pp. 146–152.
94. Wang, Y. W., Fu, Y. Y., Peng, Q. L., Guo, S. S., Liu, G., Li, J., Yang, H. H., and Chen, G. N. (2013). Dye-enhanced graphene oxide for photothermal therapy and photoacoustic imaging, *J. Mater. Chem. B,* **1**, pp. 5762–5767.
95. Sheng, Z. H., Hu, D. H., Zheng, M. B., Zhao, P. F., Liu, H. L., Gao, D. Y., Gong, P., Gao, G. H., Zhang, P. F., Ma, Y. F., et al. (2014). Smart human serum albumin-indocyanine green nanoparticles generated by programmed assembly for dual-modal imaging-guided cancer synergistic phototherapy, *ACS Nano,* **8**, pp. 12310–12322.
96. Razansky, D., Vinegoni, C., and Ntziachristos, V. (2007). Multispectral photoacoustic imaging of fluorochromes in small animals, *Opt. Lett.,* **32**, pp. 2891–2893.
97. Yao, J. J., Maslov, K., Hu, S., and Wang, L. H. V. (2009). Evans blue dye-enhanced capillary-resolution photoacoustic microscopy in vivo, *J. Biomed. Opt.,* **14**.
98. Levi, J., Kothapalli, S. R., Ma, T. J., Hartman, K., Khuri-Yakub, B. T., and Gambhir, S. S. (2010). Design, synthesis, and imaging of an activatable photoacoustic probe, *J. Am. Chem. Soc.,* **132**, pp. 11264–11269.
99. Razansky, D., Harlaar, N. J., Hillebrands, J. L., Taruttis, A., Herzog, E., Zeebregts, C. J., van Dam, G. M., and Ntziachristos, V. (2012). Multispectral optoacoustic tomography of matrix metalloproteinase activity in vulnerable human carotid plaques, *Mol. Imaging Biol.,* **14**, pp. 277–285.
100. Jain, P. K., Lee, K. S., El-Sayed, I. H., and El-Sayed, M. A. (2006). Calculated absorption and scattering properties of gold nanoparticles of different size, shape, and composition: Applications in biological imaging and biomedicine, *J. Phys. Chem. B,* **110**, pp. 7238–7248.
101. Bastus, N. G., Comenge, J., and Puntes, V. (2011). Kinetically controlled seeded growth synthesis of citrate-stabilized gold nanoparticles of up to 200 nm: Size focusing versus Ostwald ripening, *Langmuir,* **27**, pp. 11098–11105.

102. Huang, X. H., El-Sayed, I. H., Qian, W., and El-Sayed, M. A. (2006). Cancer cell imaging and photothermal therapy in the near-infrared region by using gold nanorods, *J. Am. Chem. Soc.,* **128**, pp. 2115–2120.

103. Liopo, A., Conjusteau, A., Tsyboulski, D., Ermolinsky, B., Kazansky, A., and Oraevsky, A. (2012). Biocompatible gold nanorod conjugates for preclinical biomedical research, *J. Nanomed. Nanotechnol.,* **S2**.

104. de la Zerda, A., Kim, J. W., Galanzha, E. I., Gambhir, S. S., and Zharov, V. P. (2011). Advanced contrast nanoagents for photoacoustic molecular imaging, cytometry, blood test and photothermal theranostics, *Contrast Media Mol. Imaging,* **6**, pp. 346–369.

105. Grabinski, C., Schaeublin, N., Wijaya, A., D'Couto, H., Baxamusa, S. H., Hamad-Schifferli, K., and Hussain, S. M. (2011). Effect of gold nanorod surface chemistry on cellular response, *ACS Nano,* **5**, pp. 2870–2879.

106. Yavuz, M. S., Cheng, Y. Y., Chen, J. Y., Cobley, C. M., Zhang, Q., Rycenga, M., Xie, J. W., Kim, C., Song, K. H., Schwartz, A. G., et al. (2009). Gold nanocages covered by smart polymers for controlled release with near-infrared light, *Nat. Mater.,* **8**, pp. 935–939.

107. Skrabalak, S. E., Au, L., Li, X., D., and Xia, Y. N. (2007). Facile synthesis of Ag nanocubes and Au nanocages, *Nat. Protoc.,* **2**, pp. 2182–2190.

108. Cheng, K., Kothapalli, S. R., Liu, H. G., Koh, A. L., Jokerst, J. V., Jiang, H., Yang, M., Li, J. B., Levi, J., Wu, J. C., et al. (2014). Construction and validation of nano gold tripods for molecular imaging of living subjects, *J. Am. Chem. Soc.,* **136**, pp. 3560–3571.

109. Chen, J. Y., Wiley, B., Li, Z. Y., Campbell, D., Saeki, F., Cang, H., Au, L., Lee, J., Li, X. D., and Xia, Y. N. (2005). Gold nanocages: Engineering their structure for biomedical applications, *Adv. Mater.,* **17**, pp. 2255–2261.

110. Chen, Y. S., Frey, W., Kim, S., Homan, K., Kruizinga, P., Sokolov, K., and Emelianov, S. (2010). Enhanced thermal stability of silica-coated gold nanorods for photoacoustic imaging and image-guided therapy, *Opt. Express,* **18**, pp. 8867–8877.

111. Chen, Y. S., Frey, W., Kim, S., Kruizinga, P., Homan, K., and Emelianov, S. (2011). Silica-coated gold nanorods as photoacoustic signal nanoamplifiers, *Nano Lett.,* **11**, pp. 348–354.

112. Jokerst, J. V., Thangaraj, M., Kempen, P. J., Sinclair, R., and Gambhir, S. S. (2012). Photoacoustic imaging of mesenchymal stem cells in living mice via silica-coated gold nanorods, *ACS Nano,* **6**, pp. 5920–5930.

113. Niidome, T., Yamagata, M., Okamoto, Y., Akiyama, Y., Takahashi, H., Kawano, T., Katayama, Y., and Niidome, Y. (2006). PEG-modified gold nanorods with a stealth character for in vivo applications, *J. Control Release,* **114**, pp. 343–347.

114. Kwag, D. S., Park, K., Oh, K. T., and Lee, E. S. (2013). Hyaluronated fullerenes with photoluminescent and antitumoral activity, *Chem. Commun.,* **49**, pp. 282–284.

115. Raoof, M., Mackeyev, Y., Cheney, M. A., Wilson, L. J., and Curley, S. A. (2012). Internalization of C60 fullerenes into cancer cells with accumulation in the nucleus via the nuclear pore complex, *Biomaterials,* **33**, pp. 2952–2960.

116. Shim, M., Kam, N. W. S., Chen, R. J., Li, Y. M., and Dai, H. J. (2002). Functionalization of carbon nanotubes for biocompatibility and biomolecular recognition, *Nano Lett.,* **2**, pp. 285–288.

117. Leeuw, T. K., Reith, R. M., Simonette, R. A., Harden, M. E., Cherukuri, P., Tsyboulski, D. A., Beckingham, K. M., and Weisman, R. B. (2007). Single-walled carbon nanotubes in the intact organism: Near-IR imaging and biocompatibility studies in Drosophila, *Nano Lett.,* **7**, pp. 2650–2654.

118. Zhou, F. F., Xing, D., Ou, Z. M., Wu, B. Y., Resasco, D. E., and Chen, W. R. (2009). Cancer photothermal therapy in the near-infrared region by using single-walled carbon nanotubes, *J. Biomed. Opt.,* **14**.

119. Xiang, L. Z., Yuan, Y., Xing, D., Ou, Z. M., Yang, S. H., and Zhou, F. F. (2009). Photoacoustic molecular imaging with antibody-functionalized single-walled carbon nanotubes for early diagnosis of tumor, *J. Biomed. Opt.,* **14**.

120. Kim, J. W., Galanzha, E. I., Shashkov, E. V., Moon, H. M., and Zharov, V. P. (2009). Golden carbon nanotubes as multimodal photoacoustic and photothermal high-contrast molecular agents, *Nat. Nanotechnol.,* **4**, pp. 688–694.

121. de la Zerda, A., Bodapati, S., Teed, R., May, S. Y., Tabakman, S. M., Liu, Z., Khuri-Yakub, B. T., Chen, X. Y., Dai, H. J., and Gambhir, S. S. (2012). Family of enhanced photoacoustic imaging agents for high-sensitivity and multiplexing studies in living mice, *ACS Nano,* **6**, pp. 4694–4701.

122. Yang, K., Hu, L. L., Ma, X. X., Ye, S. Q., Cheng, L., Shi, X. Z., Li, C. H., Li, Y. G., and Liu, Z. (2012). Multimodal imaging guided photothermal therapy using functionalized graphene nanosheets anchored with magnetic nanoparticles, *Adv. Mater.,* **24**, pp. 1868–1872.

123. Qin, H., Zhou, T., Yang, S., and Xing, D. (2015). Fluorescence quenching nanoprobes dedicated to in vivo photoacoustic imaging and high-efficient tumor therapy in deep-seated tissue, *Small,* **11**, pp. 2675–2686.

124. Tian, Q. W., Tang, M. H., Sun, Y. G., Zou, R. J., Chen, Z. G., Zhu, M. F., Yang, S. P., Wang, J. L., Wang, J. H., and Hu, J. Q. (2011). Hydrophilic flower-like cus superstructures as an efficient 980 nm laser-driven photothermal agent for ablation of cancer cells, *Adv. Mater.,* **23**, pp. 3542–3547.

125. Li, Y. B., Lu, W., Huang, Q. A., Huang, M. A., Li, C., and Chen, W. (2010). Copper sulfide nanoparticles for photothermal ablation of tumor cells, *Nanomedicine (UK),* **5**, pp. 1161–1171.

126. Liu, T., Wang, C., Gu, X., Gong, H., Cheng, L., Shi, X. Z., Feng, L. Z., Sun, B. Q., and Liu, Z. (2014). Drug delivery with PEGylated MoS2 nano-sheets for combined photothermal and chemotherapy of cancer, *Adv. Mater.,* **26**, pp. 3433–3440.

127. Tian, Q. W., Jiang, F. R., Zou, R. J., Liu, Q., Chen, Z. G. Zhu, M. F., Yang, S. P., Wang, J. L., Wang, J. H., and Hu, J. Q. (2011). Hydrophilic Cu9S5 nanocrystals: a photothermal agent with a 25.7% heat conversion efficiency for photothermal ablation of cancer cells in vivo, *ACS Nano,* **5**, pp. 9761–9771.

128. Hessel, C. M., Pattani, V. P., Rasch, M., Panthani, M. G., Koo, B., Tunnell, J. W., and Korgel, B. A. (2011). Copper selenide nanocrystals for photothermal therapy, *Nano Lett.,* **11**, pp. 2560–2566.

129. Zhou, M., Ku, G., Pageon, L., and Li, C. (2014). Theranostic probe for simultaneous in vivo photoacoustic imaging and confined photothermolysis by pulsed laser at 1064 nm in 4T1 breast cancer model, *Nanoscale,* **6**, pp. 15228–15235.

130. Wang, Z. T., Huang, P., Jacobson, O., Wang, Z., Liu, Y. J., Lin, L. S., Lin, J., Lu, N., Zhang, H. M., Tian, R., et al. (2016). Biomineralization-inspired synthesis of copper sulfide-ferritin nanocages as cancer theranostics, *ACS Nano,* **10**, pp. 3453–3460.

131. Chou, S. S., Kaehr, B., Kim, J., Foley, B. M., De, M., Hopkins, P. E., Huang, J., Brinker, C. J., and Dravid, V. P. (2013). Chemically exfoliated MoS2 as near-infrared photothermal agents, *Angew. Chem. Int. Edit.,* **52**, pp. 4160–4164.

132. Sheeran, P. S., Luois, S., Dayton, P. A., and Matsunaga, T. O. (2011). Formulation and acoustic studies of a new phase-shift agent for diagnostic and therapeutic ultrasound, *Langmuir,* **27**, pp. 10412–10420.

133. Lindner, J. R. (2004). Microbubbles in medical imaging: Current applications and future directions, *Nat. Rev. Drug Discov.,* **3**, pp. 527–532.

134. Reznik, N., Seo, M., Williams, R., Bolewska-Pedyczak, E., Lee, M., Matsuura, N., Gariepy, J., Foster, F. S., and Burns, P. N. (2012). Optical studies of vaporization and stability of fluorescently labelled perfluorocarbon droplets, *Phys. Med. Biol.,* **57**, pp. 7205–7217.

135. Akers, W. J., Kim, C., Berezin, M., Guo, K., Fuhrhop, R., Lanza, G. M., Fischer, G. M., Daltrozzo, E., Zumbusch, A., Cai, X., et al. (2011). Noninvasive photoacoustic and fluorescence sentinel lymph node identification using dye-loaded perfluorocarbon nanoparticles, *ACS Nano,* **5**, pp. 173–182.

136. Qu, M., Mallidi, S., Mehrmohammadi, M., Truby, R., Homan, K., Joshi, P., Chen, Y. S., Sokolov, K., and Emelianov, S. (2011). Magneto-photo-acoustic imaging, *Biomed. Opt. Express,* **2**, pp. 385–395.

137. Hong, J. Y., Yoon, H., and Jang, J. (2010). Kinetic study of the formation of polypyrrole nanoparticles in water-soluble polymer/metal cation systems: A light-scattering analysis, *Small,* **6**, pp. 679–686.

138. Au, K. M., Lu, Z. H., Matcher, S. J., and Armes, S. P. (2011). Polypyrrole nanoparticles: A potential optical coherence tomography contrast agent for cancer imaging, *Adv. Mater.,* **23**, pp. 5792–5795.

139. Fan, Q. L., Cheng, K., Hu, X., Ma, X. W., Zhang, R. P., Yang, M., Lu, X. M., Xing, L., Huang, W., Gambhir, S. S., et al. (2014). Transferring biomarker into molecular probe: Melanin nanoparticle as a naturally active platform for multimodality imaging, *J. Am. Chem. Soc.,* **136**, pp. 15185–15194.

140. Zhang, R. P., Fan, Q. L., Yang, M., Cheng, K., Lu, X. M., Zhang, L., Huang, W., and Cheng, Z. (2015). Engineering melanin nanoparticles as an efficient drug-delivery system for imaging-guided chemotherapy, *Adv. Mater.,* **27**, pp. 5063–5069.

141. Lin, J., Wang, M., Hu, H., Yang, X. Y., Wen, B., Wang, Z. T., Jacobson, O., Song, J. B., Zhang, G. F., Niu, G., et al. (2016). Multimodal-imaging-guided cancer phototherapy by versatile biomimetic theranostics with UV and gamma-irradiation protection, *Adv. Mater.,* **28**, pp. 3273–3279.

142. Kumar, A., Kumar, S., Rhim, W. K., Kim, G. H., and Nam, J. M. (2014). Oxidative nanopeeling chemistry-based synthesis and photodynamic and photothermal therapeutic applications of plasmonic core-petal nanostructures, *J. Am. Chem. Soc.,* **136**, pp. 16317–16325.

143. Lee, H., Dellatore, S. M., Miller, W. M., and Messersmith, P. B. (2007). Mussel-inspired surface chemistry for multifunctional coatings, *Science,* **318**, pp. 426–430.

144. Liu, X. S., Cao, J. M., Li, H., Li, J. Y., Jin, Q., Ren, K. F., and Ji, J. (2013). Mussel-inspired polydopamine: A biocompatible and ultrastable coating for nanoparticles in vivo, *ACS Nano,* **7**, pp. 9384–9395.

145. Kiriy, A., Senkovskyy, V., and Sommer, M. (2011). Kumada catalyst-transfer polycondensation: Mechanism, opportunities, and challenges, *Macromol. Rapid Commun.,* **32**, pp. 1503–1517.

146. Lee, H., Rho, J., and Messersmith, P. B. (2009). Facile conjugation of biomolecules onto surfaces via mussel adhesive protein inspired coatings, *Adv. Mater.,* **21**, pp. 431–434.

147. Xu, L. Q., Yang, W. J., Neoh, K. G., Kang, E. T., and Fu, G. D. (2010). Dopamine-induced reduction and functionalization of graphene oxide nanosheets, *Macromolecules,* **43**, pp. 8336–8339.

148. Yang, S. H., Kang, S. M., Lee, K. B., Chung, T. D., Lee, H., and Choi, I. S. (2011). Mussel-inspired encapsulation and functionalization of individual yeast cells, *J. Am. Chem. Soc.,* **133**, pp. 2795–2797.

149. Ham, H. O., Liu, Z. Q., Lau, K. H. A., Lee, H., and Messersmith, P. B. (2011). Facile DNA immobilization on surfaces through a catecholamine polymer, *Angew. Chem. Int. Edit.,* **50**, pp. 732–736.

150. Liu, Y. L., Ai, K. L., Liu, J. H., Deng, M., He, Y. Y., and Lu, L. H. (2013). Dopamine-melanin colloidal nanospheres: An efficient near-infrared photothermal therapeutic agent for in vivo cancer therapy, *Adv. Mater.,* **25**, pp. 1353–1359.

151. Heeger, A. J. (2001). Semiconducting and metallic polymers: The fourth generation of polymeric materials (Nobel lecture), *Angew. Chem. Int. Edit.,* **40**, pp. 2591–2611.

152. Bidez, P. R., Li, S. X., MacDiarmid, A. G., Venancio, E. C., Wei, Y., and Lelkes, P. I. (2006). Polyaniline, an electroactive polymer, supports adhesion and proliferation of cardiac myoblasts, *J. Biomat. Sci. Polym. E,* **17**, pp. 199–212.

153. Yang, J., Choi, J., Bang, D., Kim, E., Lim, E. K., Park, H., Suh, J. S., Lee, K., Yoo, K. H., Kim, E. K., et al. (2011). Convertible organic nanoparticles for near-infrared photothermal ablation of cancer cells, *Angew. Chem. Int. Edit.,* **50**, pp. 441–444.

154. Hsiao, C. W., Chen, H. L., Liao, Z. X., Sureshbabu, R., Hsiao, H. C., Lin, S. J., Chang, Y., and Sung, H. W. (2015). Effective photothermal killing of pathogenic bacteria by using spatially tunable colloidal gels with nano-localized heating sources, *Adv. Funct. Mater.,* **25**, pp. 721–728.

155. Shi, J., Chen, Y., Wang, Q. A., and Liu, Y. (2010). Construction and efficient radical cation stabilization of cyclodextrin/aniline polypseudorotaxane and its conjugate with carbon nanotubes, *Adv. Mater.*, **22**, pp. 2575–2578.

156. Wang, J. P., Yan, R., Guo, F., Yu, M., Tan, F. P., and Li, N. (2016). Targeted lipid-polyaniline hybrid nanoparticles for photoacoustic imaging guided photothermal therapy of cancer, *Nanotechnology*, **27**, p. 285102.

157. Wu, C. F., and Chiu, D. T. (2013). Highly fluorescent semiconducting polymer dots for biology and medicine, *Angew. Chem. Int. Edit.*, **52**, pp. 3086–3109.

158. Zhu, C. L., Liu, L. B., Yang, Q., Lv, F. T., and Wang, S. (2012). Water-soluble conjugated polymers for imaging, diagnosis, and therapy, *Chem. Rev.*, **112**, pp. 4687–4735.

159. Pu, K. Y., Shuhendler, A. J., Jokerst, J. V., Mei, J. G., Gambhir, S. S., Bao, Z. N., and Rao, J. H. (2014). Semiconducting polymer nanoparticles as photoacoustic molecular imaging probes in living mice, *Nat. Nanotechnol.*, **9**, pp. 233–239.

160. Pu, K. Y., Shuhendler, A. J., and Rao, J. H. (2013). Semiconducting polymer nanoprobe for in vivo imaging of reactive oxygen and nitrogen species, *Angew. Chem. Int. Edit.*, **52**, pp. 10325–10329.

161. Lovell, J. F., Jin, C. S., Huynh, E., Jin, H. L., Kim, C., Rubinstein, J. L., Chan, W. C. W., Cao, W. G., Wang, L. V., and Zheng, G. (2011). Porphysome nanovesicles generated by porphyrin bilayers for use as multimodal biophotonic contrast agents, *Nat. Mater.*, **10**, pp. 324–332.

162. Misteli, T., and Spector, D. L. (1997). Applications of the green fluorescent protein in cell biology and biotechnology, *Nat. Biotechnol.*, **15**, pp. 961–964.

163. Greer, L. F., and Szalay, A. A. (2002). Imaging of light emission from the expression of luciferases in living cells and organisms: A review, *Luminescence*, **17**, pp. 43–74.

164. Razansky, D., Distel, M., Vinegoni, C., Ma, R., Perrimon, N., Koster, R. W., and Ntziachristos, V. (2009). Multispectral opto-acoustic tomography of deep-seated fluorescent proteins in vivo, *Nat. Photonics*, **3**, pp. 412–417.

165. Paproski, R. J., Heinmiller, A., Wachowicz, K., and Zemp, R. J. (2014). Multi-wavelength photoacoustic imaging of inducible tyrosinase

reporter gene expression in xenograft tumors, *Sci. Rep. (UK)*, **4**, p. 5329.

166. Li, L., Zemp, R. J., Lungu, G., Stoica, G., and Wang, L. H. V. (2007). Photoacoustic imaging of lacZ gene expression in vivo, *J. Biomed. Opt.*, **12**.
167. Li, L., Zhang, H. F., Zemp, R. J., Maslov, K., and Wang, L. H. V. (2008). Simultaneous imaging of a lacz-marked tumor and microvasculature morphology in vivo by dual-wavelength photoacoustic microscopy, *J. Innov. Opt. Heal. Sci.*, **1**, pp. 207–215.
168. Oetting, W. S. (2000). The tyrosinase gene and oculocutaneous albinism type 1 (OCA1): A model for understanding the molecular biology of melanin formation, *Pigm. Cell Res.*, **13**, pp. 320–325.
169. Zonios, G., Dimou, A., Bassukas, I., Galaris, D., Tsolakidis, A., and Kaxiras, E. (2008). Melanin absorption spectroscopy: New method for noninvasive skin investigation and melanoma detection, *J. Biomed. Opt.*, **13**.
170. Paproski, R. J., Forbrich, A., Harrison, T., Hitt, M., and Zemp, R. J. (2011). Photoacoustic imaging of gene expression using tyrosinase as a reporter gene, *Photons Plus Ultrasound: Imaging Sens.*, **7899**.
171. Weissleder, R., Simonova, M., Bogdanova, A., Bredow, S., Enochs, W. S., and Bogdanov, A. (1997). MR imaging and scintigraphy of gene expression through melanin induction, *Radiology*, **204**, pp. 425–429.
172. Krumholz, A., VanVickle-Chavez, S. J., Yao, J. J., Fleming, T. P., Gillanders, W. E., and Wang, L. H. V. (2011). Photoacoustic microscopy of tyrosinase reporter gene in vivo, *J. Biomed. Opt.*, **16**.
173. Qin, C. X., Cheng, K., Chen, K., Hu, X., Liu, Y., Lan, X. L., Zhang, Y. X., Liu, H. G., Xu, Y. D., Bu, L. H., et al. (2013). Tyrosinase as a multifunctional reporter gene for photoacoustic/MRI/PET triple modality molecular imaging, *Sci. Rep.-Uk*, **3**.
174. Ponomarev, V., Doubrovin, M., Serganova, I., Vider, J., Shavrin, A., Beresten, T., Ivanova, A., Ageyeva, L., Tourkova, V., Balatoni, J., et al. (2004). A novel triple-modality reporter gene for whole-body fluorescent, bioluminescent, and nuclear noninvasive imaging, *Eur. J. Nucl. Med. Mol. I*, **31**, pp. 740–751.
175. Stritzker, J., Kirscher, L., Scadeng, M., Deliolanis, N. C., Morscher, S., Symvoulidis, P., Schaefer, K., Zhang, Q., Buckel, L., Hess, M., et al. (2013). Vaccinia virus-mediated melanin production allows MR and optoacoustic deep tissue imaging and laser-induced thermotherapy of cancer, *P. Natl. Acad. Sci. U. S. A.*, **110**, pp. 3316–3320.

176. Jathoul, A. P., Laufer, J., Ogunlade, O., Treeby, B., Cox, B., Zhang, E., Johnson, P., Pizzey, A. R., Philip, B., Marafioti, T., et al. (2015). Deep in vivo photoacoustic imaging of mammalian tissues using a tyrosinase-based genetic reporter, *Nat. Photonics,* **9**, pp. 239–246.

177. Urabe, K., Aroca, P., Tsukamoto, K., Mascagna, D., Palumbo, A., Prota, G., and Hearing, V. J. (1994). The inherent cytotoxicity of melanin precursors: A revision, *Biochim. Biophys. Acta,* **1221**, pp. 272–278.

178. Homan, K. A., Souza, M., Truby, R., Luke, G. P., Green, C., Vreeland, E., and Emelianov, S. (2012). Silver nanoplate contrast agents for in vivo molecular photoacoustic imaging, *ACS Nano,* **6**, pp. 641–650.

179. Grootendorst, D. J., Jose, J., Fratila, R. M., Visscher, M., Velders, A. H., Ten Haken, B., Van Leeuwen, T. G., Steenbergen, W., Manohar, S., and Ruers, T. J. M. (2013). Evaluation of superparamagnetic iron oxide nanoparticles (Endorem (R)) as a photoacoustic contrast agent for intra-operative nodal staging, *Contrast Media. Mol. I,* **8**, pp. 83–91.

180. Shashkov, E. V., Everts, M., Galanzha, E. I., and Zharov, V. P. (2008). Quantum dots as multimodal photoacoustic and photothermal contrast agents, *Nano Lett.,* **8**, pp. 3953–3958.

181. Asharani, P. V., Yi, L. W., Gong, Z. Y., and Valiyaveettil, S. (2011). Comparison of the toxicity of silver, gold and platinum nanoparticles in developing zebrafish embryos, *Nanotoxicology,* **5**, pp. 43–54.

182. Asharani, P. V., Wu, Y. L., Gong, Z. Y., and Valiyaveettil, S. (2008). Toxicity of silver nanoparticles in zebrafish models, *Nanotechnology,* **19**.

183. Kim, C., Cho, E. C., Chen, J., Song, K. H., Au, L., Favazza, C., Zhang, Q., Cobley, C. M., Gao, F., Xia, Y., et al. (2010). In vivo molecular photoacoustic tomography of melanomas targeted by bioconjugated gold nanocages, *ACS Nano,* **4**, pp. 4559–4564.

184. Huang, G. J., Si, Z., Yang, S. H., Li, C., and Xing, D. (2012). Dextran based pH-sensitive near-infrared nanoprobe for in vivo differential-absorption dual-wavelength photoacoustic imaging of tumors, *J. Mater. Chem.,* **22**, pp. 22575–22581.

185. Kang, B., Yu, D. C., Dai, Y. D., Chang, S. Q., Chen, D., and Ding, Y. T. (2009). Cancer-cell targeting and photoacoustic therapy using carbon nanotubes as "bomb" agents, *Small,* **5**, pp. 1292–1301.

186. Cheng, Z. L., Al Zaki, A., Hui, J. Z., Muzykantov, V. R., and Tsourkas, A. (2012). Multifunctional nanoparticles: Cost versus benefit of adding targeting and imaging capabilities, *Science,* **338**, pp. 903–910.

187. Peer, D., Karp, J. M., Hong, S., FaroKHzad, O. C., Margalit, R., and Langer, R. (2007). Nanocarriers as an emerging platform for cancer therapy, *Nat. Nanotechnol.,* **2**, pp. 751–760.

188. Alazraki, N. P., Eshima, D., Eshima, L. A., Herda, S. C., Murray, D. R., Vansant, J. P., and Taylor, A. T. (1997). Lymphoscintigraphy, the sentinel node concept, and the intraoperative gamma probe in melanoma, breast cancer, and other potential cancers, *Semin. Nucl. Med.,* **27**, pp. 55–67.

189. McMasters, K. M., Tuttle, T. M., Carlson, D. J., Brown, C. M., Noyes, R. D., Glaser, R. L., Vennekotter, D. J., Turk, P. S., Tate, P. S., Sardi, A., et al. (2000). Sentinel lymph node biopsy for breast cancer: A suitable alternative to routine axillary dissection in multi-institutional practice when optimal technique is used, *J. Clin. Oncol.,* **18**, pp. 2560–2566.

190. Song, K. H., Kim, C. H., Cobley, C. M., Xia, Y. N., and Wang, L. V. (2009). Near-infrared gold nanocages as a new class of tracers for photoacoustic sentinel lymph node mapping on a rat model, *Nano Lett.,* **9**, pp. 183–188.

191. Song, K. H., Kim, C., Maslov, K., and Wang, L. V. (2009). Noninvasive in vivo spectroscopic nanorod-contrast photoacoustic mapping of sentinel lymph nodes, *Eur. J. Radiol.,* **70**, pp. 227–231.

192. Galanzha, E. I., Kokoska, M. S., Shashkov, E. V., Kim, J. W., Tuchin, V. V., and Zharov, V. P. (2009). In vivo fiber-based multicolor photoacoustic detection and photothermal purging of metastasis in sentinel lymph nodes targeted by nanoparticles, *J. Biophotonics,* **2**, pp. 528–539.

193. Qin, H., Zhou, T., Yang, S., Chen, Q., and Xing, D. (2013). Gadolinium(III)-gold nanorods for MRI and photoacoustic imaging dual-modality detection of macrophages in atherosclerotic inflammation, *Nanomedicine (Lond),* **8**, pp. 1611–1624.

194. Newby, A. C. (2007). Metalloproteinases and vulnerable atherosclerotic plaques, *Trends. Cardiovasc. Med.,* **17**, pp. 253–258.

195. Galis, Z. S., Sukhova, G. K., Lark, M. W., and Libby, P. (1994). Increased expression of matrix metalloproteinases and matrix degrading activity in vulnerable regions of human atherosclerotic plaques, *J. Clin. Invest.,* **94**, pp. 2493–2503.

196. Schafers, M., Schober, O., and Hermann, S. (2010). Matrix-metalloproteinases as imaging targets for inflammatory activity in atherosclerotic plaques, *J. Nucl. Med.,* **51**, pp. 663–666.

197. Jose, J., Grootendorst, D. J., Vijn, T. W., Wouters, M. W., van Boven, H., van Leeuwen, T. G., Steenbergen, W., Ruers, T. J. M., and

Manohar, S. (2011). Initial results of imaging melanoma metastasis in resected human lymph nodes using photoacoustic computed tomography, *J. Biomed. Opt.*, **16**, p. 096021.

198. McCormack, D., Al-Shaer, M., Goldschmidt, B. S., Dale, P. S., Henry, C., Papageorgio, C., Bhattacharyya, K., and Viator, J. A. (2009). Photoacoustic detection of melanoma micrometastasis in sentinel lymph nodes, *J. Biomech. Eng. T Asme.*, **131**.
199. Olszewski, W. L., and Tarnok, A. (2008). Photoacoustic listening of cells in lymphatics: Research art or novel clinical noninvasive lymph test, *Cytom. Part A*, **73A**, pp. 1111–1113.
200. Zharov, V. P., Galanzha, E. I., and Tuchin, V. V. (2006). In vivo photothermal flow cytometry: Imaging and detection of individual cells in blood and lymph flow, *J. Cell. Biochem.*, **97**, pp. 916–932.
201. Galanzha, E. I., Shashkov, E. V., Tuchin, V. V., and Zharov, V. P. (2008). In vivo multispectral, multiparameter, photoacoustic lymph flow cytometry with natural cell focusing, label-free detection and multicolor nanoparticle probes, *Cytom. Part A*, **73A**, pp. 884–894.
202. Galanzha, E. I., Tuchin, V. V., and Zharov, V. P. (2007). Advances in small animal mesentery models for in vivo flow cytometry, dynamic microscopy, and drug screening, *World J. Gastroentero*, **13**, pp. 192–218.
203. Thomas, T. S., Dale, P. S., Weight, R. M., Atasoy, U., Mageee, J., and Viator, J. A. (2008). Photoacoustic detection of breast cancer cells in human blood, article no. 685609, *Photons Plus Ultrasound: Imaging and Sensing 2008: The Ninth Conference on Biomedical Thermoacoustics, Optoacoustics, and Acoustic-Optics*, **6856**, pp. 85609–85609.
204. Viator, J. A., Gupta, S., Goldschmidt, B. S., Bhattacharyya, K., Kannan, R., Shukla, R., Dale, P. S., Boote, E., and Katti, K. (2010). Gold nanoparticle mediated detection of prostate cancer cells using photoacoustic flowmetry with optical reflectance, *J. Biomed. Nanotechnol.*, **6**, pp. 187–191.
205. Kessenbrock, K., Plaks, V., and Werb, Z. (2010). Matrix metalloproteinases: Regulators of the tumor microenvironment, *Cell*, **141**, pp. 52–67.
206. Harris, A. L. (2002). Hypoxia–A key regulatory factor in tumour growth, *Nat. Rev. Cancer*, **2**, pp. 38–47.
207. Fukumura, D., and Jain, R. K. (2007). Tumor microenvironment abnormalities: Causes, consequences, and strategies to normalize, *J. Cell. Biochem.*, **101**, pp. 937–949.

208. Webb, B. A., Chimenti, M., Jacobson, M. P., and Barber, D. L. (2011). Dysregulated pH: A perfect storm for cancer progression, *Nat. Rev. Cancer,* **11**, pp. 671–677.

209. Tannock, I. F., and Rotin, D. (1989). Acid pH in tumors and its potential for therapeutic exploitation, *Cancer Res.,* **49**, pp. 4373–4384.

210. Li, C., Xia, J. A., Wei, X. B., Yan, H. H., Si, Z., and Ju, S. H. (2010). pH-activated near-infrared fluorescence nanoprobe imaging tumors by sensing the acidic microenvironment, *Adv. Funct. Mater.,* **20**, pp. 2222–2230.

211. Swietach, P., Vaughan-Jones, R. D., and Harris, A. L. (2007). Regulation of tumor pH and the role of carbonic anhydrase 9, *Cancer Metast. Rev.,* **26**, pp. 299–310.

212. Thews, O., Gassner, B., Kelleher, D. K., Schwerdt, G., and Gekle, M. (2006). Impact of extracellular acidity on the activity of P-glycoprotein and the cytotoxicity of chemotherapeutic drugs, *Neoplasia,* **8**, pp. 143–152.

213. Wojtkowiak, J. W., Verduzco, D., Schramm, K. J., and Gillies, R. J. (2011). Drug resistance and cellular adaptation to tumor acidic pH microenvironment, *Mol. Pharmaceut.,* **8**, pp. 2032–2038.

214. Chen, Q., Liu, X. D., Chen, J. W., Zeng, J. F., Cheng, Z. P., and Liu, Z. (2015). A self-assembled albumin-based nanoprobe for in vivo ratiometric photoacoustic pH imaging, *Adv. Mater.,* **27**, pp. 6820–6827.

215. Chen, Q., Liu, X. D., Zeng, J. F., Cheng, Z. P., and Liu, Z. (2016). Albumin-NIR dye self-assembled nanoparticles for photoacoustic pH imaging and pH-responsive photothermal therapy effective for large tumors, *Biomaterials,* **98**, pp. 23–30.

216. Shuhendler, A. J., Pu, K. Y., Cui, L., Uetrecht, J. P., and Rao, J. H. (2014). Real-time imaging of oxidative and nitrosative stress in the liver of live animals for drug-toxicity testing, *Nat. Biotechnol.,* **32**, pp. 373–380.

217. Li, H., Zhang, P., Smaga, L. P., Hoffman, R. A., and Chan, J. (2015). Photoacoustic probes for ratiometric imaging of copper(II), *J. Am. Chem. Soc.,* **137**, pp. 15628–15631.

218. Cash, K. J., Li, C. Y., Xia, J., Wang, L. H. V., and Clark, H. A. (2015). Optical drug monitoring: Photoacoustic imaging of nanosensors to monitor therapeutic lithium in vivo, *ACS Nano,* **9**, pp. 1692–1698.

219. Zhong, J., Wen, L., Yang, S., Xiang, L., Chen, Q., and Xing, D. (2015). Imaging-guided high-efficient photoacoustic tumor therapy with targeting gold nanorods, *Nanomedicine (UK),* **11**, pp. 1499–1509.

220. Lewinski, N., Colvin, V., and Drezek, R. (2008). Cytotoxicity of nanoparticles, *Small,* **4**, pp. 26–49.

221. Brewer, G. J. (2010). Copper toxicity in the general population, *Clin. Neurophysiol.,* **121**, pp. 459–460.

Index